T0136880

Studies in Systems, Decision and Control

Volume 203

Series Editor

Janusz Kacprzyk, Systems Research Institute, Polish Academy of Sciences, Warsaw, Poland

The series "Studies in Systems, Decision and Control" (SSDC) covers both new developments and advances, as well as the state of the art, in the various areas of broadly perceived systems, decision making and control–quickly, up to date and with a high quality. The intent is to cover the theory, applications, and perspectives on the state of the art and future developments relevant to systems, decision making, control, complex processes and related areas, as embedded in the fields of engineering, computer science, physics, economics, social and life sciences, as well as the paradigms and methodologies behind them. The series contains monographs, textbooks, lecture notes and edited volumes in systems, decision making and control spanning the areas of Cyber-Physical Systems, Autonomous Systems, Sensor Networks, Control Systems, Energy Systems, Automotive Systems, Biological Systems, Vehicular Networking and Connected Vehicles, Aerospace Systems, Automation, Manufacturing, Smart Grids, Nonlinear Systems, Power Systems, Robotics, Social Systems, Economic Systems and other. Of particular value to both the contributors and the readership are the short publication timeframe and the world-wide distribution and exposure which enable both a wide and rapid dissemination of research output.

** Indexing: The books of this series are submitted to ISI, SCOPUS, DBLP, Ulrichs, MathSciNet, Current Mathematical Publications, Mathematical Reviews, Zentralblatt Math: MetaPress and Springerlink.

More information about this series at http://www.springer.com/series/13304

Yuriy P. Kondratenko · Arkadii A. Chikrii ·
Vyacheslav F. Gubarev ·
Janusz Kacprzyk
Editors

Advanced Control Techniques in Complex Engineering Systems: Theory and Applications

Dedicated to Professor Vsevolod M. Kuntsevich

Editors
Yuriy P. Kondratenko
Department of Intelligent Information Systems
Petro Mohyla Black Sea National University
Mykolaiv, Ukraine

Vyacheslav F. Gubarev
Space Research Institute of NAS and SSA of Ukraine
Kyiv, Ukraine

Arkadii A. Chikrii
Glushkov Cybernetics Institute of NAS of Ukraine
Kyiv, Ukraine

Janusz Kacprzyk
Systems Research Institute
Polish Academy of Sciences
Warsaw, Poland

ISSN 2198-4182 ISSN 2198-4190 (electronic)
Studies in Systems, Decision and Control
ISBN 978-3-030-21929-1 ISBN 978-3-030-21927-7 (eBook)
https://doi.org/10.1007/978-3-030-21927-7

This Springer imprint is published by the registered company Springer Nature Switzerland AG
The registered company address is: Gewerbestrasse 11, 6330 Cham, Switzerland

To Academican Vsevolod Mikhailovich Kuntsevich

Academician Vsevolod M. Kuntsevich

Professor Vsevolod Mikhailovich Kuntsevich is an outstanding scientist in the field of theory and practice of control systems, Academician of the National Academy of Sciences of Ukraine (1992), Honored Worker of Science and Technology of Ukraine, Winner of the State Prizes of the Ukrainian SSR (1979) and Ukraine in the field of science and technology (1991, 2000), Director of the Space Research Institute of the National Academy of Sciences and the National Space Agency of Ukraine (1996–2007), Editor-in-chief of the Journal "Problemy Upravleniya I Informatiki" published in English in the USA under the title "Journal of Automation and Information Sciences" (since 1988).

Born March 15, 1929. In 1952, he graduated from the Kyiv Polytechnic Institute. He began his professional career as an engineer at the Institute of Mining, Academy of Sciences of the Ukrainian SSR. In 1957, he became a Ph.D.-student at the Institute of Electrical Engineering of the Academy of Sciences of the Ukrainian SSR, where, after defending his Ph.D. thesis, he worked as a junior and then as a senior researcher at the laboratory of automatic control of production processes. During these years, Vsevolod Mikhailovich investigated the problems of the theory and application of extremal systems, which were only born as a new scientific direction in the general theory of control.

In 1963, V. M. Kuntsevich moved to the Institute of Cybernetics of the Academy of Sciences of the Ukrainian SSR, where he defended his thesis for the degree of Doctor of Technical Sciences and from 1966 headed the department of discrete control systems. His scientific studies of this period are devoted to the development of the theory of a special class of nonlinear frequency-pulse modulated control systems. The results were included in the monograph, which was the first work in this direction both in the former USSR and beyond.

The problem of the synthesis of control systems using the apparatus of Lyapunov functions is highlighted in the scientific works of the scientist, published in 1973–1978. During this period, the problem of synthesizing optimal control systems of a wide class of nonlinear objects was investigated and solved.

V. M. Kuntsevich and his students have developed a new approach to the construction of adaptive control systems, based on methods for solving parametric identification problems, which provide guaranteed estimates, as well as new algorithms for solving optimal control problems under conditions of uncertainty.

Since 1986, Vsevolod Mikhailovich has been paying attention to the modern problem of control theory—robustness, in the process of studying which he obtained a number of significant results in the field of robust stability.

V. M. Kuntsevich led several projects related to the development of automatic control systems that are implemented in the aviation industry, as well as in the process control system in the oil refining industry.

In 1988, Vsevolod Mikhailovich was elected as a corresponding member of the Academy of Sciences of Ukraine, and in 1992 as an Academician of the National Academy of Sciences (NAS) of Ukraine.

In 1995 V. M. Kuntsevich became deputy director of the V. M. Glushkov Institute of Cybernetics of the NAS of Ukraine, and in May 1996 headed the Institute for Space Research, NAS of Ukraine and the National Space Agency of Ukraine. As director, Vsevolod Mikhailovich directed his activities towards the establishment of this Institute as the leading institution of space science in Ukraine, developing fundamental and applied scientific research and coordinating it with the main areas of the National Space Program of Ukraine. Much attention V. M. Kuntsevich gives basic questions of future experiments at the embedded Ukrainian research module of the International Space Station.

Professor V. M. Kuntsevich is the author of more than 250 scientific works, among which 8 monographs. In Ukraine, he created a scientific school in the field of discrete control systems, prepared more than 30 candidates (Ph.D.) and 11 doctors of science (Dr.Sc.). Over 25 years, Vsevolod Mikhailovich gave a course of lectures on control theory at the Kyiv Polytechnic Institute.

V. M. Kuntsevich received the State Prize of the Ukrainian SSR in the field of science and technology in 1979 as a co-author of the first in the world "Encyclopedia of Cybernetics"; in 1991, for his work on the theory of invariance and its applications, together with other scientists, he was awarded the title of laureate of the State Prize of Ukraine. In 1987, for a series of studies on the theory of digital control systems, Vsevolod Mikhailovich was awarded by S. A. Lebedev's Prize of the Presidium of the Academy of Sciences of the Ukrainian SSR, and in

1995—by the V. M. Glushkov Prize for the development and application of systems analysis methods.

Scientific achievements of V. M. Kuntsevich are also marked by government awards—the Order "Badge of Honor", "For Merits" (III degree), and the medals "For Labor Excellence" and "For Valiant Labor". In 1999 V. M. Kuntsevich was awarded by the title "Honored Worker of Science and Technology of Ukraine", and in 2000 he was awarded by the State Prize of Ukraine in the field of science and technology for a series of scientific works on multi-channel control systems for transport facilities and rolling mills.

In 1988, Vsevolod Mikhailovich became the chief editor of the journal Avtomatika (since 1994, "Problemy Upravleniya i Informatiki"), and since 2009—the International Scientific and Technical Journal "Problemy Upravleniya I Informatiki".

For a long time, Prof. V. Kuntsevich was a member of the USSR National Committee on Automatic Control, and in 1992 he was elected as Chairman of the National Committee of the Ukrainian Association on Automatic Control, the national organization of the International Federation for Automatic Control (IFAC).

Now, V. M. Kuntsevich is a Honorary Director of the Space Research Institute of the National Academy of Sciences and the National Space Agency of Ukraine and Head of the Department "Control of Dynamic Systems", Deputy Academician-Secretary of the Informatics Division of the NAS of Ukraine.

The 2019 year marks a remarkable event for the National Academy of Science of Ukraine and, more generally, for our entire community, that is, the 90th anniversary of Prof. Vsevolod M. Kuntsevich.

On the one hand, Prof. Kuntsevich has been for years the driving force behind the use of modern scientific control models and methods in the analysis of broadly perceived complex systems, notably space and complex industrial systems. He is one of the originators of a comprehensive study of control theory for uncertainty and imprecision of information in such system, and has proposed many original tools and techniques.

On the other hand, as President of the National Committee of the Ukrainian Association on Automatic Control (UAAC), he has greatly contributed to the activities of the UAAC and its great stature and visibility. Thanks to his vision and constant efforts to convince all kinds of relevant parties, notably the scientific and scholarly community, both in Ukraine and all over the world, the UAAC has played a constantly increasing role as one of the world's leading scientific and scholarly associations.

Third, in addition to his illustrious scientific and scholarly career, Prof. Vsevolod M. Kuntsevich has been always aware that theory should be complemented with practice. This is true for virtually all fields of science, in particular engineering. The contributions of Prof. V. Kuntsevich in this respect, i.e. at the crossroads of science and engineering have been remarkable too.

Finally, one should also mention the role which Prof. V. Kuntsevich has played as a teacher and mentor, and also a loyal and supportive friend to so many people, including all of us.

In view of that special role that Prof. V. Kuntsevich has played in so many fields of activities, from a purely scientific and scholarly level, though organizational activities, to real engineering application, we have decided—as a token of appreciation of the entire community—to publish this special volume *Advanced Control Techniques in Complex Engineering Systems: Theory and Applications* dedicated to him, his influence and legacy.

Mykolaiv, Ukraine — Yuriy P. Kondratenko
Kyiv, Ukraine — Arkadii A. Chikrii
Kyiv, Ukraine — Vyacheslav F. Gubarev
Warsaw, Poland — Janusz Kacprzyk
December 2018

Preface

This volume is dedicated to Prof. Vsevolod M. Kuntsevich, Academician of the National Academy of Sciences of Ukraine (NASU), President of the National Committee of the Ukrainian Association on Automatic Control, Honorary Director of the Space Research Institute of the National Academy of Sciences and the National Space Agency.

By dedicating this book to Prof. Vsevolod M. Kuntsevich, we, the editors, on behalf of the entire research and professional community, wish to present a small token of appreciation for his great scientific and scholarly achievements, long-time service to many scientific and professional communities, notably those involved in automation, cybernetics, control, management and, more specifically, the foundations and applications of tools and techniques for dealing with uncertain information, robustness, non-linearity, extremal systems, discrete control systems, adaptive control systems and others.

At a more personal level, we also wish to thank him for a constant support of what we and many of our collaborators and students have been undertaking for so many years.

In terms of the structure of the volume, the 15 chapters, presented by authors from 16 different countries from all over the world—Austria, Chile, Georgia, Germany, Mexico, Norway, P.R. of China, Poland, F.Y.R. Macedonia, Romania, Russia, Spain, Turkey, the United States, Ukraine and the United Kingdom, are grouped into four parts: (1) Fundamental Theoretical Issues in Controlled Complex Engineering Systems, (2) Artificial Intelligence and Soft Computing in Control and Decision Making Systems, (3) Advanced Control Techniques for Industrial and Collaborative Automation, and (4) Modern Applications for Management and Information Processing in Complex Engineering Systems.

The chapters have been thought out to provide an easy to follow introduction to the topics that are addressed, including the most relevant references, so that anyone interested in them can start their introduction to the topic through these references. At the same time, all of them correspond to different aspects of work in progress being carried out in various research groups throughout the world and, therefore, provide information on the state of the art of some of these topics.

Part I: Fundamental Theoretical Issues in Controlled Complex Engineering Systems, includes seven contributions. Chapter "Method of Resolving Functions in the Theory of Conflict—Controlled Processes", by A. Chikrii, R. Petryshyn, I. Cherevko and Y. Bigun, is devoted to the investigation of game problems on bringing a trajectory of a dynamic system to a cylindrical terminal set. The authors proceed with the representation of a trajectory of dynamic system in the form in which the block of initial data is separated from the control block. This makes it feasible to consider a wide spectrum of functional-differential systems. The method of resolving functions, based on the use of the inverse Minkowski functionals, serves as a conceptual tool for the study. Attention is focused on the case when Pontryagin's condition does not hold. In this case the upper and lower resolving functions of two types are introduced. With their help the sufficient conditions for reaching a terminal set in a finite time are deduced. Various method schemes are provided and a comparison with Pontryagin's first direct method is given. The efficiency of the suggested mathematical scheme is illustrated with a model example.

The paper "Impact of Average-Dwell-Time Characterizations for Switched Nonlinear Systems on Complex Systems Control" by G. Dimirovski, J. Wang, H. Yue and J. Stefanovski focuses on the theory of switched systems which is largely based on assuming a certain small but finite time interval termed an average dwell time. Thus it appears to be dominantly characterized by some slow switching condition with the average dwell time satisfying a certain lower bound which implies a constraint nonetheless. In the cases of nonlinear systems there may well appear non-expected complexity phenomena of a particular different nature when switching no longer occurs. A fast switching condition with the average dwell time satisfying an upper bound is explored and established. A comparison analysis of these innovated characterizations via a slightly different overview yields new results on the transient behaviour of switched nonlinear systems while preserving the system stability. The approach based on multiple Lyapunov functions is used in the analysis and switched systems framework and is extended shedding new light on the underlying, switching caused system complexities.

A. Kurzhanski, in his Chapter "On the Problem of Optimization in Group Control", considers a team of several controlled motions with two types of members, a target set and an array of external obstacles. The problem is for both types to simultaneously reach the target, avoiding the obstacles and the possible mutual collisions, while performing the overall process in minimal time. The problem solution is described in terms of the Hamiltonian formalism.

In Chapter "Model Predictive Control for Discrete MIMO Linear Systems", V. Gubarev, M. Mishchenko and B. Snizhko consider the mathematical model in the form of discrete linear system equations with multiple inputs and multiple outputs (MIMO) which can describe different dynamic processes which important for practice. Some realizations of model predictive control for the discrete MIMO linear system are considered. The approach developed is based on the Cauchy formula for the discrete linear system. The problem of control design is reduced to solving a series of extremal problems including the optimization with constrains. The authors present the final results in an analytical form appropriate for solving different problems including control analysis and design.

V. Ushakov, in his Chapter "Krasovskii's Unification Method and the Stability Defect of Sets in a Game Problem of Approach on a Finite Time Interval", considers a nonlinear conflict-controlled system in a finite-dimensional Euclidean space over a finite time interval in which the controls of the players are constrained by geometric restrictions. The author discusses game problem of the approach of the system to a compact target set in the phase space of the system at a fixed instant of time. The problem is investigated in the framework of the positional formalization proposed by N. N. Krasovskii. The author considers the central property of stability in the theory of positional differential games and, in particular, the generalization of this property such as the stability defect of sets in the space of game positions.

Chapter "Control of Stochastic Systems Based on the Predictive Models of Random Sequences", by I. Atamanyuk, J. Kacprzyk, Y. Kondratenko and M. Solesvik, is devoted to the development of a mathematical models of stochastic control systems based on the predictive models of random sequences. In particular, the linear algorithms of the forecast of a control object state for an arbitrary number of states of an investigated object and the values of a control parameter are obtained. In the mean-square sense the algorithms make it possible to obtain an optimal estimation of the future values of a forecasted parameter, in the case when true known values for an observation interval are used and provided that the measurements are made with errors. Using an arbitrary number of non-linear stochastic relations, a predictive model of a control system is obtained as well. The schemes that reflect the peculiarities of ways to determine the parameters of a nonlinear algorithm and its functioning regularities are introduced in the work. The models developed make it possible to take into consideration the peculiarities of the sequence of the change of control object parameters and also to make full use of all known a priori and a posteriori information about the random sequence that has been investigated. The algorithms obtained in this chapter can be used in different areas of human activity to solve a wide range of problems of the control of objects of a stochastic nature.

A. Chentsov and D. Khachai, in Chapter "Program Iterations Method and Relaxation of a Pursuit-Evasion Differential Game", consider a special case of the non-linear zero-sum pursuit-evasion differential game. This game is defined by two closed sets—a target set and a set defining state constraints. The authors find an optimal non-anticipating strategy for the first player (the pursuer). Namely, the authors construct the pursuer's successful solvability set specified by a limit function of an iterative procedure in the space of positions. For the positions outside of the successful solvability set, they consider a relaxation of their game by determining the smallest size of neighborhoods of two mentioned sets for which the pursuer can solve his problem. Then, they construct a successful solvability set in terms of those neighborhoods.

Part II: Artificial Intelligence and Soft Computing in Control and Decision Making Systems, includes 3 contributions. In Chapter "Fuzzy Real-Time Multi-objective Optimization of a Prosthesis Test Robot Control System",

Y. Kondratenko, P. Khalaf, H. Richter and D. Simon investigate the fuzzy real-time multi-objective optimization of a combined test robot/transfemoral prosthesis system with three degrees of freedom. The impedance control parameters are optimized with respect to the two objectives of a ground force and vertical hip position tracking. The control parameters are first optimized off-line with an evolutionary algorithm at various values of the walking speed, surface friction, and surface stiffness. These control parameters comprise a gait library of Pareto-optimal solutions for various walking scenarios. The user-preferred Pareto point for each walking scenario can be selected either by expert decision makers or by using an automated selection mechanism such as the point that is at the minimum distance to the ideal point. Then, given a walking scenario that has not yet been optimized, a fuzzy logic system is used to interpolate in real time control parameters. This approach enables an automated real-time multi-objective optimization. Simulation results confirm the effectiveness and efficiency of the proposed approach.

O. Castillo, in Chapter "Bio-inspired Optimization of Type-2 Fuzzy Controllers in Autonomous Mobile Robot Navigation", performs a comparison of using type-2 fuzzy logic with two different bio-inspired methods: the Ant Colony Optimization and Gravitational Search Algorithm. Each of these methods is enhanced with a methodology for dynamic parameter adaptation using interval type-2 fuzzy logic, where based on some metrics about the algorithm, like the percentage of iterations elapsed or the diversity of the population, the author aims at controlling its behavior and therefore control its abilities to perform a global or a local search. To test these methods two benchmark control problems are used in which a fuzzy controller is optimized to minimize the error in the simulation with nonlinear complex plants.

Chapter "A Status Quo Biased Multistage Decision Model for Regional Agricultural Socioeconomic Planning Under Fuzzy Information", by J. Kacprzyk, Y. Kondratenko, J. M. Merigo, J. H. Hormazabal, G. Sirbiladze, A. M. Gil-Lafuente, presents a new multistage control model for agricultural regional development planning which is analyzed in terms of some life quality indicators covering various social and economic aspects. The model follows a novel philosophy to decision making, which has been for some time advocated by some prominent economists, sociologists, psychologists, etc., notably Daniel Kahneman, the 2002 winner of the Nobel Prize in economics, that in the modeling of all kinds of the so called human centric system, i.e. systems in which the human beings or their groups are crucial stakeholders, the very specifics of the human being should be accounted for. In considered case, this concerns the so called cognitive biases, specifically the so called status quo bias which states that humans usually prefer to stay with situations they know, and do not want large changes. A new multistage control planning model is proposed based on this general attitude, and its analysis in the context of agricultural regional development planning is discussed.

Part III: Advanced Control Techniques for Industrial and Collaborative Automation includes three contributions. Chapter "Holonic Hybrid Supervised Control of Semi-continuous Radiopharmaceutical Production Processes", by T. Borangiu, S. Răileanu, E. V. Oltean and A. Silişteanu, discusses the application of the holonic paradigm to the supervised hybrid control of semi-continuous processes which are exemplified by the production of radiopharmaceutical substances. The supervisor of the control system fulfils two main functionalities: the optimization of the global process planning that includes all client orders most recently received (the values of process parameters and operations timing are initially computed to maximize the number of accepted orders—the optimal state trajectory); the reconfiguration of parameters of the optimal state trajectory whenever unexpected events occur (in the production subprocesses or in the environment parameters), providing thus the robustness to disturbances. The implementation of the holonic supervised hybrid control system uses a multi-agent framework in a semi-heterarchical topology. Two scenarios validating the optimization of planning and experimental results are discussed.

B. Katalinic, D. Haskovic, I. Kukushkin and I. Zec, in the paper "Hybrid Control Structure and Reconfiguration Capabilities in Bionic Assembly System", present their research results focused on the investigation of working scenarios and the efficiency of the next generation of modern assembly systems—the Bionic Assembly System (BAS). It is based on biologically inspired principles of self-organisation, reduced centralized control, networking between units and natural parallel distribution of processes. The BAS control system combines two principles: the subordination from factory level to the BAS control structure, and the self-organization at the shop floor level. The BAS is a human centric system which promotes the integration of workers in the working process. The main goal of the BAS is to increase the system efficiency and robustness which can be attained by reconfiguration. Various reconfiguration capabilities are described in the chapter where the investigation is limited to a normal working mode. The results show that production systems with a high technical similarity with the BAS can increase their efficiency using the proposed concepts. This represents a promising direction in development of future modern assembly systems.

In Chapter "A Flatness-Based Approach to the Control of Distributed Parameter Systems Applied to Load Transportation with Heavy Ropes", by A. Irscheid, M. Konz, J. Rudolph, the cooperative transport of a rigid body carried by multiple heavy ropes that are suspended by tricopters is used as an example for flatness-based trajectory planning for distributed-parameter systems. At first, the load-side ends of the ropes are parametrized by a flat output of the nonlinear boundary system that describes the motion of the rigid body. This parametrization is then used to express the solution of the linearized partial differential equations for the heavy ropes by the means of operational calculus. Evaluating the derived parametrization at the controllable ends yields the desired trajectories for the tricopter positions. A position controller is used to track the reference trajectories of the tricopters. This control can be used to realize a desired transition of the load position and orientation in finite time. Experimental results of the introduced method are presented for validation.

Part IV: Modern Applications for Management and Information Processing in Complex Engineering Systems, includes two contributions. R. E. Hiromoto, M. Haney, A. Vakanski and B. Shareef, in Chapter "Toward a Secure IoT Architecture", discuss the design of a cyber-secure, the Internet of Things (IoT), supply chain risk management architecture. The purpose of the architecture is to reduce vulnerabilities of malicious supply chain risks by applying machine learning (ML), cryptographic hardware monitoring (CHM), and distributed network coordination (DNC) techniques to guard against unforeseen hardware component failures and malicious attacks. These crosscutting technologies are combined into an Instrumentation-and-Control/Operator-in-the-Loop (ICOL) architecture that learns normal and abnormal system behaviors. In the event that the ICOL detects possible abnormal system-component behaviors, an ICOL network alert is triggered that requires an operator verification-response action before any control parameters can affect the operation of the system. The operator verification-response is fed back into the ML systems to recalibrate the classification of normal and abnormal states of the system. As a consequence, this proposal adheres to the notion of an Operator-in-the-Loop strategy that combines DNC, ML, and CHM with the creative problem-solving capabilities of human intelligence.

In Chapter "Formal Concept Analysis for Partner Selection in Collaborative Simulation Training", M. Solesvik, Y. Kondratenko, I. Atamanyuk and O. J. Borch propose the application of formal concept analysis technique to partner selection for joint simulator training. The simulator training is an important tool for the preparation of sailors. However, the cost of the simulator centers is high. That is why the educational institutions and centers can only afford a limited number of simulators. Though the price for the simulator classes is quite high, it still more advantageous to train sailors in a class than offshore. New cloud-based technologies make it possible to connect simulators situated in different places, including a cross-country communication. This makes it possible to carry out a joint training of different simulator types. The authors develop the approach using the formal concept analysis to facilitate partner selection from the pool of potential collaborating institutions.

The chapters selected for this book provide an overview of some up to date and important problems in the area of control engineering and controlled complex systems design as well as the advanced control techniques that relevant research groups within this area are employing to try to solve them.

We would like to express our deep appreciation to all authors for their contributions as well as to the peer reviewers for their relevant comments and suggestions. We certainly look forward to working with all contributors again in nearby future.

Mykolaiv, Ukraine — Prof. Dr. Sc. Yuriy P. Kondratenko
Kyiv, Ukraine — Acad. NASU, Prof. Dr. Sc. Arkadii A. Chikrii
Kyiv, Ukraine — Corr.-Acad. NASU, Prof. Dr. Sc. Vyacheslav F. Gubarev
Warsaw, Poland — Acad. PAN, Prof. Dr. Sc. Janusz Kacprzyk
January 2019

Contents

Part I Fundamental Theoretical Issues in Controlled Complex Engineering Systems

Method of Resolving Functions in the Theory of Conflict—Controlled Processes 3
Arkadii A. Chikrii, R. Petryshyn, I. Cherevko and Y. Bigun

Impact of Average-Dwell-Time Characterizations for Switched Nonlinear Systems on Complex Systems Control 35
Georgi Dimirovski, Jiqiang Wang, Hong Yue and Jovan Stefanovski

On the Problem of Optimization in Group Control 51
Alexander B. Kurzhanski

Model Predictive Control for Discrete MIMO Linear Systems 63
Vyacheslav F. Gubarev, M. D. Mishchenko and B. M. Snizhko

Krasovskii's Unification Method and the Stability Defect of Sets in a Game Problem of Approach on a Finite Time Interval 83
Vladimir N. Ushakov

Control of Stochastic Systems Based on the Predictive Models of Random Sequences 105
Igor Atamanyuk, Janusz Kacprzyk, Yuriy P. Kondratenko and Marina Solesvik

Program Iterations Method and Relaxation of a Pursuit-Evasion Differential Game 129
Alexander Chentsov and Daniel Khachay

Part II Artificial Intelligence and Soft Computing in Control and Decision Making Systems

Fuzzy Real-Time Multi-objective Optimization of a Prosthesis Test Robot Control System 165
Yuriy P. Kondratenko, Poya Khalaf, Hanz Richter and Dan Simon

Bio-inspired Optimization of Type-2 Fuzzy Controllers in Autonomous Mobile Robot Navigation 187
Oscar Castillo

A Status Quo Biased Multistage Decision Model for Regional Agricultural Socioeconomic Planning Under Fuzzy Information 201
Janusz Kacprzyk, Yuriy P. Kondratenko, Jos'e M. Merigó, Jorge Hernandez Hormazabal, Gia Sirbiladze and Ana Maria Gil-Lafuente

Part III Advanced Control Techniques for Industrial and Collaborative Automation

Holonic Hybrid Supervised Control of Semi-continuous Radiopharmaceutical Production Processes 229
Theodor Borangiu, Silviu Răileanu, Ecaterina Virginia Oltean and Andrei Silişteanu

Hybrid Control Structure and Reconfiguration Capabilities in Bionic Assembly System 259
Branko Katalinic, Damir Haskovic, Ilya Kukushkin and Ilija Zec

A Flatness-Based Approach to the Control of Distributed Parameter Systems Applied to Load Transportation with Heavy Ropes 279
Abdurrahman Irscheid, Matthias Konz and Joachim Rudolph

Part IV Modern Applications for Management and Information Processing in Complex Engineering Systems

Toward a Secure IoT Architecture 297
Robert E. Hiromoto, Michael Haney, Aleksandar Vakanski and Bryar Shareef

Formal Concept Analysis for Partner Selection in Collaborative Simulation Training 325
Marina Solesvik, Yuriy P. Kondratenko, Igor Atamanyuk and Odd Jarl Borch

Part I
Fundamental Theoretical Issues in Controlled Complex Engineering Systems

Method of Resolving Functions in the Theory of Conflict—Controlled Processes

Arkadii A. Chikrii, R. Petryshyn, I. Cherevko and Y. Bigun

Abstract The paper is devoted to investigation of game problems on bringing a trajectory of dynamic system to a cylindrical terminal set. We proceed with representation of a trajectory of dynamic system in the form, in which the block of initial data is separated from the control block. This makes it feasible to consider a wide spectrum of functional-differential systems. The method of resolving functions, based on use of the inverse Minkovski functionals, serves as ideological tool for study. Attention is focused on the case when Pontryagin's condition does not hold. In this case the upper and lower resolving functions of two types are introduced. With their help sufficient conditions of approach a terminal set in a finite time are deduced. Various method schemes are provided and comparison with Pontryagin's first direct method is given. Efficiency of suggested mathematical scheme is illustrated with a model example.

1 Introduction

There are a number of fundamental methods, which reveal structure of game problems and provide mathematical constructions for analysis of conflict situations. First of all, these are Pontryagin's methods (first direct method and method of alternated integral) [1], the extreme targeting rule of Krasovskii [2], the method of Isaacs [3], related with the dynamic programming and expressed in the form of functional relationships, the method of T_ε-operators of Pshenichnyi [4], which employs ideas of alternated integral, and is, in essence, is a reflection of dynamic programming ideas in the form of set-valued mappings. The method of resolving functions [5],

A. A. Chikrii (✉)
Glushkov Cybernetics Institute of NAS of Ukraine, Kiev 03187, Ukraine
e-mail: g.chikrii@gmail.com

A. A. Chikrii · R. Petryshyn · I. Cherevko · Y. Bigun
Yuriy Fedkovych Chernivtsi National University, Chernivtsi 58012, Ukraine
e-mail: yaroslav.bigun@gmail.com

Y. P. Kondratenko et al. (eds.), *Advanced Control Techniques in Complex Engineering Systems: Theory and Applications*, Studies in Systems, Decision and Control 203, https://doi.org/10.1007/978-3-030-21927-7_1

created in recent years, can be also assigned to these methods. Various problems of control in conditions of uncertainty are explored in [6–12].

It should be noted that the extreme targeting rule and method of resolving functions theoretically substantiate, known from practice, the method of pursuit along the Euler line of sight and the rule of pursuit along a ray, in particular, "parallel pursuit" rule, respectively. The game methods have found wide application [13–15]. The method of resolving functions can be applied to study of the impulse processes on the basis of results of the monograph [16] and of functional-differential systems [17].

This paper is dedicated to development of the method of resolving functions. An attractive feature of this method is that it allows efficiently apply modern technique of set-valued mappings and their selections to substantiate the game constructions and to obtain on their basis rich in content results. This method was originated during analysis of the game problems of pursuit-evasion, in particular, as a consequence of solving the problem of escape of single controlled object from a group of pursuers.

In 1969 L. S. Pontryagin formulated and solved in linear case the problem of avoiding meeting one controlled object with another one, starting from arbitrary initial states and on all half-infinite interval of time [1]. Later on, at the conference on game theory (1974) the statement and solution of the problem of avoiding meeting single player with a group of pursuers was independently presented by Gusyatnikov and Chikrii [18]. In so doing, the different methods were used: the manoeuvre of turning movement [19]—by the former and the method of escape along direction by the latter [20].

In the papers [18, 20] the following function was introduced:

$$\omega(n, \nu) = \min_{\|p_i\|=1} \max_{\|\nu\|=1} \min_{i=1,\dots,k} |(p_i, \nu)|, \; p_i, \nu \in R^n, n, k \geq 2, 0 \leq \omega(n, \nu) \leq 1 \quad (1)$$

It was destined to play a key role in creating the method of resolving functions. Function $\omega(n, \nu)$ determines an advantage of the evader in control resources upon each of the pursuers, which is suffice to rectilinearly escape, starting from arbitrary initial states.

Consider the game of group pursuit with exact capture as a goal:

$$\dot{x}_i = u_i, \; \|u_i\| \leq 1, i = 1, \dots, k, \; x_i(0) = x_i^0, \; x_i \in R^n,$$
$$\dot{y} = \nu, \; \|\nu\| \leq 1, \; y(0) = y^0, \; y \in R^n, n \geq 2. \quad (2)$$

Removing external minimum and modulo and setting $p_i = \frac{x_i^0 - y_0}{\|x_i^0 - y_0\|}$ in (1) we come to the assertion: $y_0 \in \mathrm{int}co\{x_i^0\}$ if and only if $\max_{\|\nu\|=1} \min_{i=1,\dots k} (p_i, \nu) > 0$, whence follows the main result: for feasibility of capture in the group pursuit problem (2) it is necessary and sufficient that

$$y_0 \in \mathrm{int}co\{x_i^0\}$$

This fact was noticed by B. N. Pshenichnyi when seeing the function (1) [21]. In the proof of this result the pursuers apply the rule of parallel pursuit and their controls appear as counter-controls [2] which depend on the initial players' states and certain scalar function, the latter being the greatest positive root of special quadratic equation. This function is a prototype of the resolving function.

The full description of the method of resolving functions is contained in the monograph [5] and it's modification for more general dynamics—in the paper [22]. The problems of group pursuit are also considered in the books [23, 24].

This paper is devoted to the problem of pursuit with single pursuer and a single evader in the case when Pontryagin's condition (the condition for advantage of the pursuer over the evader) does not hold. The upper and the lower resolving functions of two types are introduced that makes it possible to essentially broaden the class of problems solvable on the basis of the above mentioned ideology.

2 Problem Statement and Players Goals

Let us consider in finite-dimensional Euclidian space R^n, $n \geq 2$, conflict-controlled process whose evolution is given by the equation

$$z(t) = g(t) + \int_0^t \Omega(t, \tau)\phi(u(\tau), v(\tau))d\tau, t \geq 0 \tag{3}$$

where $z(t) \in R^n$, the function of initial data $g(t)$, $g : R_+ \to R^n$, $\mathrm{R}_+ = \{t : t \geq 0\}$, is the Lebesgue measurable and bounded for $t > 0$, the matrix function $\Omega(t, \tau)$, $t \geq \tau \geq 0$, is bounded almost everywhere, measurable in t and summable in τ for every $t \in R'_+$. Control unit includes function $\phi(u, v)$, $\varphi : U \times V \to R^n$, which is assumed to be jointly continuous in its variables on direct product of nonempty compacts U and V, $U \in K(R^m)$, $V \in K(R^l)$, n, m, l are natural integers.

Admissible controls of the players, $u(\tau)$, $u : R_+ \to U$, $v(\tau)$, $v : \mathrm{R}_+ \to V$, are measurable functions of time.

Let the terminal set M^*, having the cylindrical form be given

$$M^* = M_0 + M \tag{4}$$

where M_0 is linear subspace from R^n, a $M \in K(L)$, $L = M_0^{\perp}$ is orthogonal complement to M_0 in R^n.

Goals of the players are opposite, moreover, the performance criterion is the time. The first one (u) tries in the shortest time to bring a trajectory (3) to the terminal set (4), and the second one (v) tries to delay maximally the instant of trajectory hitting the set M^* or avoid the hitting. General representation of solution of dynamical system in the form (3) makes it possible to consider in the frames of unified scheme a wide range of quasilinear functional-differential systems, functioning under the condition of conflict, in particular, systems of ordinary differential, integral-differential and differential-difference equations as well as systems of equations with classical fractional derivatives of Rienmann–Liouville, regularized fractional derivatives of Dzhirbashian–Nersesian–Caputo, sequential fractional derivatives of Miller-Ross, the generalized Hilfer derivatives [25], impulse systems [26]. Similar representation in discrete case makes it feasible to study multistep processes, in particular, the discrete systems of fractional order by Grunwald–Letnikov [27] and the hybrid systems.

Specific forms of the function $g(t)$ and the matrix function $\Omega(t, \tau)$ define the type of conflict-controlled process.

Let us take the side of the first player. If the game (3)–(4) is evolving on the interval $[0, T]$, then we select control at time instant t in the form of measurable function

$$u(t) = u(g(T), v_t(\cdot)), u(t) \in U, t \in [0, T] \tag{5}$$

where $v_t(\cdot) = \{v(s) : s \in [0, t]\}$ is a prehistory of admissible control of the second player until the time instant t or, in the form of counter-control,

$$u(t) = u(g(T), v(t)), u(t) \in U, t \in [0, T]. \tag{6}$$

In the case (5) we speak about quasistrategy of the first player, in the case (6)—about Hajek stroboscopic strategy [11], which prescribes counter-control by Krasovskii [2].

Objective of the paper is to establish sufficient conditions for termination of approach game (3)–(4) in certain guaranteed time, using the second player control prehistory, as well as in the class of stroboscopic strategies by comparing guaranteed time of game termination with guaranteed time of the Pontryagin's first direct method [1, 12].

3 Pontryagin's First Direct Method

Let us denote by π the operator of orthogonal projecting from R^n into L. We set

$$\varphi(U, v) = \{\varphi(u, v) : u \in U\}$$

and introduce set-valued mapping

$$W(t, \tau, v) = \pi\Omega(t, \tau)\varphi(U, v)$$

on the set $\Delta \times V$, where

$$\Delta = \{(t, \tau) : 0 \leq \tau \leq t < +\infty\}$$

is a plane cone.

Condition 1 Mapping $W(t, \tau, v)$ is closed-valued on the direct product of cone Δ and compact V.

Let us consider

$$W(t, \tau) = \bigcap_{v \in V} W(t, \tau, v), 0 \leq \tau \leq t < +\infty,$$

and denote [28]

$$\text{dom} W = \{(t, \tau) \in \Delta : W(t, \tau) \neq \emptyset\}.$$

Pontryagin's condition. $\text{dom} W = \Delta$.

If Condition 1 and Pontryagin's condition hold, then mapping $W(t, \tau)$ is closed-valued and measurable in τ, $\tau \in [0, t]$. Therefore [29], there exists at least one measurable in τ selection, namely, Pontryagin's selection.

Let us consider Pontryagin's function, under the above mentioned conditions,

$$p(g(\cdot)) = \inf\{t \geq 0 : t \in P(g(\cdot))\},$$

where

$$P(g(\cdot)) = \{t \geq 0 : \pi g(t) \in M - \int_0^t W(t, \tau) d\tau\}. \tag{7}$$

Here by integral of set-valued mapping is meant Aumann's integral [30]. If in (7) the inclusion in braces does not hold for all t, $t \geq 0$, then we set

$$P(g(\cdot)) = \emptyset, \text{and } p(g(\cdot)) = +\infty.$$

Theorem 1 *Let for conflict-controlled process* (3)–(4) Condition 1 and *Pontryagin's condition hold, and* $P \in P(g(\cdot)) \neq \emptyset$.

Then a trajectory of the process (3) *can be brought into the terminal set* (4) *at instant* P *by means of the Hajek stroboscopic strategy, which prescribes the corresponding Krasovskii' counter-control.*

Proof From the assumptions of Theorem 1 and the inclusion in the relation (7) there follows that

$$\pi g(P) \in M - \int_0^P W(P, \tau) d\tau.$$

This means that there exist a point m, $m \in M$, and, by definition of Aumann's integral, Pontryagin's measurable selection $\gamma(P,\tau)$, $\gamma(P,\tau) \in W(P,\tau)$, $\tau \in [0, P]$, such that

$$\pi g(P) = m - \int_0^P \gamma(P,\tau)d\tau. \tag{8}$$

Let us consider the set-valued mapping

$$U_0(\tau, v) = \{u \in U : \pi\Omega(P,\tau)\phi(u, v) - \gamma(P,\tau) = 0\}, \tau \in [0, P], v \in V$$

The mapping $U_0(\tau, v)$ is closed-valued and $L \times B$-measurable [22, 31]. Therefore, by the theorem on measurable choice [29], there exists at least one $L \times B$-measurable selection $u_0(\tau, v)$, $u_0(\tau, v) \in U_0(\tau, v)$, which appears as superpositionally measurable function [22]. Let us select control of the first player in the form of measurable function $u(\tau) = u_0(\tau, v(\tau))$, $\tau \in [0, P]$, where $v(\tau)$ is an admissible control of the second player.

From the representation (3), in view of the relation (8) and equality in the expression for $U_0(\tau, v)$, it immediately follows it that $\pi g(P) = m \in M$.

Remark The above provided scheme is usually called the first Pontryagin direct method and the time $p(g(\cdot))$ is the minimal guaranteed time of this scheme. The suggested proof differs from the original one [1] and is a generalization of the result of P. B. Gusyatnikov and M. S. Nikolskij, dealing with the mentioned method for linear differential games and using Filippov-Castaing theorem on measurable choice [32].

In next section we do not assume that Pontryagin condition holds and focus our investigations on development of the ideology of resolving functions [5, 22] in this case.

4 Schemes of the Method of Resolving Functions

Let $\gamma(t,\tau)$, $\gamma : \Delta \to L$, be some function, which is almost everywhere bounded, measurable in t and summable in τ, $\tau \in [0, t]$ for each $t \in R_+$. In the sequel it will be called the shift function.

Let us denote

$$\xi(t) = \xi(t, g(t), \gamma(t,\cdot)) = \pi g(t) + \int_0^t \gamma(t,\tau)d\tau$$

and consider the set-valued mapping

$$\mathsf{A}(t,\tau,v)=\{\alpha\geq 0:[\pi\Omega(t,\tau)\phi(u,v)-\gamma(t,\tau)]\cap\alpha[M-\xi(t)]\neq\emptyset\},\qquad(9)$$

$$\mathsf{A}:\Delta\times V\rightarrow 2^{R_+}.$$

Condition 2 Set-valued mapping $\mathsf{A}(t,\tau,v)$ has nonempty images on the set $\Delta\times V$.

Under this condition, we consider scalar upper and lower resolving functions of the first type

$$\alpha^*(t,\tau,v)=sup\{\alpha:\alpha\in \mathrm{A}(t,\tau,v)\},$$
$$\alpha_*(t,\tau,v)=inf\{\alpha:\alpha\in \mathrm{A}(t,\tau,v)\}.$$

If, in addition, Pontryagin's condition holds, then, to emphasize the role of the Minkowski functional [28, 33] and its inverse in the method scheme, we represent the upper resolving function $\alpha^*(t,\tau,v)$ in another way. To this end, we introduce the inverse Minkowski functional [5] for closed set X, $X\subset R^n$, $0\in X$:

$$\alpha_X(p)=sup\{\alpha\geq 0:\alpha p\in X\},\ p\in R^n.$$

Then [22] $\alpha^*(t,\tau,v)=\underset{m\in M}{sup}\,\alpha_{W(t,\tau,v)-\gamma(t,\tau)}(m-\xi(t)).$

Since image of the mapping $\mathsf{A}(t,\tau,v)$ present itself numerical sets on the semi-axis R_+, then the upper resolving function is the support function of this mapping in the direction $+1$. Taking into account properties of the conflict-controlled process (3), (4), Conditions 1 and 2, and the theorem on characterization and inverse image [29], we can show [22, 31] that closed-valued mapping $\mathsf{A}(t,\tau,v)$ is jointly $L\times B$-measurable in (τ,v), $\tau\in[0,t]$, $v\in V$, and the upper and the lower resolving functions are jointly $L\times B$-measurable in (τ,v) by virtue of the theorem on support function [29].

Let $V(\cdot)$ be a set of measurable functions $v(\tau)$, $\tau\in[0,+\infty)$, taking their values in V. Since at fixed t the function $\alpha^*(t,\tau,v)$ is jointly $L\times B$-measurable in (τ,v), then it is superpositionally measurable [22], i.e. $\alpha^*(t,\tau,v(\tau))$ is measurable in τ, $\tau\in[0,t]$, for each measurable function $v(\cdot)\in V(\cdot)$. The lower resolving function possesses the same property.

The upper resolving function $\alpha^*(t,\tau,v)$, in essence, reflects maximal gain of the first player in the game at the time instant τ on interval $[0,t]$ under instantaneous counter-action v. We associate with this function the set

$$T(g(\cdot),\gamma(\cdot,\cdot))=\{t>0:\inf_{v(\cdot)\in V(\cdot)}\int_0^t\alpha^*(t,\tau,v(\tau))d\tau\geq 1\}\qquad(10)$$

and its least element

$$t(g(\cdot),\gamma(\cdot,\cdot))=\inf\{t:t\in T(g(\cdot),\gamma(\cdot,\cdot))\}.$$

If for some t, $t > 0$, $\alpha^*(t, \tau, v) \equiv +\infty$, $\tau \in [0, t]$, $v \in V$, then in this case it is natural to set the value of integral in the relation (10) to be equal to $+\infty$. Then the corresponding inequality is readily satisfied and $t \in T(g(\cdot), \gamma(\cdot, \cdot))$. In case the inequality in (10) does not hold for all $t > 0$, we set $T(g(\cdot), \gamma(\cdot, \cdot)) = \emptyset$ and correspondingly $t(g(\cdot), \gamma(\cdot, \cdot)) = +\infty$.

Let us denote

$$\mathsf{A}(t, \tau) = \bigcap_{v \in V} \mathsf{A}(t, \tau, v), \; (t, \tau) \in \Delta.$$

Condition 3 The set-valued mapping $\mathsf{A}(t, \tau)$ has nonempty image on the cone Δ.

Under this condition we introduce upper and lower resolving functions of the second type

$$\alpha^*(t, \tau) = \sup\{\alpha : \alpha \in \mathsf{A}(t, \tau)\},$$
$$\alpha_*(t, \tau) = \inf\{\alpha : \alpha \in \mathsf{A}(t, \tau)\}.$$

Lemma Let conditions 1 and 3 hold for the conflict-controlled process (3)–(4), the mapping $\mathsf{A}(t, \tau, v)$ be compact-valued, and the upper resolving function $\alpha^*(t, \tau, v)$ be bounded on the set $\Delta \times V$. Then the following inequality is true:

$$\inf_{v \in V} \alpha^*(t, \tau, v) \geq \alpha^*(t, \tau), \; (t, \tau) \in \Delta. \tag{11}$$

If, in addition, the mapping $\mathsf{A}(t, \tau, v)$, $(t, \tau) \in \Delta$, $v \in V$, is convex-valued, then (11) turns into equality.

Proof By construction, the studied functions have the forms

$$\inf_{v \in V} \alpha^*(t, \tau, v) = \inf_{v \in V} \sup\{\alpha : \alpha \in \mathsf{A}(t, \tau, v)\},$$

$$\alpha^*(t, \tau) = \sup\{\alpha : \alpha \in \bigcap_{v \in V} \mathsf{A}(t, \tau, v)\}, \; 0 \leq \tau \leq t < +\infty.$$

Let us denote $\alpha^* = \alpha^*(t, \tau)$. Since the mapping $\mathsf{A}(t, \tau, v)$ is compact-valued, then $\mathsf{A}(t, \tau)$ is compact-valued in α^*, $\alpha^* \in \mathsf{A}(t, \tau, v)$, for each $v \in V$. This implies

$$\alpha^* \leq \sup\{\alpha : \alpha \in \mathsf{A}(t, \tau, v)\} \quad \forall v \in V,$$

therefore

$$\alpha^* \leq \inf_{v \in V} \sup\{\alpha : \alpha \in \mathsf{A}(t, \tau, v)\} = \inf_{v \in V} \alpha^*(t, \tau, v).$$

From the assumption that mapping $\mathsf{A}(t, \tau, v)$ is convex-valued and compact-valued there follows that on the set $\Delta \times V$

$$\mathsf{A}(t, \tau, v) = [\alpha_*(t, \tau, v), \alpha^*(t, \tau, v)].$$

Then the nonemptiness of the images of mapping $\mathsf{A}(t, \tau)$, $(t, \tau) \in \Delta$, means that $\mathsf{A}(t, \tau) = [\alpha_*(t, \tau), \alpha^*(t, \tau)]$, moreover

$$\alpha_*(t, \tau) = \sup_{v \in V} \alpha_*(t, \tau, v) \leq \inf_{v \in V} \alpha^*(t, \tau, v) = \alpha^*(t, \tau), (t, \tau) \in \Delta.$$

Thereby equality in the relation (11) is proved.

Let us introduce into consideration the numerical functions

$$\alpha^*(t) = \int_0^t \alpha^*(t, \tau) d\tau, \alpha_*(t) = \int_0^t \alpha_*(t, \tau) d\tau,$$

$$\delta(g(\cdot), \gamma(\cdot, \cdot)) = inf\{t : t \in \Theta(g(\cdot), \gamma(\cdot, \cdot))\},$$

where

$$\Theta(g(\cdot), \gamma(\cdot, \cdot)) = \{t > 0 : \alpha^*(t) \geq 1\}.$$

Theorem 2 *Let for the conflict-controlled process* (3)–(4) Conditions 1, 3 *hold, in addition $M = \text{co}M$, moreover, for the given function $g(\cdot)$ and the shift $\gamma(\cdot, \cdot)$*

$$T \in T(g(\cdot), \gamma(\cdot, \cdot)) \neq \emptyset.$$

Then, if $\alpha_*(T) < 1$ then a trajectory of the process (3) *can be brought to the terminal set* (4) *at the time instant T, by means of control of the form* (5)*, if, otherwise, $\alpha^*(T) \geq 1$, then by means of counter-control, for arbitrary admissible counteractions of the second player.*

Proof Let $v(\tau)$, $v : [0, T] \to V$ be a measurable function. Let us assume that $\alpha^*(T, \tau, v) \neq +\infty, \tau \in [0, T], v \in V$.

We consider the testing function

$$h(t) = 1 - \int_0^t \alpha^*(T, \tau, v(\tau)) d\tau - \int_t^T \alpha_*(T, \tau) d\tau, \tau \in [0, T].$$

As it was mentioned above, the function $\alpha^*(t, \tau, v)$ is $L \times B$-measurable in (τ, v), $\tau \in [0, T]$, $v \in V$, therefore, it is superpositionally measurable and, consequently, $\alpha^*(T, \tau, v(\tau))$ is a function, measurable in τ, $\tau \in [0, T]$. The function $\alpha_*(T, \tau)$ is also measurable in τ. Hence, $h(t)$ is absolutely continuous on the interval $[0, T]$. Since

$$h(0) = 1 - \int_0^T \alpha_*(T, \tau)d\tau = 1 - \alpha_*(T) > 0,$$

and, by definition of moment of T,

$$h(T) = 1 - \int_0^T \alpha^*(T, \tau, v(\tau))d\tau \leq 0,$$

then, by the well-known theorem of mathematical analysis, there exists a time instant $t_*, t_* = t(v(\cdot)), t_* \in [0, T]$, such that $h(t_*) = 0$. Note that the instant of switching t_* depends on control prehistory of the second player $v_{t_*}(\cdot)$.

We shall call the time intervals $[0, t_*)$ and $[t_*, T]$ "active" and "passive", respectively. Let us describe how the first player chooses its control on each of them. To this end, let us consider the compact-valued mappings

$$\begin{aligned} U_1(\tau, v) = \{u \in U : \pi\Omega(T, \tau)\phi(u, v) - \gamma(T, \tau) \in \\ \in \alpha^*(T, \tau, v)[M - \xi(T)]\}, \quad \tau \in [0, t_*) \\ U_2(\tau, v) = \{u \in U : \pi\Omega(T, \tau)\phi(u, v) - \gamma(T, \tau) \in \\ \in \alpha_*(T, \tau)[M - \xi(T)]\}, \quad \tau \in [t_*, T] \end{aligned} \tag{12}$$

From construction of the mappings $\mathsf{A}(T, \tau, v)$ and $\mathsf{A}(T, \tau)$ there follows that $U_i(\tau, v), i = 1, 2$, have nonempty images.

By virtue of the theorem on inverse image [29] the set-valued mappings $U_1(\tau, v)$, $U_2(\tau, v)$ are $L \times B$-measurable [30], therefore, by the theorem on measurable choice [29], in both there exist at least one $L \times B$- measurable selection $u_1(\tau, v), u_2(\tau, v)$, which are superpositionally measurable functions.

Let us denote $u_1(\tau) = u_1(\tau, v(\tau)), u_2(\tau) = u_2(\tau, v(\tau))$.

We set control of the first player on the "active" interval to be equal to $u_1(\tau)$, and on the "passive" one-to $u_2(\tau)$. Thus, despite the fact that the first player on each of the interval does not use directly prehistory of control of the second player but only its instantaneous control, in order to determine the switching instant t_* the prehistory is required .

Under chosen controls, we have from representation (3)

$$\begin{aligned} \pi z(T) &= \pi g(T) + \int_0^T \pi\Omega(T, \tau)\phi(u(\tau), v(\tau))d\tau \\ &= \pi g(T) + \int_0^{t_*} \pi\Omega(T, \tau)\phi(u_1(\tau), v(\tau))d\tau \end{aligned}$$

$$+ \int_{t_*}^{T} \pi\Omega(T,\tau)\phi(u_2(\tau), v(\tau))d\tau \pm \int_{0}^{T} \gamma(T,\tau)d\tau$$

Using the relations (12), from above formula we deduce

$$\pi z(T) \in \pi g(T) + \int_{0}^{t_*} \alpha^*(T,\tau,v(\tau))[M - \xi(T)]d\tau$$

$$+ \int_{t_*}^{T} \alpha_*(T,\tau)[M - \xi(T)]d\tau + \int_{0}^{T} \gamma(T,\tau)d\tau$$

$$= \xi(T)(1 - \int_{0}^{t_*} \alpha^*(T,\tau,v(\tau))d\tau - \int_{t_*}^{T} \alpha_*(T,\tau)d\tau)$$

$$+ \int_{0}^{t_*} \alpha^*(T,\tau,v(\tau))Md\tau + \int_{t_*}^{T} \alpha_*(T,\tau)Md\tau$$

$$= [\int_{0}^{t_*} \alpha^*(T,\tau,v(\tau))d\tau + \int_{t_*}^{T} \alpha_*(T,\tau)d\tau)]M = M.$$

Here the equality $h(t_*) = 0$ is twice taken into account, and transition of set-valued mappings with set M can be substituted by usage of the support functions technique [28, 34].

As follows from the expression (10), the case $\alpha^*(T,\tau,v) = +\infty$ for some $\tau \in [0,T]$, $v \in V$, is possible only under the conditions

$$0 \in M - \xi(T), 0 \in \Omega(T,\tau)\phi(U,v) - \gamma(T,\tau).$$

It is evident that in this case

$$\mathsf{A}(T,\tau,v) = [0,+\infty), \tau \in [0,T], v \in V,$$

and $\alpha_*(T,\tau) = 0$, $\tau \in [0,T]$.

This makes it possible to select as resolving functions at the points $\tau \in [0,T]$, where $\alpha^*(T,\tau,v(\tau)) = +\infty$, an arbitrary finite, superpositionally measurable function, which takes its values on the semi-infinite interval $[0,+\infty)$ with the only condition that final resolving function on the interval $[0,T]$ provides the relation $h(t_*) = 0$ for certain switching instant t_*, $t_* \in [0,T]$. Thereby construction of control on "active" and "passive" intervals is reduced to the previous case.

The case when $\alpha^*(T, \tau, v) = +\infty$ for all $\tau \in [0, T]$, $v \in V$, corresponds to the first Pontryagin method [1, 12]. Indeed, inclusion

$$0 \in \Omega(T, \tau)\phi(U, v) - \gamma(T, \tau) \quad \forall \tau \in [0, T], \quad v \in V,$$

provides non-emptiness of the set of images of Pontryagin's mapping $W(T, \tau)$ on the game interval $[0, T]$, and the shift function $\gamma(T, \tau)$ appears as measurable selection of the mapping $W(T, \tau)$, i.e., the Pontryagin selection.

The inclusion $0 \in M - \xi(T)$ implies the relation

$$\pi g(T) \in M - \int_0^T W(T, \tau)d\tau,$$

from which, by virtue of Theorem 1, there follows that the game (3)–(4) can be terminated in the class of stroboscopic strategies.

Let us consider separately the case $\alpha^*(T) \geq 1$, $\alpha_*(T) < 1$. We introduce the testing function

$$h_1(t) = 1 - \int_0^t \alpha^*(T, \tau)d\tau - \int_t^T \alpha_*(T, \tau)d\tau.$$

It is natural to examine only the case $\alpha^*(T, \tau) \neq +\infty$, $\tau \in [0, T]$. Then

$$h_1(0) = 1 - \alpha_*(T) > 0, h_1(T) = 1 - \alpha^*(T) \leq 0,$$

and, by virtue of continuity of the function $h_1(t)$, there exists a moment $t_*^1, t_*^1 \in [0, T]$, such that $h_1(t_*^1) = 0$. It should be noted that the time instant t_*^1 does not depend on $v(\cdot)$ in this case.

We consider set-valued mappings (12) on "active" and "passive" intervals $[0, t_*^1)$ and $[t_*^1, T]$, with the function $\alpha^*(T, \tau)$ replaced for $\alpha^*(T, \tau, v)$ in the expression for $U_1(\tau, v)$. Using the property for $L \times B$-measurability of compact-valued mappings $U_1(\tau, v)$, $U_2(\tau, v)$, we select in them $L \times B$-measurable selections, which determine admissible controls on both intervals. Final considerations are similar to the conclusions in previous case.

Let us introduce into consideration the functions

$$\alpha(t) = \int_0^t \inf_{v \in V} \alpha^*(t, \tau, v)d\tau,$$

$$\alpha(t, \tau) = \frac{1}{\alpha(t)} \inf_{v \in V} \alpha^*(t, \tau, v), 0 \leq \tau \leq t < +\infty.$$

Condition 4 For the chosen shift function $\gamma(t,\tau)$, $(t,\tau) \in \Delta$, the function $\inf_{v \in V} \alpha^*(t,\tau,v)$ is measurable in τ, $\tau \in [0,t]$, $t > 0$, and

$$\inf_{v(\cdot)\in V(\cdot)} \int_0^t \alpha^*(t,\tau,v)d\tau = \int_0^t \inf_{v\in V} \alpha^*(t,\tau,v)d\tau, t > 0.$$

Assumptions on the function $\alpha^*(t,\tau,v)$, ensuring fulfillment of the above equality are analyzed, for example, in [31].

Theorem 3 *Let for the game problem* (3)–(4) *with some shift function* $\gamma(t,\tau)$, $(t,\tau) \in \Delta$, $T \in T(g(\cdot), \gamma(\cdot,\cdot)) \neq \emptyset$, Conditions 1, 3, 4 *hold and the mapping* $\mathsf{A}(t,\tau,v)$ *be convex-valued on the set* $\Delta \times V$, $M = \text{co}M$. *Let us suppose that the following inequality is true:*

$$\alpha(T,\tau) \geq \sup_{v\in V} \alpha_*(T,\tau,v), \tau \in [0,T]. \tag{13}$$

Then a trajectory of the process (3) can be brought to the set (4) at the time-instant T with help of certain counter-control.

Proof It suffices to consider the case $\alpha^*(T,\tau,v) < +\infty$, $\tau \in [0,T]$, $v \in V$. Since, by virtue of the inequality in (4.2), $\alpha(T) \geq 1$, then

$$\alpha(T,\tau) = \frac{1}{\alpha(T)} \inf_{v\in V} \alpha^*(T,\tau,v) \leq \inf_{v\in V} \alpha^*(T,\tau,v), \tau \in [0,T],$$

and, thereby,

$$\alpha(T,\tau) \leq \alpha^*(T,\tau,v) \quad \forall v \in V, \tau \in [0,T].$$

Taking into account the inequality (13), one can draw conclusion that $\alpha(T,\tau) \in \mathsf{A}(T,\tau,v)$ for $v \in V$, $\tau \in [0,T]$, and, consequently, $\alpha(T,\tau) \in \mathsf{A}(T,\tau)$, $\tau \in [0,T]$.

Let us consider the set-valued mapping

$$\begin{aligned} U(\tau,v) = \{u \in U : \pi\Omega(T,\tau)\phi(u,v) - \gamma(T,\tau) \in \\ \in \alpha(T,\tau) \in \alpha(T,\tau)[M - \xi(T)]\}, \tau \in [0,T], v \in V \end{aligned} \tag{14}$$

Analogously to the previous case, the mapping $U(t,\tau)$ is compact-valued and $L \times B$-measurable, therefore, by the theorem on measurable choice [29], there exists a measurable selection $u(t,\tau)$, $u(t,\tau) \in U(t,\tau)$, which appears as superpositionally measurable function. Then, if $v(\cdot)$, $v(\cdot) \in V(\cdot)$, is an arbitrary admissible control of the second player, then we set control of the first player to be equal to

$$u(\tau) = u(\tau, v(\tau)), \tau \in [0,T].$$

From representation (3), with account of the inclusion in (14), we obtain

$$\pi z(T) \in \xi(T)[1 - \int_0^T \alpha(T,\tau)d\tau] + \int_0^T \alpha(T,\tau)Md\tau.$$

Since M is a convex compact, and $\alpha(t,\tau)$, $\tau \in [0,T]$, is a nonnegative function, moreover, $\int_0^T \alpha(T,\tau)d\tau = 1$, then $\int_0^T \alpha(T,\tau)Md\tau = M$ and, therefore, $\pi z(T) \in M$.

5 Scheme with Fixed Points of the Terminal Set Solid Part

One of the assumptions of the above mentioned theorems, concerning the method of resolving functions, is convexity of the set M (solid part of the terminal set). Now we provide one scheme of the method of resolving functions without this assumption, in which the targeting point of the set remains the same as time goes on.

For simplicity sake, we assume that Pontryagin's condition holds. It immediately follows that the lower resolving function equals zero and the upper one coincides with the ordinary resolving function [22]. As previously, in this case $\gamma(t,\tau)$ is a selection of the set-valued mapping $W(t,\tau)$. We fix some point m, $m \in M$, and set $\eta(t,m) = \xi(t) - m$, $t \geq 0$. Let us introduce a set-valued mapping

$$\mathrm{A}(t,\tau,v,m) = \{\alpha \geq 0 : -\alpha\eta(t,m) \in \pi\Omega(t,\tau)\varphi(U,v) - \gamma(t,\tau)\}$$

and its support function in the direction $+1$

$$\alpha(t,\tau,v,m) = \sup\{\alpha : \alpha \in \mathrm{A}(t,\tau,v,m)\}, t \geq \tau \geq 0, v \in V, m \in M.$$

Note that if $\eta(t,m) = 0$ then $\mathrm{A}(t,\tau,v,m) = [0,+\infty)$ for $\tau \in [0,t]$, $v \in V$, and, consequently, $\alpha(t,\tau,v,m) = +\infty$.

Let us consider the set

$$T(g(\cdot), m, \gamma(\cdot,\cdot)) = \{t > 0 : \inf_{v(\cdot)} \int_0^t \alpha(t,\tau,v(\tau),m)d\tau \geq 1\}$$

and its least element

$$t(g(\cdot), m, \gamma(\cdot,\cdot)) = \inf\{t > 0 : t \in T(g(\cdot), m, \gamma(\cdot,\cdot))\}.$$

under assumption that this element exists. If the function $t(g(\cdot), m, \gamma(\cdot,\cdot))$ is lower semi-continuous in m, $m \in M$, then it generates the marginal set

$$M(g(\cdot), \gamma(\cdot, \cdot)) = \{m \in M : t(g(\cdot), m, \gamma(\cdot, \cdot) = \bar{t}(g(\cdot), \gamma(\cdot, \cdot))\},$$

where

$$\bar{t}(g(\cdot), \gamma(\cdot, \cdot)) = \min_{m \in M} t(g(\cdot), m, \gamma(\cdot, \cdot)).$$

For the sake of comparison, note that [22]

$$t(g(\cdot), \gamma(\cdot, \cdot)) = \inf\{t > 0 : \inf_{v(\cdot)} \int_0^t \sup_{m \in M} \alpha(t, \tau, v(\tau), m) d\tau \geq 1\},$$

$$\bar{t}(g(\cdot), m, \gamma(\cdot, \cdot)) = \inf\{t > 0 : \sup_{m \in M} \inf_{v(\cdot)} \int_0^t \alpha(t, \tau, v(\tau), m) d\tau \geq 1\},$$

where $\sup_{m \in M} \alpha(t, \tau, v, m) = \alpha(t, \tau, v)$.

Theorem 4 *Let for the conflict-controlled process* (3)–(4) *Pontryagin's condition hold, and for given function $g(t)$, some selection $\gamma(t, \tau), \gamma(t, \tau) \in W(t, \tau)$ and point m, $m \in M$,*

$$T_m \in T(g(\cdot), m, \gamma(\cdot, \cdot)) \neq \emptyset.$$

Then the projection of a trajectory (3) *on the subspace L can be brought into the point m at the moment T_m with the help of certain control, prescribed by the quasi-strategy. If, in addition, $m \in M(g(\cdot), \gamma(\cdot, \cdot))$, then it will occur at the moment $\bar{t}(g(\cdot), \gamma(\cdot, \cdot))$.*

Without going into details of the proof we only describe the procedure of control construction. It should be mentioned that the line of reasoning is analogous, for example, to that of the proof of Theorem 2.

If $\eta(T_m, m) \neq 0$, then we divide the interval $[0, T_m]$ into the active and passive parts $[0, t_m)$ and $[t_m, T_m]$, where the moment of switching t_m is a root of the testing function

$$h_m(t) = 1 - \int_0^t \alpha(T_m, \tau, v(\tau), m) d\tau.$$

Then control of the first player on the active part $[0, t_m)$ is chosen in the form of superpositionally measurable selection of the set-valued mapping

$$\text{U}(\tau, v, m) = \{u \in U : \pi\Omega(T_m, \tau)\varphi(u, v) - \gamma(T_m, \tau) = -\alpha(T_m, \tau, v, m)\eta(T_m, m)\},$$

and on the passive part $[t_m, T_m]$—analogously, but with $\alpha(T_m, \tau, v, m) = 0$.

If, otherwise, $\eta(T_m, m) = 0$, then on all interval $[0, T_m]$ control of the first player is chosen in the same way as on the passive part.

6 Connection of the First Direct Method with the Method of Resolving Functions

Let us deduce relations between guaranteed times of the above mentioned methods.

Proposition 1 *Let the game problem* (3)–(4) *be given. Then for fulfillment of Pontryagin's condition: $W(t, \tau) \neq \emptyset$, $(t, \tau) \in \Delta$, it is necessary and sufficient that there exists a shift function $\gamma(t, \tau)$ such that*

$$0 \in \mathsf{A}(T, \tau, v) \quad \forall (t, \tau) \in \Delta, v \in V.$$

Proof Let the condition $W(t, \tau) \neq \emptyset$, $(t, \tau) \in \Delta$, hold. Then, by virtue of closed-valuedness and measurability of the mapping $W(t, \tau)$, from the theorem on measurable choice [29] there follows that there exists a measurable selection $\gamma(t, \tau)$, $\gamma(t, \tau) \in W(t, \tau)$. This implies that

$$0 \in W(t, \tau) - \gamma(t, \tau) \quad \forall (t, \tau) \in \Delta$$

or

$$0 \in W(T, \tau, v) - \gamma(t, \tau) \quad \forall (t, \tau) \in \Delta, v \in V.$$

Thereby, zero value of α in expression (9) provides non-emptiness of the sets intersection and, therefore,

$$0 \in \mathsf{A}(T, \tau, v) \quad \forall (t, \tau) \in \Delta, v \in V. \tag{15}$$

Consideration in inverse order results in the required conclusion.

Thus, within the framework of Proposition 1 the shift function $\gamma(t, \tau)$ appears as the Pontryagin selection. We denote the set of such selections by Γ. In addition, we have that

$$0 \in \mathsf{A}(t, \tau) \quad \forall (t, \tau) \in \Delta$$

and the corresponding lower resolving functions are

$$\alpha_*(t, \tau, v) = \alpha_*(t, \tau) = 0 \quad \forall (t, \tau) \in \Delta, v \in V. \tag{16}$$

Proposition 2 *Let for some* $t > 0$ $W(t,\tau) \neq \emptyset$, $\tau \in [0, t]$. *In this case the inclusion*

$$\pi g(t) \in M - \int_0^t W(t,\tau)d\tau \tag{17}$$

holds if and only if there exists a measurable in τ *Pontryagin's selection, such that*

$$\xi(t, g(t), \gamma(t,\cdot)) \in M.$$

Proof Let the inclusion (17) hold. Then, by definition of the Auman integral there exists a Pontryagin selection, such that

$$\pi g(t) + \int_0^t \gamma(t,\tau)d\tau = \xi(t) \in M. \tag{18}$$

Conversely, if for some Pontryagin's selection the relation (18) is true, then by transferring the integral of selection into the right-hand side of (18) we, all the more, get the inclusion (17).

Thus, if for some time instant t and the chosen Pontryagin's selection the inclusion (18) holds, then

$$\mathsf{A}(t,\tau,v) = [0, +\infty) \quad \forall (t,\tau) \in \Delta, v \in V.$$

Thereby

$$\mathsf{A}(t,\tau) = [0, +\infty), (t,\tau) \in \Delta.$$

Hence, in this case the corresponding upper resolving functions of both types coincide:

$$\alpha^*(t,\tau,v) = \alpha^*(t,\tau) = +\infty, (t,\tau) \in \Delta, v \in V.$$

and the corresponding lower resolving functions equal zero.

Before comparing guaranteed times of the considered methods let us deduce so-called functional form [22] of the first direct method, expressed through special resolving functions.

Let for conflict-controlled process (3), (4) Pontryagin's condition hold. We analyze the set-valued mapping

$$\mathrm{B}(t,\tau) = \{\beta \geq 0 : [W(t,\tau) - \gamma(t,\tau)] \cap \beta[M - \xi(t)] \neq \emptyset\},$$

where $\gamma(t, \tau)$ is Pontryagin's selection which is measurable in τ, $\tau \in [0, t]$. Its support function in the direction $+1$ is

$$\beta^*(t, \tau) = sup\{\beta : \beta \in \mathrm{B}(t, \tau)\}, t \geq \tau \geq 0.$$

Since $0 \in \mathrm{B}(t, \tau)$ for $(t, \tau) \in \Delta$, by virtue of the Pontryagin condition, then

$$\beta_*(t, \tau) = \inf\{\beta : \beta \in \mathrm{B}(t, \tau)\} = 0, t \geq \tau \geq 0.$$

If $\xi(t) \bar{\in} M$, then, by the theorem on characterization and inverse image [29], the mapping $\mathrm{B}(t, \tau)$ is closed-valued and measurable in τ, $\tau \in [0, t]$. Correspondingly, on the basis of the theorem on support function [29], the function $\beta^*(t, \tau)$ is also measurable in τ, $\tau \in [0, t]$. In the case $\xi(t) \in M$ for some $t > 0$

$$\mathrm{B}(t, \tau) = [0, +\infty), \tau \in [0, t],$$

and $\beta^*(t, \tau) = +\infty$ for all $\tau \in [0, t]$.

The resolving function $\beta^*(t, \tau)$ generates the guaranteed time

$$p_1(g(\cdot), \gamma(\cdot, \cdot)) = \inf\{t \geq 0 : \int_0^t \beta^*(t, \tau) d\tau \geq 1\}. \quad (19)$$

Theorem 5 *Let for conflict-controlled process* (3)–(4) *the Pontryagin condition hold,* $M = \mathrm{co}M$ *and let for the function* $g(\cdot)$ *and the Pontryagin selection* $\gamma(\cdot, \cdot)$

$$p_1 = p_1(g(\cdot), \gamma(\cdot, \cdot)) < +\infty,$$

with infinum in the relation (19) *attained.*

Then a trajectory of the process (3) can be brought into the terminal set (4) at the time instant p_1 by some counter-control, therewith

$$\inf_{\gamma(\cdot,\cdot) \in \Gamma} p_1(g(\cdot), \gamma(\cdot, \cdot)) = p(g(\cdot)). \quad (20)$$

Proof of the first part of statement follows from the proof of Theorem 2, and the equality (20) is obtained in the case of differential games in [34, Theorem 6]. In the case under study the proof is similar.

From the above provided constructions, under assumptions of Theorem 1, 2, it readily follows the inequality

$$\inf_{\gamma(\cdot,\cdot) \in \Gamma} t(g(\cdot), \gamma(\cdot, \cdot)) \leq p(g(\cdot)). \quad (21)$$

The case of equality (21) is highlighted by the following statement.

Theorem 6 *Let for the conflict-controlled process* (3), (4) *Pontryagin's condition hold and let us suppose that for some shift function the mapping* $\mathsf{A}(t, \tau, v), (t, \tau) \in \Delta$, $v \in V$, *is convex-valued, besides set M is convex.*

Then, if for some T, $T \in T(g(\cdot), \gamma(\cdot, \cdot)) \neq \emptyset$ *the condition*

$$\bigcap_{v \in V} [W(T, \tau, v) - \alpha(T, \tau)M] = W(T, \tau) - \alpha(T, \tau)M, 0 \leq \tau \leq T, \tag{22}$$

holds, then

$$\inf_{\gamma(\cdot,\cdot)} t(g(\cdot), \gamma(\cdot, \cdot)) = \inf_{\gamma(\cdot,\cdot)} \delta(g(\cdot), \gamma(\cdot, \cdot)) = p(g(\cdot)).$$

Proof It will suffice to prove the inequality

$$p(g(\cdot)) \leq \inf_{\gamma(\cdot,\cdot) \in \Gamma} t(g(\cdot), \gamma(\cdot, \cdot)). \tag{23}$$

Since $\alpha(T, \tau) \in \mathsf{A}(T, \tau, v), (t, \tau) \in \Delta, v \in V$, then, by virtue of the property for convex-valuedness of the mapping $\mathsf{A}(T, \tau, v)$, we have from the relation (9)

$$[W(T, \tau, v) - \gamma(T, \tau)] \cap \alpha(T, \tau)[M - \xi(T)] \neq \emptyset, 0 \leq \tau \leq T, v \in V.$$

The latter is equivalent to the inclusion

$$0 \in W(T, \tau, v) - \alpha(T, \tau)M + \alpha(T, \tau)\xi(T) - \gamma(T, \tau), \tau \in [0, T], v \in V,$$

or

$$0 \in \bigcap_{v \in V} [W(T, \tau, v) - \alpha(T, \tau)M] + \alpha(T, \tau)\xi(T) - \gamma(T, \tau), \tau \in [0, T].$$

Taking into account the condition (22) we get

$$0 \in W(T, \tau) - \alpha(T, \tau)M + \alpha(T, \tau)\xi(T) - \gamma(T, \tau), \tau \in [0, T].$$

Upon integration from 0 to T both parts of this inclusion, taking into account the condition $M = \text{co}M$, we have

$$\pi g(T) \in M - \int_0^T W(T, \tau) d\tau.$$

This implies the inequality (23) and, therefore, the theorem is proved.

Applications of approach methods for conflict-controlled processes to design of control systems are presented, for example, in [13, 14].

Illustrative example

Dynamics of conflict-controlled process is defined by the system of equations

$$\dot{z} = \lambda z + u - v, z \in R^n, z(0) = z_0, \lambda \in R^n, \tag{24}$$

and control parameters are subject to the constraints

$$\|u\| \le a \ge 1, \quad \|v\| \le 1.$$

This means that in the general scheme $A = \lambda E$, E- unit quadratic matrix of order n, control block is $\varphi(u, v) = u - v$, and control domains are $U = aS = S_a$, $V = S$, where $S = \{z : \|z\| \le 1\}$ is a unit ball centered at the origin.

Terminal set is given by the relationship $\|z\| \le \varepsilon$. It presents itself the ε—neighbourhood of the origin and in the case of separated motions of the players, corresponds to the approach of the pursuer with the evader at a distance of ε. Hence,

$$M^* = M = \varepsilon S, M_0 = \{0\}.$$

The orthogonal complement L to M_0 in R^n is all space R^n, $L = R^n$, and the operator of orthogonal projecting $\pi : R^n \to L$, is the operator of identical transformation, given by the unit matrix E. The fundamental matrix $e^{At} = e^{\lambda t} E$. Then the set-valued mapping $W(t) = e^{\lambda t} \, [aS \underline{*} S] = e^{\lambda t}(a - 1)S \neq \emptyset$.

Since $a \ge 1$ Pontryagin's condition holds and W(t) is a ball of radius $e^{\lambda t}(a - 1)$ centered at the origin. The time, provided by the first direct method, is determined by the inclusion

$$e^{\lambda t} z_0 \in \varepsilon S - \int_0^t e^{\lambda t}(a - 1)S d\tau \tag{25}$$

In view of the properties of Aumann integral [30] of set-valued mapping with spherical images,

$$\int_0^t e^{\lambda \tau}(a - 1)S d\tau = \begin{cases} \frac{1}{\lambda}(e^{\lambda t} - 1)(a - 1)S, \lambda \neq 0, \\ t(a - 1)S, \lambda = 0. \end{cases}.$$

Since the images enjoy the property for central symmetry, the minus sign before integral in (25) must be replaced by the plus sign. Then the inclusion (25) takes the form ($\lambda \neq 0$)

$$e^{\lambda t} z_0 \in [\varepsilon + \frac{1}{\lambda}(e^{\lambda t} - 1)(a - 1)]S. \tag{26}$$

Of particular interest is the least time, at which the inclusion (26) holds. By virtue of continuity of the left-hand part and the radius of ball in the right-hand part of inclusion (26), this time is determined by the equation

$$\left\| e^{\gamma t} z_0 \right\| = \varepsilon + \frac{1}{\lambda}(e^{\lambda t} - 1)(a - 1), \tag{27}$$

that corresponds to the first instant of the vector $e^{\lambda t} z_0$ hitting the boundary of the ball $[\varepsilon + \frac{1}{\lambda}(e^{\lambda t} - 1)(a - 1)]S$. From Eq. (27) we deduce that

$$e^{\lambda t} = \frac{\varepsilon - \frac{1}{\lambda}(a-1)}{\|z_0\| - \frac{1}{\lambda}(a-1)} = \frac{\lambda\varepsilon - (a-1)}{\lambda\|z_o\| - (a-1)},$$

hence

$$t = \frac{1}{\lambda} \ln \frac{\lambda\varepsilon - (a-1)}{\lambda\|z_0\| - (a-1)} = \min\{t > 0 : t \in P(z_0)\} \tag{28}$$

Since $\|z_0\| > \varepsilon$, then this time is positive only for $\lambda < 0$. Even in the case of equal resources ($a = 1$) for $\lambda < 0$ the time (28) is finite if $\varepsilon \neq 0$. If, otherwise $\varepsilon = 0$, then the trajectory hits the origin only at the time $+\infty$. Note that for $\lambda < 0$ the constraint $0 < \frac{\lambda\varepsilon-(a-1)}{\lambda\|z_0\|-(a-1)} < 1$ should hold that imposes condition that the numerator and the denominator should be of the same sign. Because the denominator is negative, the numerator should be negative too: $\lambda\varepsilon - (a - 1) < 0$. But this is always the case if $\lambda < 0$.

If $\lambda = 0$ from (25) there follows the inclusion

$$z_0 \in \varepsilon S + t(a-1)S = [\varepsilon + t(a-1)]S.$$

The first moment of hitting is defined by the equation $\|z_0\| = \varepsilon + t(a-1)$, whence we have $t = \frac{\|z_0\|-\varepsilon}{a-1}$.

Note, that when $\lambda > 0$, in order time t be positive the inequality in (28) $\frac{\lambda\varepsilon-(a-1)}{\lambda\|z_0\|-(a-1)} > 1$ should be fulfilled that leads to the inequality $\varepsilon > \|z_0\|$, that contradicts the initial assumption $\|z_0\| > \varepsilon$. Thus, in this case it is impossible to terminate the game (24) in a finite time. The selection of Pontryagin's mapping $W(t, \tau)$ has the form

$$\gamma(t, \tau) = -e^{\lambda(t-\tau)}(a-1)\frac{z_0}{\|z_0\|}.$$

It defines the counter-control of the pursuer

$$u(\tau) = v(\tau) - (a-1)\frac{z_0}{\|z_0\|},$$

which is performing parallel ε-approach at the moment t in the point $\varepsilon \frac{z_0}{\|z_0\|}$.

In fact, upon substitution this control into representation of the trajectory in the form of Cauchy formula we obtain

$$z(t) = e^{\lambda t} z_0 - \int_0^t e^{\lambda t}(a-1)\frac{z_0}{\|z_0\|}d\tau$$
$$= \frac{\lambda\varepsilon - (a-1)}{\lambda\|z_0\| - (a-1)} z_0 - \frac{1}{\lambda}(\frac{\lambda\varepsilon - (a-1)}{\lambda\|z_0\| - (a-1)} - 1)(a-1)\frac{z_0}{\|z_0\|}$$
$$= \frac{z_0}{\|z_0\|}[\frac{\lambda\varepsilon - (a-1)}{\lambda\|z_0\| - (a-1)}\|z_0\| - \frac{1}{\lambda}(\frac{\lambda\varepsilon - \lambda\|z_0\|}{\lambda\|z_0\| - (a-1)})](a-1)$$
$$= \frac{z_0}{\|z_0\|}[\frac{(\lambda\varepsilon - (a-1))\|z_0\| - (\varepsilon - \|z_0\|)(a-1)}{\lambda\|z_0\| - (a-1)}]$$
$$= \frac{z_0}{\|z_0\|}\frac{\varepsilon(\lambda\|z_0\| - (a-1))}{\lambda\|z_0\| - (a-1)} = \varepsilon\frac{z_0}{\|z_0\|}.$$

Parallelism of the ε- approach follows from the trajectory representation

$$z(t) = [e^{\lambda t} + \int_0^t e^{\lambda\tau}\frac{a-1}{\|z_0\|}d\tau]z_0, \forall t \geq 0.$$

Let us apply the apparatus of resolving functions to solving example (24). Since the mapping $W(t,\tau)$ has the form of a ball centered at the zero then Pontryagin's condition is fulfilled. That is why, we choose in $W(t,\tau)$ a selection coinciding with the origin, i.e. $\gamma(t) \equiv 0$. Then $\xi(t) = e^{\lambda t} z_0$, the set-valued mapping A has the form

$$\mathrm{A}(t,\tau,v) = \{\alpha \geq 0 : e^{\lambda(t-\tau)}(aS - v) \cap \alpha[\varepsilon S - e^{\lambda t} z_0] \neq \emptyset\},$$

and the resolving function

$$\alpha(t,\tau,v) = \sup\{\alpha \geq 0 : ae^{\lambda t} z_0 - e^{\lambda(t-\tau)} v \in [\alpha\varepsilon + e^{\lambda(t-\tau)} a]S\},$$

i.e. $\mathrm{A}(t,\tau,v) = [0, \alpha(t,\tau,v)]$ is interval on the half-axis R_+.

The resolving function satisfies the equality

$$\left\|\alpha e^{\lambda t} z_0 - e^{\lambda(t-\tau)} v\right\| = \alpha\varepsilon + e^{\lambda(t-\tau)} a,$$

that can be reduced to the quadratic equation for α,

$$(\alpha e^{\lambda t} z_0 - e^{\lambda(t-\tau)} v, \alpha e^{\lambda t} z_0 - e^{\lambda(t-\tau)} v) = [\alpha\varepsilon + e^{\lambda(t-\tau)} a]^2,$$

or, in conventional form,

$$[e^{2\lambda t}\|z_0\|^2 - \varepsilon^2]\alpha^2 - 2[(e^{\lambda t}z_0, e^{\lambda(t-\tau)}v) \\ + \varepsilon a e^{\lambda(t-\tau)}]\alpha + e^{2\lambda(t-\tau)}(\|v\|^2 - a^2) = 0.$$

Its greatest positive root appears as the resolving function

$$\alpha(t,\tau,v) = \frac{(e^{\lambda t}z_{0,} e^{\lambda(t-\tau)}v)}{e^{2\lambda t}\|z_0\|^2 - \varepsilon^2} \\ + \frac{\sqrt{[(e^{\lambda t}z_0, e^{\lambda(t-v)}v]^2 + (e^{2\lambda t}\|z_0\|^2 - \varepsilon^2)e^{2\lambda(t-\tau)}(a^2 - \|v\|^2)}}{e^{2\lambda t}\|z_0\|^2 - \varepsilon^2}.$$

Taking into account that $\min\limits_{\|v\|\le 1} \alpha(t,\tau,v)$ is attained at $v = -\frac{z_0}{\|z_0\|}$, we obain

$$\min_{\|v\|\le 1} \alpha(t,\tau,v) = \frac{-e^{\lambda(2t-\tau)}\|z_0\| + \varepsilon a e^{\lambda(t-\tau)} + \left|e^{\lambda(2t-\tau)}\|z_0\|a - \varepsilon e^{\lambda(t-\tau)}\right|}{e^{2\lambda t}\|z_0\|^2 - \varepsilon^2}. \quad (29)$$

Let us consider the function $e^{\lambda t}\|z_0\| - \varepsilon = f(t)$—for $\lambda < 0$. It is evident that $f(0) > 0$. As time grows this function is decreasing and turns into zero at $\bar{t} = \frac{1}{\lambda}\ln\frac{\varepsilon}{\|z_0\|} > 0$.

This time guarantees the game termination, if $a = 1$, according to the first direct method, with $\alpha(t,\tau,v) \equiv +\infty$. On the interval $[0,\bar{t}]$ $e^{\lambda t}\|z_0\|a - \varepsilon > 0$ if $a > 1$, that is why, removing the modulo sing in (29), we obtain

$$\min_{\|v\|\le 1} \alpha(t,\tau,v) = \frac{e^{\lambda(2t-\tau)}\|z_0\|(a-1) + \varepsilon e^{\lambda(t-\tau)}(a-1)}{e^{2\lambda t}\|z_0\|^2 - \varepsilon^2} = \frac{(a-1)e^{\lambda(t-\tau)}}{e^{\lambda t}\|z_0\| - \varepsilon}$$

in view of the above function continuity in t. Hence, the minimal time of game termination, according to the method of resolving functions, is defined by the relationship

$$\int_0^t (a-1)e^{\lambda\tau}d\tau = e^{\lambda t}\|z_0\| - \varepsilon.$$

Then

$$\frac{a-1}{\lambda}(e^{\lambda t} - 1) = e^{\lambda t}\|z_0\| - \varepsilon, e^{\lambda t} = \frac{\lambda\varepsilon - (a-1)}{\lambda\|z_0\| - (a-1)},$$

$$t = \frac{1}{\lambda}\ln\frac{\lambda\varepsilon - (a-1)}{\lambda\|z_0\| - (a-1)} = t(z_0, 0). \quad (30)$$

For $\lambda < 0$ it is positive, for $\lambda = 0$ $t = \frac{\|z_0\| - \varepsilon}{a-1}$, just like in Pontryagin's first direct method. Taking into account the expression for $B(t,\tau,v)$, we find the targeting points in the set εS and through them the control of pursuer in an explicit form. Let us consider the set-valued mapping

$$\mathrm{M}(t,\tau,v)=\{m\in\varepsilon S:\alpha(t,\tau,v)(m-e^{\lambda t}z_0)\in e^{\lambda(t-\tau)}(aS-v)\},$$

$$t\geq\tau\geq 0,\, v\in S.$$

In this example it consists of a unique selection

$$m(t,\tau,v)=\varepsilon\frac{\alpha(t,\tau,v)e^{\lambda t}z_0-e^{\lambda(t-\tau)}v}{\left\|\alpha(t,\tau,v)e^{\lambda t}z_0-e^{\lambda(t-\tau)}v\right\|}. \tag{31}$$

In fact, the set-valued mappings $\alpha(t,\tau,v)(\varepsilon S-e^{\lambda t}z_0)$ and $e^{\lambda(t-\tau)}(aS-v)$ are closed and convex-valued. That their images intersect means that their difference contains zero. We evaluate the support functions of the left-hand and the right-hand part of this inclusion and obtain following inequality

$$\begin{aligned}&\alpha(t,\tau,v)\varepsilon\|p\|+(e^{\lambda(t-\tau)}v-e^{\lambda t}\alpha(t,\tau,v)z_0,p)+e^{\lambda(t-\tau)}a\|p\|\geq 0,\\&\forall p,\|p\|=1.\end{aligned} \tag{32}$$

Minimum in p of its left-hand side is attained at the element

$$p=\frac{e^{\lambda t}\alpha(t,\tau,v)z_0-e^{\lambda(t-\tau)}v}{\left\|e^{\lambda t}\alpha(t,\tau,v)z_0-e^{\lambda(t-\tau)}v\right\|}.$$

From this it follows that the selection of the mapping $\mathrm{M}(t,\tau,v)$, at which maximum of support function of mapping $\alpha(t,\tau,v)\varepsilon S$ is attained, with account of the inequality (32), has the form (31).

Then the control, realizing the guaranteed time (30), has the form

$$u(\tau)=e^{-\lambda(T-\tau)}\{e^{\lambda(T-\tau)}v(\tau)+\bar{\alpha}(T,\tau,v(\tau))[m(T,\tau,v(\tau))-e^{\lambda T}z_0]\},$$

$$T=t(z_0,0),$$

where

$$\bar{\alpha}(T,\tau,v(\tau))=\begin{cases}\alpha(T,\tau,v),\,\tau\in[0,t_*),\\0,\,\tau\in[t_*,T],\end{cases}$$

and

$$m(T,\tau,v)=\varepsilon\frac{\bar{\alpha}(T,\tau,v)e^{\lambda T}z_0-e^{\lambda(T-\tau)}v}{\left\|\bar{\alpha}(T,\tau,v)e^{\lambda T}z_0-e^{\lambda(T-\tau)}v\right\|}.$$

Here the time t_* divides the active and passive parts and appears as a root of the testing function

$$h(t) = 1 - \int_0^t \alpha(T, \tau, v(\tau))d\tau, t > 0.$$

Since there exists a moment of switching, then a prehistory of control of the evader is used and, generally, the pursier applies control prescribed by a quasi-strategy. This control appears as counter-control both on the active and passive parts.

On other hand, because the mapping $\mathrm{A}(t, \tau, v)$ is convex-valued, then the game can be terminated in the class of counter-controls. Really, since

$$\alpha(T) = \int_0^T \inf_{\|v\| \leq 1} \alpha(T, \tau, v)d\tau = \int_0^T \frac{(a-1)e^{\lambda(T-\tau)}}{e^{\lambda T}\|z_0\| - \varepsilon} d\tau = 1.$$

then

$$\alpha^*(T, \tau) = \inf_{\|v\| \leq 1} \alpha(T, \tau, v) = \frac{(a-1)e^{\lambda(T-\tau)}}{e^{\lambda T}\|z_0\| - \varepsilon}.$$

Thus, there does not exist a moment of switching if the testing function is treated in the form

$$h_0(t) = 1 - \int_0^t \alpha^*(T, \tau)d\tau,$$

because $h_0(T) = 1$.

Control of the pursuer has the form

$$u(\tau) = e^{-\lambda(T-\tau)}\{e^{\lambda(T-\tau)}v(\tau) + \alpha^*(T, \tau)[m_1(T, \tau, v(\tau)) - e^{-\lambda T}z_0]\},$$

where

$$m_1(T, \tau, v) = \varepsilon \frac{\alpha^*(T, \tau)e^{\lambda T}z_0 - e^{\lambda(T-\tau)}v}{\left\|\alpha^*(T, \tau)e^{\lambda T}z_0 - e^{\lambda(T-\tau)}v\right\|}.$$

It is easy to see that

$$\mathrm{A}(t, \tau) = \bigcap_{v \in S} \mathrm{A}(t, \tau, v) = [0, \alpha(t, \tau)] = [0, \alpha^*(t, \tau)],$$

since

$$\alpha(t, \tau) = \alpha^*(t, \tau) = \frac{(a-1)e^{\lambda(t-\tau)}}{e^{\lambda t}\|z_0\| - \varepsilon}$$

and, thus, Theorem 2 is illustrated by the above provided calculations.

Now we proceed to analysis of the method of resolving functions with fixed points of the set εS. As before, selection $\gamma(t) \equiv 0$. Then $\eta(t, m) = e^{\lambda t} z_0 - m$ and the set-valued mapping

$$\mathrm{A}(t, \tau, v, m) = \{\alpha \geq 0 : -\alpha(e^{\lambda t} z_0 - m) \in e^{\lambda(t-\tau)}(aS - v)\}.$$

The support function of its images in direction $+1$, (the resolving function) satisfies the equality

$$\left\| e^{\lambda(t-\tau)} v - \alpha(m - e^{\lambda t} z_0) \right\| = e^{\lambda(t-\tau)} a,$$

that implies the corresponding quadratic equation for α. Its greatest positive root has the form

$$\alpha(t, \tau, v, m) = \frac{(e^{\lambda(t-\tau)} v, e^{\lambda t} z_0 - m)}{\left\| e^{\lambda t} z_0 - m \right\|^2} + \frac{\sqrt{((e^{\lambda(t-\tau)} v, e^{\lambda t} z_0 - m)^2 + \left\| e^{\lambda t} - m \right\|^2 e^{2\lambda(t-\tau)}(a^2 - \|v\|^2)}}{\left\| e^{\lambda t} z_0 - m \right\|^2}.$$

From this it follows that

$$\min_{\|v\| \leq 1} \alpha(t, \tau, v, m) = \frac{e^{\lambda(t-\tau)}(a-1)}{\left\| e^{\lambda t} z_0 - m \right\|},$$

with minimum attained at the element $v = -\frac{e^{\lambda t} z_0 - m}{\left\| e^{\lambda t} z_0 - m \right\|}$.

Then, in this scheme the least time of the game termination is

$$t(z_0, 0) = \min\{t > 0 :$$

$$\max_{m \in \varepsilon S} \int_0^t \frac{e^{\lambda(t-\tau)}(a-1)}{\left\| e^{\lambda t} z_0 - m \right\|} d\tau = 1\} = \min\{t > 0 : \frac{\int_0^t e^{\lambda(t-\tau)}(a-1) d\tau}{\min_{m \in \varepsilon S} \left\| e^{\lambda t} z_0 - m \right\|} = 1\}.$$

Minimum in the denominator is attained at the element $m = \varepsilon \frac{z_0}{\|z_0\|}$.

Hitting the given point m will occur at the moment

$$t_m = \min\{t > 0 : \frac{1}{\lambda}(e^{\lambda t} - 1)(a-1) = \left\| e^{\lambda t} z_0 - m \right\|\},$$

with the help of the control

$$u(\tau) = e^{-\lambda(t_m-\tau)}\{e^{\lambda(t_m-\tau)}v(\tau) - \bar{\alpha}(t_m, \tau, v(\tau), m)\eta(t_m, m)\},$$

where

$$\bar{\alpha}(t, \tau, v, m) = \begin{cases} \alpha(t_m, \tau, v, m), \tau \in [0, t_m^*), \\ 0, \tau \in [t_m^*, t_m], \end{cases}$$

and t_m^* is a zero of the testing function $h_m(t) = 1 - \int\limits_0^t \alpha(t_m, \tau, v(\tau), m)d\tau$.

Now we examine the case in example (24), when the control domain U presents itself a boundary of the ball S_a, i.e. $U = \partial S_a$. Then Pontryagin's condition does not hold because

$$W(t, \tau) = e^{\lambda(t-\tau)}[\partial S_a \overset{*}{-} S] = \emptyset.$$

Let us set the shift function $\gamma(t, \tau)$ equal to zero, $\xi_0(t) = e^{\lambda t} z_0$.

The set-valued mapping

$$\mathrm{A}(t, \tau, v) = \{\alpha \geq 0 : e^{\lambda(t-\tau)}(\partial S_a - v) \cap \alpha[\varepsilon S - e^{\lambda t} z_0] \neq \emptyset\}$$

has non-empty images when $t \geq \tau \geq 0$, $v \in S$ and therefore Condition 3 is fulfilled.

From geometric considerations it is easy to see that the upper resolving function of the first type $\alpha^*(t, \tau, v)$ coincides with the resolving function $\alpha(t, \tau, v)$ and, therefore, the moment of the game termination is

$$T = T(z_0, 0) = \begin{cases} \frac{1}{\lambda} \ln \frac{\lambda\varepsilon-(a-1)}{\lambda\|z_0\|-(a-1)}, \lambda < 0 \\ \frac{\|z_0\|-\varepsilon}{a-1}, \lambda = 0 \end{cases} \tag{33}$$

Let us deduce formula for the resolving function of first type:

$$\begin{aligned} \alpha_*(t, \tau, v) &= \inf\{\alpha \geq 0 : e^{\lambda(t-\tau)}(\partial S_a - v) \cap \alpha[\varepsilon S - e^{\lambda t} z_0] \neq \emptyset \\ &= \sup\{\alpha \geq 0 : \alpha[\varepsilon S - e^{\lambda t} z_0] \subset e^{\lambda(t-\tau)}(aS - v)\} \\ &= \sup\{\alpha \geq 0 : \left\|\alpha e^{\lambda t} z_0 - e^{\lambda(t-\tau)} v\right\| = e^{\lambda(t-\tau)} \cdot a - \alpha\varepsilon\} \end{aligned}$$

The greatest positive root of corresponding quadratic equation

$$\begin{aligned} &[e^{2\lambda t}\|z_0\|^2 - \varepsilon^2]\alpha^2 - 2[(e^{\lambda t} z_0, e^{\lambda(t-\tau)} v) - \varepsilon a e^{\lambda(t-\tau)}]\alpha \\ &+ e^{2\lambda(t-\tau)}(\|v\|^2 - a^2) = 0 \end{aligned}$$

has the form

$$\alpha_*(t, \tau, v) = \frac{(e^{\lambda t} z_0, e^{\lambda(t-\tau)} v) - \varepsilon a e^{\lambda(t-\tau)}}{e^{2\lambda t}\|z_0\|^2 - \varepsilon^2}$$

$$+\frac{\sqrt{[(e^{\lambda t}z_0, e^{\lambda(t-\upsilon)}v) - \varepsilon a e^{\lambda(t-\tau)}]^2 + (e^{2\lambda t}\|z_0\|^2 - \varepsilon^2)e^{2\lambda(t-\tau)}(a^2 - \|v\|^2)}}{e^{2\lambda t}\|z_0\|^2 - \varepsilon^2}.$$

Then
$\text{A}(t,\tau,v) = [\alpha_*(t,\tau,v), \alpha^*(t,\tau,v)]$ and $0 \notin \text{A}(t,\tau,v)$.
The equality $\alpha_*(t,\tau,v) = 0$ means that Pountryagin's condition is fulfilled with

$$\sup_{\|v\|\le 1} \alpha_*(t,\tau,v) = \frac{(a+1)e^{\lambda(t-\tau)}}{e^{\lambda t}\|z_0\| + \varepsilon},$$

attained at $v = \frac{z_0}{\|z_0\|}$. Then

$$\text{A}(t,\tau) = \bigcap_{\|v\|\le 1} \text{A}(t,\tau,v) = [\frac{(a+1)e^{\lambda(t-\tau)}}{e^{\lambda t}\|z_0\| + \varepsilon}, \frac{(a-1)e^{\lambda(t-\tau)}}{e^{\lambda t}\|z_0\| - \varepsilon}]$$
$$= e^{\lambda(t-\tau)}[\frac{a+1}{e^{\lambda t}\|z_0\| + \varepsilon}, \frac{a-1}{e^{\lambda t}\|z_0\| - \varepsilon}]$$

and Condition 3, implying that $\text{A}(t,\tau) \neq \emptyset$, $t \ge \tau \ge 0$, is fulfilled if $\frac{(a+1)}{e^{\lambda t}\|z_0\|+\varepsilon} \le \frac{(a-1)}{e^{\lambda t}\|z_0\|-\varepsilon}$, that leads to the inequality $e^{\lambda t}\|z_0\| - a\varepsilon \le 0$.
In this case the upper and the lower resolving functions of second type are

$$\alpha^*(t,\tau) = \frac{e^{\lambda(t-\tau)}(a-1)}{e^{\lambda t}\|z_0\| - \varepsilon}, \quad \alpha_*(t,\tau) = \frac{e^{\lambda(t-\tau)}(a+1)}{e^{\lambda t}\|z_0\| + \varepsilon}.$$

Since $\alpha^*(T) = 1$ and in case $e^{\lambda T}\|z_0\| - a\varepsilon < 0$,

$$\alpha_*(T,\tau) < \alpha^*(T,\tau),$$

then $\alpha_*(T) < 1$ and from Theorem 4 there follows that the game can be terminated in the time T, given by the Eq. (33).

This process is realized with help of the pursuer control

$$u(\tau) = e^{-\lambda(T-\tau)}\{e^{\lambda(T-\tau)}v(\tau) + \overset{\approx}{\alpha}(T,\tau,v(\tau))[m(T,\tau,v(\tau)) - e^{\lambda T}z_0]\}, \tag{34}$$

where

$$\tilde{\tilde{\alpha}}(T,\tau,v) = \begin{cases} \alpha(T,\tau,v), \ \tau \in [0, t_*^1), \\ \alpha_*(T,\tau), \ \tau \in [t_*^1, T], \end{cases}.$$

and

$$m(T,\tau,v) = \varepsilon \frac{\overset{\approx}{\alpha}(T,\tau,v)e^{\lambda T}z_0 - e^{\lambda(T-\tau)}v}{\left\|\overset{\approx}{\alpha}(T,\tau,v)e^{\lambda T}z_0 - e^{\lambda(T-\tau)}v\right\|}.$$

The moment of switching t_*^1 is a zero of testing function

$$h(t) = 1 - \int_0^t \alpha^*(T, \tau, v(\tau))d\tau - \int_t^T \alpha_*(T, \tau)d\tau, t \in [0, T].$$

The functions $\alpha^*(T, \tau, v)$ and $\alpha_*(T, \tau)$ are found above in analytic form. Since $\alpha^*(T) = 1$, then the approach can be realized with help of the control of the form (7.11) with function $m(T, \tau, v)$, in which $\overset{\approx}{\alpha}(T, \tau, v)$ is replaced for function $\alpha^*(T, \tau), \tau \in [0, T]$.

In Theorem 3 the inequality (11) is provided by the relationship

$$e^{\lambda T}\|z_0\| - a\varepsilon \leq 0.$$

Guaranteed control of the pursuer is constructed analogously to the previous case and, therefore, in this case

$$u(\tau) = e^{-\lambda(T-\tau)}\{e^{\lambda(T-\tau)}v(\tau) + \alpha(T, \tau)[m(T, \tau, v(\tau)) - e^{\lambda T}z_0]\},$$

where

$$m(T, \tau, v) = \varepsilon \frac{\alpha(T, \tau)e^{\lambda T}z_0 - e^{\lambda(T-\tau)}v}{\left\|\alpha(T, \tau)e^{\lambda T}z_0 - e^{\lambda(T-\tau)}v\right\|}, \tau \in [0, T].$$

7 Conclusions

Various schemes of the method of resolving functions for investigation of the quasi-linear game dynamic problems are presented in the paper. In so doing, the case, when the classic Pontryagin's condition does not hold, is analyzed. To tackle this difficulty, we take into consideration the upper and the lower resolving functions of two types. This makes it feasible to derive sufficient conditions for approach a cylindrical terminal set by a trajectory of conflict-controlled process in the class of quasi- and stroboscopic strategies. That a solution of dynamic system is presented in a rather general form allows encompass in unified form a wide spectrum of functional-differential systems. A comparison of guaranteed times of the schemes of the first direct method and the method of resolving functions is made.

Analytical results are exemplified by the model example with simple matrix and spherical control domains.

References

1. Pontryagin, L.S.: Selected Scientific Papers, vol. 2. Nauka, Moscow (in Russian) (1988)
2. Krasovskii, N.N.: Game Problems on the Encounter of Motions. Nauka, Moscow (in Russian) (1970)
3. Isaacs, R.: Differential Games. Wiley, New York (1965)
4. Pschenichnyi, B.N., Ostapenko, V.V.: Differential Games. Naukova Dumka, Kiev (in Russian) (1992)
5. Chikrii, A.A.: Conflict Controlled Processes. Springer Science and Business Media, Boston, London, Dordrecht (2013)
6. Osipov, YuS, Kryazhimskii, A.V.: Inverse Problems for Ordinary Differential Equation: Dynamical Solutions. Gordon and Breach, Basel (1995)
7. Sybbotin, A.I., Chentsov, A.G.: Optimization of Guaranteed Result in Control Problems. Nauka, Moscow (in Russian) (1981)
8. Kuntsevich, V.M.: Control under Uncertainty: Guaranteed Results in Control and Identification Problems. Naukova Dumka, Kyiv (in Russian) (2006)
9. Kuntsevich, V.M., Lychak, M.M.: Synthesis of Optimal and Adaptive Systems of Control: Game Approach. Naukova Dumka, Kiev (in Russian) (1985)
10. Kurzhanskii, A.B.: Control and Observation in Uncertainty Conditions. Nauka, Moscow (in Russian) (1977)
11. Hajek, O.: Pursuit games, vol. 12. Academic Press, New York (1975)
12. Nikolskii, M.S., Pontryagins, L.S.: First Direct Method in Differential Games. Izdat. Gos. Univ, Moscow (in Russian) (1990)
13. Siouris, G.M.: Aerospace Avionics Systems: A Modern Synthesis. Academic Press, San Diego (1993)
14. Siouris, G.M.: Missile Guidance and Control Systems. Springer, New York (2004)
15. Locke, A.S.: Guidance. D. Van Nostrand Company, Princeton (1955)
16. Petryshyn, R.J., Sopronyuk, T.M.: Approximate Methods for Solving Differential Equations Under Impulse Impact. Ruta, Chernivtsi (in Ukrainian) (2003)
17. Bigun, Y.I., Liubarshchuk, E.A., Cherevko, I.M.: Game problems for systems with variable delay. Automat. Informat. Sci. **48**(4), 18–31 (2016)
18. Chikrii, A.A.: Problem of evading meeting a terminal set of complex structure in differential games. In: III All-Union Conference on Differential Games, Odessa, Ukraine (in Russian) (1974)
19. Gusyatnikov, P.B.: Escape of a single nonlinear object from several more inertial pursuers. Diff. Uravn. **12**(2), 213–226 (1976). (in Russian)
20. Chikrii, A.A.: Linear Evasion Problem with Several Pursuers. Izv. Akad. Nauk SSSR. Teth. Kib. **4**, 46–50 (1976). (in Russian)
21. Pschenichnyi, B.N.: Simple Pursuit by Several Objects. Kibernetika **3**, 145–146 (1976). (in Russian)
22. Chikrii AA (2010) An Analytical method in dynamic pursuit games. In: Proceedings of the Steklov Institute of Mathematics 271:69–85
23. Grigorenko, N.L.: Mathematical Methods of Control of Several Dynamic Processes. Izdat. Gos. Univ, Moscow (in Russian) (1990)
24. Blagodatskikh, A.I., Petrov, N.N.: Conflict Interaction of Groups of Controlled Objects. Udmurtskiy universitet, Izhevsk (in Russian) (2009)
25. Chikrii AA, Matychyn II, Gromaszek K, Smolarz A (2011) Control of Fractional-Order Controlled Dynamic Systems Under Uncertainty. Modell. Optimizat. 3–56
26. Krivonos, YuG, Matychyn, I.I., Chikrii, A.A.: Dynamic Games with Discontinuous Trajectories. Naukova Dumka, Kiev (in Russian) (2005)
27. Zhukovskiy, V.I., Chikrii, A.A.: On discrete conflict-controlled processes of fractional order by Grunwald–Letnicov. Problemy upravlenya i informatiki **1**, 35–42 (2015). (in Russian)
28. Rokafellar, T.: Convex Analysis. Mir, Moscow (in Russian) (1973)

29. Aubin, J., Frankowska, H.: Set-Valued Analysis. Birkhauser, Boston, Basel, Berlin (1990)
30. Aumann, R.J.: Integrals of set-valued functions. Math. Anal. Appl. **12**, 1–12 (1965)
31. YeS, Polovinkin: Set-Valued Analysis and Differential Inclusions. Fizmatlit, Moscow (in Russian) (2014)
32. Filippov, A.F.: Differential Equations with Discontinuous Right-Hand Part. Nauka, Moscow (in Russian) (1985)
33. Pschenichnyi, B.N.: Convex analysis and extremal problems. Nauka, Moscow (in Russian) (1980)
34. Chikrii, A.A., Pittsyk, M.V., Shishkina, N.B.: The first Pontryagin direct method and some efficient ways of pursuit. Kibernetika **5**, 627–636 (1986). (in Russian)

Impact of Average-Dwell-Time Characterizations for Switched Nonlinear Systems on Complex Systems Control

Georgi Dimirovski, Jiqiang Wang, Hong Yue and Jovan Stefanovski

Honoring Academician Vsevolod M. Kuntsevich for his life-long contributions to Control Systems Theory and Applications.

Abstract It is well known, present day theory of switched systems is largely based on assuming certain small but finite time interval termed average dwell time. Thus it appears dominantly characterized by some slow switching condition with the average dwell time satisfying a certain lower bound, which implies a constraint nonetheless. In cases of nonlinear systems, there may well appear certain non-expected complexity phenomena of particularly different nature when switching becomes no longer useful. A fast switching condition with average the dwell time satisfying an upper bound is explored and established. A comparison analysis of these innovated characterizations via slightly different overview yielded new results on the transient

G. Dimirovski (✉)
Faculty of Engineering, Dogus University of Istanbul, 34722 Acibadem, Istanbul, Turkey
e-mail: gdimirovski@dogus.edu.tr

FEIT – Karpos 2, St. Cyril and St. Methodius University, Skopje 1000, Republic of Macedonia

J. Wang
Jiangsu Province Key Laboratory of Aerospace Power Systems, Nanjing University of Aeronautics & Astronautics, Nanjing 210016, People's Republic of China
e-mail: jiqiang.wang@nuaa.edu.cn

H. Yue
Department of Electronic & Electrical Engineering, University of Strathclyde, Glasgow, UK
e-mail: hong.yue@strath.ac.uk

J. Stefanovski
J. P. "Strezhevo", Division of Control & Informatics, Ul. 1 Maj BB, Bitola 6000, Republic of Macedonia
e-mail: jovanstef@t.mk

Y. P. Kondratenko et al. (eds.), *Advanced Control Techniques in Complex Engineering Systems: Theory and Applications*, Studies in Systems, Decision and Control 203, https://doi.org/10.1007/978-3-030-21927-7_2

behaviour of switched nonlinear systems, while preserving the system stability. The approach of multiple Lyapunov functions is used in current analysis and the switched systems framework is believed to be extended slightly. Thus some new insight into the underlying, switching caused, system's complexities has been achieved.

1 Introduction

Inspired by the work [31] of Ye and co-authors, recently, in 2017 the authors of work [8] have extended considerably their previous findings [31] so as to open new insights and light on the underlying system complexities induced by average dwell-time switching law in conjunction with traditional feedback control strategies. Now these authors have privilege to present their most recent extension in order to honour Academician Vsevolod M. Kuntsevich (in the middle on Fig. 1) for his many contributions to Control Systems Theory and Applications [13–16]. In particular, the first author especially honors his past learning the refined usage of Lyapunov functions stability analysis and control synthesis from the 1977 monograph "Sintez sistem avtomaticheskogo upravleniya s pomoshtyu funkciy Lyaponova" by Kunstevich and Lychak [14].

The authors' innovated extension of the Average-Dwell-Time (ADT) switched systems and control theory is intuitively considerably interesting and appealing too, since it could often yield reducing the conservatism of stability results in the appli-

Fig. 1 V. Kuntsevich at the IFAC World Congress in Seoul, R. Korea in 2008, in the middle along with his colleagues Academicians S. Vassilev, to the left, and A. Kurzanskiy, to the right (courtesy by our good common friend the late S. Zemlyakov, may he rests in peace)

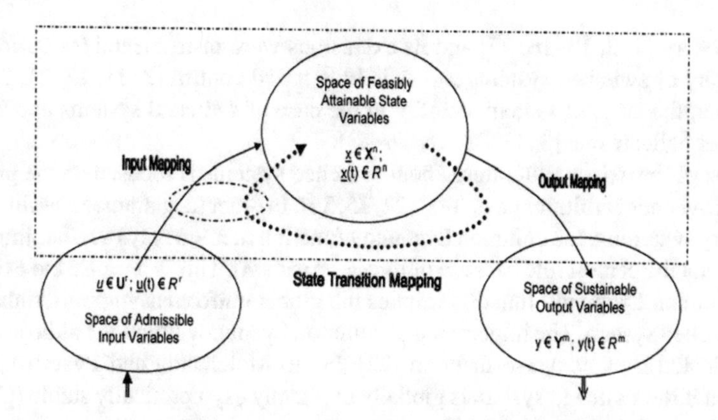

Fig. 2 Illustration of controlled, general nonlinear, systems in accordance to the fundamental laws of physics [7, 8]. Although input, state and output spaces in terms of involved classes of functions can be mathematically defined by a chosen measuring norm, all the vector-valued variables become real-valued vectors at any fixed instant of time, and norm employed is Euclidean one most often

cations. This may not be amenable even to techniques that employ the multiple Lyapunov functions approach [3]. It was shown in [8] both the slow switching and the fast switching can be studied within an appropriately redefined framework. These implications were further worked out in producing novel results on stability of switched nonlinear systems, which are further refined in here. In addition, appropriate simulations are given in this paper to illustrate those results.

It is well known [13] that the behaviour of general nonlinear dynamic systems [7], Fig. 2, and of hybrid systems in particular, to which also switched systems also belong to, may have remarkably different system dynamics from either of their components [3, 5, 6, 11, 12, 19–21]. In the case of switched systems for instance, one switched system can be stable although all its components are unstable, also, some inappropriate switching signal may destabilize the overall switched system even though all of its components are stable [3, 22, 26, 28–32]. The initial opening research [27] on switched systems and control was due to Sworder, who investigated the control of systems subject to sudden changes. This appeared as an alternative to the theory of variable structure control systems [9] due to Emalyanov. In more recent literature, following the results by Molchanov and Pyatnitsky [25] in 1989 and by Morse [26] in 1996, 1997, a switched system is defined as a rather important category of hybrid systems that consists of a family of continuous-time or discrete-time subsystems and a switching law which governs the switching between the subsystems. During the last decade, considerable attention has been paid to switched systems; in due times, many important results have been obtained since the pioneering contributions of Morse and Branicky; for instance, see [2, 4, 15, 16, 21, 22, 28–30] and [36–43] and references therein. Given the fundamental crucial property of stability in systems engineering, all approaches based on analytical studies of switched systems and switching based control have relied on Lyapunov stability theory and its various extensions; see all given references and references therein. Naturally, Lyapunov stability theory [1, 2, 13–16, 25] and its extensions were instrumental for building up the theory of switched systems and switching based control [8, 11, 12, 21, 23, 26]. However, the issue of system stability in the case of switched systems appeared to be rather delicate one [8, 11, 22, 26, 36–43].

Most of the existing literature about switched systems is focused on the problem of stability under arbitrary switching [22, 25, 36]. In order to guarantee stability under arbitrary switching, the common Lyapunov function method plays a rather important role (if not the central role because of its conservatism). This is because the existence of a common Lyapunov function implies the global uniform asymptotic stability of the switched system. The importance of common Lyapunov function has been further consolidated by a converse theorem [22] due to Molchanov and Pyatnitskiy, that asserted if the switched system is globally uniformly asymptotically stable (GUAS), then all the subsystems ought to have a common Lyapunov function.

Over time and in particular more recently, the approach exploiting multiple Lyapunov functions [3] and the associated dwell time [11, 17, 18] or the average dwell time alone [20] are recognized as rather efficient tools in stability studies of switched nonlinear systems [17, 32, 39–43]. The concept of average dwell time switching, which was introduced by Hespanha and Morse in [12], appeared more general than the standard dwell time switching for both stability analysis and related control design and synthesis problems; for instance, see [32–43]. It does imply that the number of switching actions in a finite interval is bounded from above while the average time between two consecutive switching actions is not less than a constant. It is believed the multiple Lyapunov function approach per se reduces the inherent conservatism of the common Lyapunov function approach.

In fact, when confining to dynamic systems only, some well-known design procedures for explicit construction of multiple Lyapunov functions have been developed among which the S-procedure and the methodology of Linear Matrix Inqualities (LMI) have been particularly fruitful as most of references in this paper on control synthesis design and the references therein have clearly demonstrated. In multiple Lyapunov functions approaches, it is generally assumed that each Lyapunov-like function associated for each subsystem is increasing (with the first time-derivative decreasing) with time as time elapses. For the first time, Ye and co-authors in their stability theory for hybrid systems [36] have also studied an approach allowing for a Lyapunov-like function to rise to a limited extent and thus established rather interesting property of a class of such functions called weak Lyapunov.

In order to exploit the concept and results for average-dwell time switched nonlinear systems into control synthesis design, it appeared that it is of paramount importance to keep trace with the application feasibility of Linear Matrix Inequality (LMI) methodology and the S-procedure [33–35] due to of Yakubovich. In case of other classes of switched systems for that matter the same is equally valid and important, of course. The major reason now is the availability of the respective software tools for LMI [10] and Simulink computing support by Matlab [23, 24]. However, given the well observed fact that typically control synthesis design in practical applications is usually consisted of at least a few iterative cycles of adequate synthesis-simulation-rectifying-modification [13, 14], the importance of consistency with the LMI methodology and S-procedure becomes rather crucial.

A note on notation: The notation used in this work is a fairly standard one. This may well be inferred from: $\mathbb{R}^n$ represents the n-dimensional Euclidean space; C^2 denotes the space of twice continuously differentiable functions; C^1 once differentiable piece-wise function; $\sigma : [t_0, +\infty)$ denotes a causal real-world signal; M index set of integers; $\{([t_0, t_1], \sigma([t_0, t_1])), \ldots ([t_k, t_{k+1}], \sigma([t_k, t_{k+1})), \ldots\}$ switching sequence.

2 Background and Mathematical Preliminaries

2.1 A Couple of Preliminary Notes

It is important to notice the advanced Lyapunov stability theory, the full account of which is found in Khalil's monograph [13] employing both class K and class $\mathcal{KL}$ functions (additional to the classic Lyapunov functions) as well as his Comparison Principle, is used in this study. The concepts of these functions are re-stated first for reasons of readability.

Definition 1 [13] A continuous function $k : [0, a) \to [0, \infty)$ is aid to belong to class K if it is strictly increasing and has value $k(0) = 0$. It is said to belong to class K_∞ if the upper bound of domain is $a = \infty$ and if $k(\rho) = \infty_{\rho\to\infty}$, i.e. if it is $k : [0, \infty] \to [0, \infty]$.

Definition 2 [13] A continuous mapping $k = k(\rho, \sigma)$ defined by function $k : [0, a) \times [0, \infty) \to [0, \infty)$ is said to belong to class $\mathcal{KL}$ if, for each fixed σ, the mapping $k = k(\rho, \sigma)$ belong to class K functions with respect to ρ and, for each fixed ρ, the mapping $k = k(\rho, \sigma)$ is decreasing with respect to σ such that $k(\rho, \sigma)_{\sigma\to\infty} = 0$.

2.2 On Basics of Switched Systems Theory

It is well known that, in general terms [5, 7, 14, 15], a controlled nonlinear dynamic system can be represented by means of the state transition and output measurement equations (1). In order such a system to have sustained operability functions its state transition mechanism must have the property $f(0, 0) = 0$ and output measuring mechanism must have $h(0) = 0$ on the grounds of basic natural laws.

$$\left\{\begin{array}{c} \dot{x}(t) = f(x(t), u(t)), x(t_0) = x_0 \\ y(t) = h(x(t)). \end{array}\right\} \text{ for } \forall t \in [t_0, +\infty). \quad (1)$$

In here, the system's quantities denote: the state space $x \in X^n$, the input space $u \in U^r$, $y \in Y^m$ the output space; $f: \mathbb{R}^n \times \mathbb{R}^r \to \mathbb{R}^n$, $h: \mathbb{R}^n \to \mathbb{R}^m$. Upon the synthesis design of a certain controlling infrastructure is carried out then it follows

$$u = u(t; t_0, u_0), \forall t \in [t_0, \infty). \tag{2}$$

Notice that Fig. 1 depicts a relevant illustration, which is related to general class of systems and the quoted notions of system theory. For, it depicts the controlled general nonlinear systems in accordance to the fundamental laws of physics on rigid-body energy, matter and momentum of motion, i.e. evolution of natural processes over time.

Furthermore, a controlled nonlinear dynamic system

$$\left\{ \begin{array}{c} \dot{x}(t) = f_\sigma(x(t), u(t)), x(t_0) = x_0 \\ y(t) = h_\sigma(x(t)) \end{array} \right\} \text{ for } \forall t \in [t_0, +\infty), \tag{3}$$

where subsystems $f_{\sigma(t)} = f_i,\ h_{\sigma(t)} = h_i$ with $i \in M$ in an index set M are fixed given models, and the control input $u = u(t)$ is also given (upon its synthesis) is called a *switched* system. An autonomous switched nonlinear dynamic system thus appears to be defined as follows:

$$\left\{ \begin{array}{c} \dot{x}(t) = f_\sigma(x(t)), x(t_0) = x_0 \\ y(t) = h_\sigma(x(t)). \end{array} \right\} \text{ for } \forall t \in [t_0, +\infty). \tag{4}$$

For a causal signal $\sigma : [t_0, +\infty)$, if it is

$$\sigma(t^+) = \Sigma\{[t_0, t], (\sigma[t_0, t]), \{(x([t_0, t]), y([t_0, t])\}, \text{ for } \forall t \in [t_0, +\infty), \tag{5}$$

a piece-wise constant time-sequence function such that $\sigma\left(t^+\right) = \lim_{\tau \downarrow t} \sigma(\tau)$, for $\tau \geq 0$ in continuous time case, and $\sigma\left(t^+\right) = \sigma(t+1)$, for $t \geq 0$ in discrete time case, then it is called a switching signal.

A switching signal is said to be a *switching path* if it is defined as mapping of finite, semi-open time interval into the index set M such that $\sigma : [t_0, t_1) \to M$ for every $[t_0, t_1)$ with $t_0 < t_1 < +\infty$. A switching law is called a time-driven switching law if it depends only on time and its past value $\sigma\left(t^+\right) = \sum(t, \sigma(t))$. A switching law is called a state-feedback switching law if it depends only on its past value an on the values of state variables at that time $\sigma\left(t^+\right) = \sum(\sigma(t), x(t))$ for $\forall t \in [t_0, +\infty)$. A switching law is called a output-feedback switching law if it depends only on its past value an on the values of output variables at that time $\sigma\left(t^+\right) = \sum(\sigma(t), y(t))$ for $\forall t \in [t_0, +\infty)$. At present, no other concepts and notions about feasible switching signals matter.

The concept of average dwell time is given below and the respective rather important fundamental result is cited too, both due to Hespana and Morse [9].

Definition 3 [12] For a switching signal σ and any $t_2 > t_1 > t_0$, let $N_\sigma(t_1, t_2)$ be the number of switching over the interval $[t_1, t_2)$. If the condition $N_\sigma(t_1, t_2) \leqslant N_0 + (t_2 - t_1)/\tau_a$ holds for $N_0 \geqslant 1$, $\tau_a > 0$, then N_0 and τ_a are called the *average dwell time* (ADT) and the *chatter bound*, respectively.

Theorem 1 [12] *Consider the switched system* (4), *and let* α *and* μ *be given constants. Suppose that there exist smooth functions* $V_{\sigma(t)} : \Re^N \to \Re, \sigma(t) \in \ell$, *and two* K_∞ *functions* k_1 *and* k_2 *such that for each* $\sigma(t) = i$, *the following conditions hold:*

$$k_1(|x_t|) \le V_i(x_t) \le k_2(|x_t|), \dot{V}_i(x_t) \le -\alpha V_i(x_t), \text{and} \tag{6}$$

$$\text{for any}(i, j) \in \ell \times \ell, i \ne j, V_i(x_t) \le \mu V_j(x_t). \tag{7}$$

Then the system is globally uniformly asymptotically stable for any switching signal with ADT $\tau_a > \tau_a^* = \frac{\ln \mu}{\alpha}$.

Theorem 1 in fact considers multiple Lyapunov functions with a "jump" on the switching boundary. An extension in [36], due to Ye et al., the underlying essentialities of which were further highlighted in [38] by Zhang and Gao, allows the Lyapunov-like function to rise to a limited extent, and in addition to the jump on switching boundary [8, 39, 40]. This is the so-called *weak* Lyapunov function, and it allows both the jump on the switching boundary and the increase over any time interval of interest. Now consider $\sigma(t) \in i$ and then, within the interval $[t_i, t_{i+1}]$, denote the unions of scattered subintervals during which the week Lyapunov function is increasing and decreasing by $T_r(t_i, t_{i+1})$ and $T_d(t_i, t_{i+1})$, respectively. Hence it holds $[t_i, t_{i+1}) = T_r(t_i, t_{i+1}) \cup T_d(t_i, t_{i+1})$. Further let use $T_r(t_{i+1} - t_i)$ and $T_d(t_{i+1} - t_i)$ to represent the length of $T_r(t_i, t_{i+1})$ and $T_d(t_i, t_{i+1})$, correspondingly. Then the following important result can be obtained.

Theorem 2 [36, 38] *Consider the switched system*

$$\dot{x}_t = f_\sigma(x_t), \tag{8}$$

and let $\alpha > 0$, $\beta > 0$ *and* $\mu > 1$ *are prescribed constants. If there exist smooth functions* $V_{\sigma(t)} : \Re^n \to \Re$ *and two* K_∞ *functions* k_1 *and* k_2 *such that for each* $\sigma(t) = i$, *the following conditions hold:*

$$k_1(|x_t|) \le V_i(x_t) \le k_2(|x_t|), \tag{9}$$

$$\dot{V}_i(x_t) \le \begin{cases} -\alpha V_i(x_t) & over \quad t \in T_d(t_i, t_{i+1}) \\ \beta V_i(x_t) & over \quad t \in T_r(t_i, t_{i+1}) \end{cases}, \tag{10}$$

$$V_i(x_t) \le \mu V_j(x_t) \quad \forall\big(\sigma(t) = i \quad \& \quad \sigma(t^-) = j\big). \tag{11}$$

Then the system is GUAS for any switching signal with ADT

$$\tau_a > \tau_a^s = \frac{(\alpha + \beta)T_{\max} + \ln^{\mu}}{\alpha}, T_{\max} = \max T_r(t_{i-1}, t_i), \quad \forall i. \tag{12}$$

It may well be seen that the result above actually includes Theorem 1 *as a special case. Namely,* $\beta = 0$ *implies no increase over the interval and hence* $T_{\max} = 0$, *then the ADT condition reduces to the ADT condition* $\tau_a > \tau_a^* = \frac{\ln \mu}{\alpha}$, (6) and (7) *in* Theorem 1. *It is precisely this generality of the weak Lyapunov functions that has given incentives to explore the alternatives on fast and slow switching rules* [8, 31] *in this study.*

3 Fast and Slow Switching Modes: Recent Discoveries

In the sequel the derived novel results, which are subject to further exploration in this paper, are presented first. In what follows subsequently, the following notations from work [3] of Branicky are adopted throughout. In particular, a general arbitrary switching sequence is expressed by

$$\sum = \{x_0; (i_0, t_0), (i_1, t_1), \ldots, (i_j, t_j), \ldots, |i_j \in \overline{M}, j \in N\} \tag{13}$$

in which t_0 is the initial time, x_0 is the initial state, (i_k, j_k) means that the i_k-th subsystem is activated for $t \in [t_k, t_{k+1})$. Therefore, when $t \in [t_k, t_{k+1})$, the trajectory of the switched system (1) is produced by the i_k-th subsystem. Thus, for any $j \in \overline{M}$, the set

$$\sum_t (j) = \{[t_{j_1}, t_{j_1+1}), \cdots [t_{j_n}, t_{j_n+1}) \cdots \sigma(t) = j, t_{j_k} \le t \le t_{j_k+1}, k \in N\} \tag{14}$$

denotes the sequence of switching times of the j-th subsystem, in which the j-th subsystem is switched on at t_{j_k} and switched off at t_{j_k+1}.

3.1 Novel Insights into the Complexity of Switching

It should be noted, Theorem 2 is a slow switching result in the sense that it is characterized by a lower bound on the average dwell time. Recently, the following two novel results have emanated from the novel insights into the switching complexity.

Theorem 3 [8, 31] *Consider the switched system*

$$\dot{x}_t = f_{\sigma}(x_t), \tag{15}$$

and let $\alpha > 0$, $\beta > 0$ *and* $\mu > 1$ *are prescribed constants. If there exist smooth functions* $V_{\sigma(t)} : \Re^n \to \Re$ *and two* K_∞ *functions* k_1 *and* k_2 *such that for each* $\sigma(t) = i$, *the following conditions hold:*

$$k_1(|x_t|) \leq V_i(x_t) \leq k_2(|x_t|), \tag{16}$$

$$V_i(x_t) \leq \mu V_j(x_t) \ \ \forall\big(\sigma(t) = i \ \ \& \ \ \sigma(t^-) = j\big), \tag{17}$$

$$\dot{V}_i(x_t) \leq \begin{cases} -\alpha V_i(x_t) \ over \ t \in T_d(t_i, t_{i+1}) \\ \beta V_i(x_t) \quad over \ t \in T_r(t_i, t_{i+1}). \end{cases} \tag{18}$$

Then the system is GUAS for any switching signal with ADT

$$\tau_a < \tau_a^f = \frac{(\alpha + \beta)T_{\text{min}} - \ln^{\mu}}{\beta}, \quad T_{\text{min}} = \min T_d(t_{i-1}, t_i), \forall i. \tag{19}$$

Intuitively, a switched nonlinear will achieve induced stability by arbitrary switching if the upper bound for the fast switching is larger than the lower bound for slow switching [20]. *This is the essence of the subsequent theorems.*

Theorem 4 [8, 31] *Consider the switched system*

$$\dot{x}_t = f_\sigma(x_t), \tag{20}$$

and let $\alpha > 0$, $\beta > 0$ are prescribed constants. If there exist smooth functions $V_{\sigma(t)} : \Re^n \rightarrow \Re$ and two K_∞ functions k_1 and k_2 such that for each $\sigma(t) = i$, the following conditions hold:

$$(|x_t|) \leq V_i(x_t) \leq k_2(|x_t|), \tag{21}$$

$$V_i(x_t) = V_j(x_t) \ \ \forall\big(\sigma(t) = i \ \ \& \ \ \sigma(t^-) = j\big) \tag{22}$$

$$\dot{V}_i(x_t) \leq \begin{cases} -\alpha V_i(x_t) \ over \ t \in T_d(t_i, t_{i+1}), \\ \beta V_i(x_t) \ \ over \ t \in T_r(t_i, t_{i+1}). \end{cases} \tag{23}$$

Then the system is GUAS for arbitrary switching signal if the following conditions are fulfilled:

$$\frac{T_{\text{max}}}{T_{\text{min}}} \leq \frac{\alpha}{\beta}, T_{\text{max}} \equiv \max T_r(t_{i-1}, t_i), T_{\text{min}} \equiv \min T_d(t_{i-1}, t_i), \quad \forall i. \tag{24}$$

Theorem 5 [8, 31] *Consider the switched system*

$$\dot{x}_t = f_\sigma(x_t) \tag{25}$$

and let $\alpha > 0$, $\mu > 1$ be given constants. Suppose that there exist C^1 functions $V_{\sigma(t)} : \Re^N \rightarrow \Re$, $\sigma(t) \in \ell$, and two K_∞ functions k_1 and k_2 such that $\forall \sigma(t) = i$, $k_1(|x_t|) \leq V_i(x_t) \leq k_2(|x_t|)$, $\dot{V}_i(x_t) \leqslant -\alpha V_i(x_t)$, and $\forall (i, j) \in \ell \times \ell, i \neq j$, $V_i(x_t) \leq \mu V_j(x_t)$.

Then the system is GUAS for any switching signal if and only if the ADT satisfies the condition $\tau_a > \tau_a^* = \frac{\ln \mu}{\alpha}$.
The details of proofs for the above presented results are found in [8]. *It is important to notice that the slow switching condition reduces to the average dwell time of Hespanha and Morse precisely, as they have defined it in their 1999 seminal paper* [12].

3.2 Novel Supporting Evidence via Simulation Experiments

At this point, let recall the expression of a general arbitrary switching sequence, namely:

$$\sum = \{x_0; (i_0, t_0), (i_1, t_1), \ldots, (i_j, t_j), \ldots, |i_j \in \overline{M}, j \in N\} \tag{26}$$

Also, notice that t_0 is the initial time, x_0 is the initial state, (i_k, j_k) means that the i_k-th subsystem is activated for $t \in [t_k, t_{k+1})$. Duplicate the template file by using the Save As command, and use the naming convention prescribed by your conference for the name of your paper.

Next, let recall that by means of Jacobians the basic linearization of general nonlinear systems, such as the class of time-varying nonlinear plant processes to be controlled, is as follows:

$$\dot{\vec{x}}(t) = \vec{f}(\vec{x}(t), \vec{u}(t); t), \vec{x}(t_0) = \vec{x}_0, \vec{y}(t) = \vec{g}(\vec{x}(t); t). \tag{27}$$

Upon linearization of nonlinear functions $\vec{f}, \vec{g}$ in the vicinity neighborhood of a certain operating state, e.g. such as a *steady-state* operating point *EP* $(\vec{u}_c, \vec{x}_c, \vec{y}_c)$ with the *steady-states* $\vec{x}_c, \vec{y}_c$ which may happen under some *steady control input* $\vec{u}_c$, model (27) yields

$$\begin{cases} \dot{\vec{x}}(t) = A(t)_{n\times n}\vec{x}(t) + B(t)_{n\times r}\vec{u}(t), \\ \quad \vec{y}(t) = C(t)_{m\times n}\vec{x}(t), \end{cases} \quad \vec{x}(t_0) = \vec{x}_0, \tag{28a}$$

where matrices $A(t)_{n\times n} = \partial\vec{f}/\partial\vec{x}_{|x_c,u_c}$, $B(t)_{n\times r} = \partial\vec{f}/\partial\vec{u}_{|x_c,u_c}$, $C(t)_{n\times r} = \partial\vec{g}/\partial\vec{x}_{|x_c,u_c}$. Should the *steady-state* operating point *EP* $(\vec{u}_c, \vec{x}_c, \vec{y}_c)$ is a *desired equilibrium* (which usually is in practice) that can be achieved under a certain *synthesized equilibrium control* vector $\vec{u}_c = \vec{u}_c^e = const$, defining the desired *equilibrium state* vector $\vec{x}_c = \vec{x}_c^e = const$ and hence the desired steady-state *equilibrium output* vector $\vec{y}_c = \vec{y}_c^e = const$, then model (27) yield

$$\dot{\vec{x}} = A_{n\times n}\vec{x}(t) + B_{n\times r}\vec{u}(t), \vec{x}(t_0) = x_0; y(t) = C_{n\times m}\vec{x}(t). \tag{28b}$$

It is therefore that the *Lyapunov asymptotic stability* requirement on the system's steady-state equilibrium is indispensable and rigorous requirement. It is this requirement precisely which is being enhanced by involving a switching law in addition to the synthesized feedback control. The presented illustrations obtained by means of simulated time-responses further below, along with the above theoretical results, highlight the concepts essence of both the fast and slow switching as well as the average dwell time. Also, they emphasize the importance of achieving asymptotic Lyapunov stability under arbitrary switching as well as under average-dwell time switching. In any case, these findings guarantee in the close loop the ultimately uniform bounded operation of the plant at desired equilibrium steady-state shall be reached despite the possible uncertainties of plant model.

For this purpose let now consider the application of the arbitrary switching sequence (26) according to average-dwell time principle to a second-order two-input-two-output, uncertain nonlinear plant system of class (27), whose states are detectable and measurable. (If they are not detectable and measurable., then state estimator ought to be employed in composite control strategy combining estimated state feedback and average-dwell time principle.) Further, let assume the respective Jacobian state-transition and output subsystems represent a controllable and observable linearized system model in the vicinity of the desired steady-state operating equilibrium.

Thus, let consider the following example [8] for which an uncertain plant of class (27) yields system (28b) having the system matrices:

$$A_{11} = \begin{bmatrix} -5 & 4 \\ 0 & 2 \end{bmatrix}, A_{12} = \begin{bmatrix} -15 & 1 \\ 0 & -10 \end{bmatrix}, A_{21} = \begin{bmatrix} 2 & 0 \\ 1 & -5 \end{bmatrix}, A_{22} = \begin{bmatrix} -15 & 0 \\ -5 & -4 \end{bmatrix}, \tag{29a}$$

$$B_{11} = B_{12} = \begin{bmatrix} 1 & 1 \\ 1 & 1 \end{bmatrix}, B_{21} = B_{22} = \begin{bmatrix} 3 & 1 \\ 1 & 5 \end{bmatrix}, \tag{29b}$$

$$C_{11} = C_{12} = \begin{bmatrix} 1 & 1 \\ 1 & 1 \end{bmatrix}, C_{11} = C_{12} = \begin{bmatrix} 1 & 1 \\ 1 & 1 \end{bmatrix}. \tag{29c}$$

Next, let define region subsets

$$\Omega_1 \cup \Omega_2 \equiv \left\{ \begin{array}{l} \Omega_1 = \{x(t) \in R^n \big| x(t)^{\mathrm{T}}(P_1 - P_2)x(t) \geq 0, x(t) \neq 0\}, \\ \Omega_2 = \{x(t) \in R^n \big| x(t)^{\mathrm{T}}(P_2 - P_1)x(t) \geq 0, x(t) \neq 0\}. \end{array} \right\} \tag{30}$$

where matrices P_1, P_2 are designed in the course of stabilizing state feedback control synthesis using LMI as in most of previous works on switched systems and control, which were surveyed here in the introduction. In particular, see work [43]. It is therefore that system's state space is generated by $\Omega_1 \cup \Omega_2 = R^n \backslash \{0\}$. The closed-loop system therefore should be asymptotically stable or uniformly bounded at least due to the switching law

$$\sigma(t) = \left\{ \begin{array}{ll} 1 & x(t) \in \Omega_1, \\ 2 & x(t) \in \Omega_2 \backslash \Omega_1. \end{array} \right\} \tag{31}$$

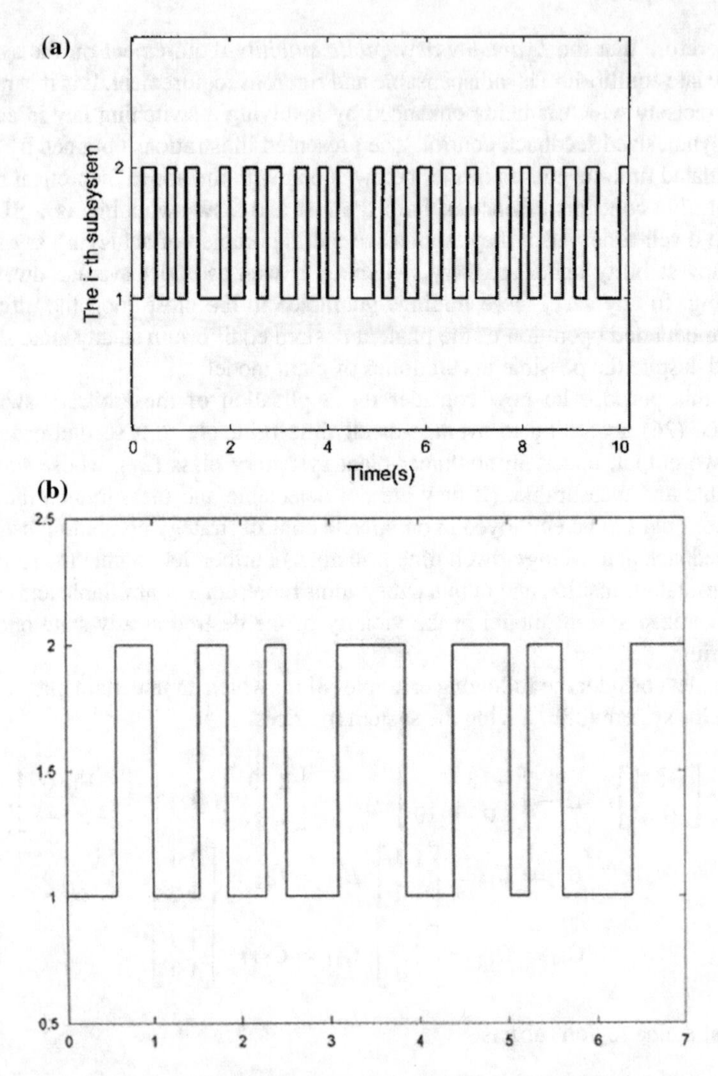

Fig. 3 Evolution of the time responses for the considered class of switching sequences demonstrating the fast (**a**) and the slow (**b**) switching modes with the average-dwell-time strategy, in conjunction with the state-feedback controlled transitions in Fig. 3, which converge to the equilibrium state in step-wise mode with continuously convergent trend

regardless of the plant uncertainty. The obtained computer simulation results are depicted in Figs. 3 and 4. In the former figure there are presented the switching sequences employed, while in the latter figure there are presented the state responses of the controlled plant process.

In fact, Fig. 3 depicts two typical switching sequences combining consistently fast and slow switching of two different modes, which thus demonstrate asymptotic stability can be achieved under arbitrary switching law over two consistent state zones.

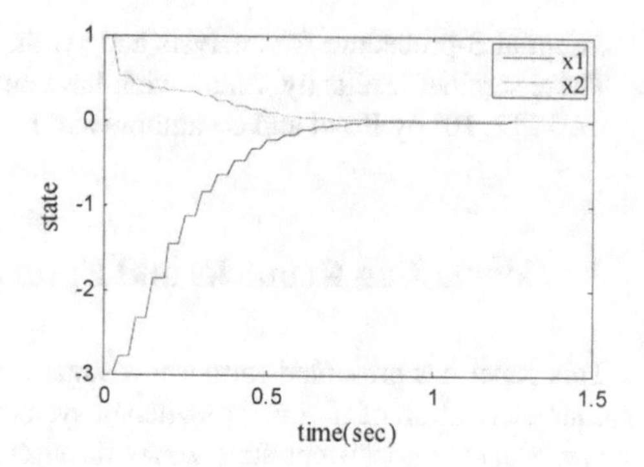

Fig. 4 Fast step-wise stable evolution of the time responses of both states towards the plant's equilibrium state in closed loop under switching based state feedback control with switching laws in Fig. 3b, which combine consistently fast and slow switching modes stability

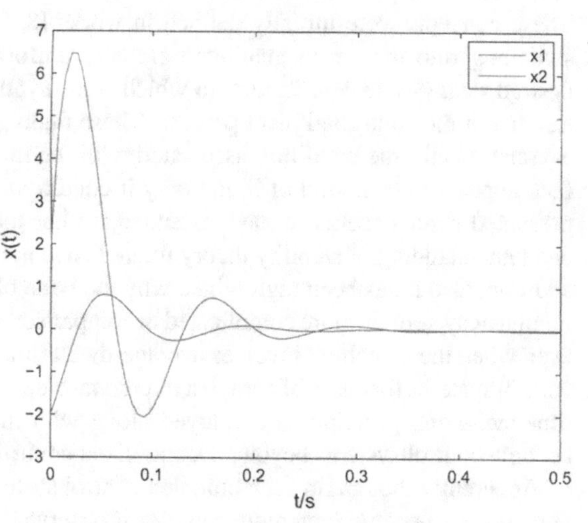

Fig. 5 Fast step-wise stable evolution of the time responses of both states towards the plant's equilibrium state in closed loop under switching based state feedback control with switching laws in Fig. 3b, which combine consistently fast and slow switching modes stability along with the filter's smoothing effect

The corresponding simulation results for the plant's state vector response, when the initial conditions of states are $x(0) = [-3, 1]^T$, have been found as depicted in Fig. 4.

Before summarizing this study, let observe another similar system with uncertainties handled by means of the same procedure employing region subsets (30) and the switching law (31). Again, P_1, P_2 were designed in the course of stabilizing state feedback control synthesis using LMI and a smoothing filter was associated with it. The system's state space is thus generated by $\Omega_1 \cup \Omega_2 = R^n \backslash \{0\}$. The simulation results for the plant's state vector response in this case, when the initial states were chosen as $x(0) = [+2, -2]^T$, have been found as depicted in Fig. 5.

The computer simulation results have been obtained via Matlab-Simulink computing tools of MathWorks Co [23, 24] on the grounds of using Linear Matrix Inequalities [2, 10]. At this point again it should be noted, as Brocket has emphasized in article [4], the methodology of Linear Matrix Inequalities (LMI) is based on the seminal mathematical and systems-theoretical discoveries by Yakubovich for nonlinear control systems. These include not only matrix inequalities [28, 29] but also the rather essential S-procedure for analysis and synthesis of nonlinear control systems [30]. These seminal results by Yakubovich have enabled the emergence of the respective works [2, 10] by Boyd and co-authors and by Gahinet and coauthors.

4 Concluding Remarks and Future Research

This paper has presented further investigation and supportive evidence to the novel characterization of nonlinear switched systems via adopting constrained switching law over two consistent state zones through concepts of slow and fast switching. These concepts were initially defined in works [8, 31]. It should be noted, a fast switching rule may even guarantee globally uniform asymptotic stability of the desired steady-state equilibrium, to which some synthesized state-feedback control has driven the controlled plant process. These findings thus prove that the standard average dwell time condition associated with the multiple Lyapunov functions, in fact, appears to be a kind of if and only if condition. It should be noted, the above presented results seems to have extended further the existing insight, knowledge and understanding of stability theory for switched nonlinear systems and control. In addition, also it has been highlighted why the issue of system stability for switched nonlinear systems is more complicated in comparison with the non-switched systems even when the switching involves average-dwell time. It is therefore that inevitably control systems for cases of complex interconnected plants even when average-dwell time switching principle is employed along with more traditional state feedback become control systems beyond decentralized control.

Apparently though, in such complex control systems, the system dynamics phenomenology become even more complex if observer-based output feedback strategy ought to be employed along with average-dwell time switching law feedbacks. Thus this brief outline tackling of these issues seems to have defined a new direction for future research in which there several paths to follow. However, proper definition of tasks to be explored in the near future is yet to be elaborated on and precisely defined in mathematical terms since it involves not only uncertainties but also information constraints. For this purpose, notice the recent work [29] by Sun and co-authors, where an idea on how to approach the state-constrained case was developed within context of switched nonlinear systems. Another idea is found in works [17, 18] by Li and co-authors in which impulsive observers for switched nonlinear systems were instrumental. On the other hand, it is also instrumental to notice recent work [1] by Andriyevsky and co-authors who suggested a way towards unified theory of control, computation and communications.

Acknowledgements This research has been generously supported by China Postdoctoral Science Foundation (grant No. 20100471336), Beijing, and NUAA Research Funding Projects (grants No. S0985-021 and No. 1002-56YAH11011), Nanjing, P. R. of China. In part, it was also supported by a special Fund for China-Macedonia co-operation of Ministry of Education & Science of R. of Macedonia (grant No. 14-3154/1), a special Fund for Turkey-Russia co-operation of Turkish Agency TUBITAK and Russian Foundation for Basic Research (grant TUBITAK-RFBR 05.09-

06.09.2012/1) as well as by Fund for Science of Dogus University. Here Georgi Dimirovski and Jiqiang Wang want to acknowledge their former doctoral students, respectively, for carrying out computer simulations. They would like also to mention some inspiring question that emerged while discussing on the exciting area of switched systems and switching based control for nonlinear dynamic plants.

References

1. Andriyevsky, B.R., Matveev, A.S., Fradkov, A.L.: Control and estimation under information constraints: toward a unified theory of control, computation and communications. Avtomatika i Telemechanika **71**(4), 34–99 (2010) (in Russian)
2. Boyd, S.P., El Ghaoui, L., Feron, E., Balakrishnan, V.: Linear Matrix Inequalities in Systems and Control Theory, SIAM Studies in Applied Mathematics, vol. 15. The SIAM, Philadelphia, PA, USA (1994)
3. Branicky, M.S.: Multiple Lyapunov functions and other analysis tools for switched and hybrid systems. IEEE Trans. Autom. Control **43**(4), 475–482 (1998)
4. Brockett, R.W.: The wider influence of the work of V. A. Yakubovoch. Int. J. Robust Nonlinear Syst. **17**(5–6), 363–368 (2007)
5. Dayawansa, W.P., Martin, C.F.: A converse Laypunov theorem for a class of dynamical systems which undergo switching. IEEE Trans. Autom. Control **44**(4), 751–760 (1999)
6. DeCarlo, R.A., Branicky, M.S., Petterson, S., Lenartsson, B.: Perspectives and results on the stability and stabilizability of hybrid systems. Proc. IEEE **88**(7), 1069–1082 (2000)
7. Dimirovski, G.M., Gough, N.E., Barnett, S.: Categories in systems and control theory. Int. J. Syst. Sci. **8**(9), 1081–1090 (1977)
8. Dimirovski, G.M., Wang, J., Yue, H.: Complexity of constrained switching for switched nonlinear systems with average dwell time: novel characterization. In: Proceedings SMC2016 of the IEEE International Conference on Systems, Man, and Cybernetics (General Chair Imre J. Rudas, Technical Program Chair Shun-Feng Su), Budapest, HU, 9–12 October, pp. 2376–2381. The IEEE, Piscataway, NJ, USA (2017)
9. Emelyanov, S.V.: Variable Structure Control Systems. Nauka, Moscow, USSR (1967) (in Russian)
10. Gahinet, P., Nemirovskii, A., Laub, A. J., Chilali, M.: The LMI Tool Box. The MathWorks, Inc., Natick, NJ (1995)
11. Hespanha, J.P.: Uniform stability of switched linear systems: extensions of Lassalle's invariance principle. IEEE Trans. Autom. Control **49**(4), 470–482 (2004)
12. Hespanha, J.P., Morse, A.S.: Stability of switched systems with average dwell time. In: Proceedings of 38th Conference on Decision and Control (General Chair E. W. Kamen; Program Chair Ch. Casandras), Phoenix, AZ, 7–10 December, pp. 2655–2660. The IEEE, Piscataway, NJ, USA (1999)
13. Khalil, H.K.: Nonlinear Systems, 3rd edn. Prentice Hall, Upper Saddle River, NJ, USA (2002)
14. Kuntsevich, V.M., Lychak, M.M.: Synthesis of Systems of Automatic Control with the Aid of Lyapunov Functions. Nauka, Moscow, USSR (1977) (in Russian)
15. Kuntsevich, V.M.: Control Under Uncertainty: Guaranteed Results in Control and Identification. Naukova Dumka, Kyiev, UA (2006) (in Russian)
16. Kuntsevich, V.M., Gubarev, V.F., Kondratenko, Y.P., Lebedev, D.V., Lysenko, V.P. (eds.): Control Systems Theory and Applications. Rover Publishers, Gistrup, DK–Delft, NL (2018)
17. Li, J., Ma, R., Dimirovski, G.M., Fu, J.: Dwell-time based stabilization of switched linear singular systems with all unstable modes. J. Frankl. Inst. Eng. Appl. Math. **354**(7), 2712–2724 (2017)
18. Li, J., Ma, R., Dimirovski, G.M.: Adaptive impulsive observers for a class of switched nonlinear systems with unknown parameter. Asian J. Control **19**(3), 1153–1163 (2017)

19. Li, L.-L., Zhao, J., Dimirovski, G.M.: Observer-based reliable exponential stabilization and H_{inf} approach for switched systems with faulty actuators: an average dwell-time approach. Nonlinear Anal. Hybrid Syst. **5**, 479–491 (2011)
20. Li, T.-F., Dimirovski, G.M., Liu, Y., Zhao, J.: Improved stability of a class of switched neutral systems via Lyapunov-Krasovskii functionals and an average dwell time scheme. Int. J. Syst. Sci. **44**(6), 1076–1088 (2013)
21. Liberson, D., Morse, A.S.: Basic problems in stability and design of switched systems. Control Syst. Mag. **19**(1), 59–70 (1999)
22. Liberzon, D.: Switching in Systems and Control. Birkhauser, Boston, MA (2003)
23. MathWorks: MATLAB. The MathWorks, Inc., Natick, NJ (1994)
24. MathWorks: SIMULINK. The MathWorks, Inc., Natick, MA (1999)
25. Molchanov, A.P., Pyatnitskiy, Y.S.: Criteria for asymptotic stability of differential and difference inclusions encountered in control theory. Syst. Control Lett. **13**(1), 59–64 (1989)
26. Morse, A.S.: Supervisory control of families of linear set-point controllers: Pt.1 Exact matching; and Pt. 2 Robustness. IEEE Trans. Autom. Control. **41**(10), 1413–1431 (1996); **42**(10), 1500–1515 (1997)
27. Sworder, D.D.: Control of systems subject to sudden changes. Proc. IEEE **64**(8), 1219–1225 (1976)
28. Sun, X.M., Zhao, J., Dimirovski, G.M.: State feedback control for discrete delay systems with controller failures based on average-dwell time method. IET Control Theory Appl. **2**(2), 126–132 (2008)
29. Sun, Y., Zhao, J., Dimirovski, G.M.: Adaptive control for a class of state-constrained high-order switched nonlinear systems with unstable subsystems. Nonlinear Anal. Hybrid Syst. **32**, 91–105 (2019)
30. Sun, Z., Ge, S.S.: Analysis and synthesis of switched linear control systems. Automatica **41**(2), 181–185 (2005)
31. Wang, J., Hu, Z., Yue, H., Dimirovski, G.M.: Two characterizations of switched nonlinear systems with average dwell time. Int. J. Appl. Math. Mach. Learn. **4**(2), 81–91 (2016)
32. Xiang, J., Xiao, W.: Stabilization of switched continuous-time systems with all modes unstable via dwell-time switching. Automatica **50**(3), 940–945 (2014)
33. Yakubovich, V.A.: Solution to some matrix inequalities encountered in the theory of automatic control. Dokladi Akademii Nauk SSSR **143**(6), 1304–1307 (1962) (in Russian)
34. Yakubovich, V.A.: Solution to some matrix inequalities encountered in the nonlinear theory of automatic control. Dokladi Akademii Nauk SSSR **156**(2), 278–281 (1964) (in Russian)
35. Yakubovich, V.A.: S-procedure in the nonlinear theory of automatic control. Vestnik Leningradskogo Gosudarstvenogo Universsiteta, Seriya Matematika, Mechanika, Astronomiya **1**, 62–77 (1971) (in Russian)
36. Ye, H., Michel, A.N., Hou, L.: Stability theory for hybrid dynamical systems. IEEE Trans. Autom. Control **43**(4), 461–474 (1998)
37. Zhai, G.S., Hu, B., Yasuda, K., Michel, A.N.: Piecewise Laypunov functions for switched systems with average dwell time. Asian J. Control **2**(3), 192–197 (2000)
38. Zhang, L., Gao, H.: Asynchronously switched control of switched linear systems with average dwell time. Automatica **46**(9), 953–958 (2010)
39. Zhao, J., Nie, H.: Sufficient conditions for input-to-state stability of switched systems. Acta Autom. Sin. **29**(2), 252–257 (2003)
40. Zhao, J., Dimirovski, G.M.: Quadratic stability of switched nonlinear systems. IEEE Trans. Autom. Control **49**(4), 574–578 (2004)
41. Zhao, J., Hill, D.J.: Passivity and stability of switched nonlinear systems. Syst. Control Lett. **57**(2), 158–164 (2008)
42. Zhao, J., Hill, D.J.: Dissipativity theory for switched systems. IEEE Trans. Autom. Control **53**(4), 941–953 (2008)
43. Zhao, S., Dimirovski, G.M., Ma, R.: Robust H_{inf} control for non-minimum phase switched cascade systems with time delay. Asian J. Control **17**(4), 1–10 (2015)

On the Problem of Optimization in Group Control

Alexander B. Kurzhanski

Abstract Given is a team of m controlled motions with two types of members, a target set $\mathcal{M}$ and an array $\mathbf{E}_k(t)$ of external obstacles. The problem is for both types to simultaneously reach the target, avoiding the obstacles and the possible mutual collisions, while performing the overall process in minimal time. The problem solution is described in terms of the Hamiltonian formalism.

Keywords Coordinated control · Closed-loop control strategies · Non-linear feedback · Hamiltonian formalism · Group motion

1 Introduction

The problems of designing feedback (closed-loop) control strategies under uncertainty and their optimization are at the core of research in mathematical theory of control, having yielded a vast literature that includes topics related to this paper (see [1, 2, 4, 5, 9, 13]). At the same time the directions of recent research begun to involve control problems for groups of motions motivated by numerous examples from applied areas, such as [3, 7, 15, 16]. However the problem of optimizing group controls was beyond serious attention. Continuing previous research of the author (see [10–12]), the present chapter deals with the problem of optimal control for group motions. Namely, given a team of m controlled motions, a target set $\mathcal{M}$ and an array $\mathbf{E}_i$ of external obstacles, the team is to simultaneously reach the target while avoiding these obstacles and also the possible mutual internal collisions. The further point is that the time for such performance *should be minimal*.

The specifics of the team is that each member is described by a ball with its own *safety radius* where no other team member would be present. And the overall team consists of two types of members—those with a larger safety radius R and those with a smaller one r. On the other hand it is desirable to keep all the identical team

A. B. Kurzhanski (✉)
Lomonosov Moscow State University, GSP-2, Moscow 119992, Russia
e-mail: kurzhans@mail.ru; kurzhans@cs.msu.ru

Y. P. Kondratenko et al. (eds.), *Advanced Control Techniques in Complex Engineering Systems: Theory and Applications*, Studies in Systems, Decision and Control 203, https://doi.org/10.1007/978-3-030-21927-7_3

members near each other for communication reasons. To cope with such constraints, we require the team to move within two nonintersecting virtual containers described by ellipsoidal-valued tubes with cross-sections of two different sizes. The distance between the obstacles are assumed such that only some of them allow the passage of the larger team members.

Hence the overall problem is therefore one of motion planning aimed at simultaneously reaching the target $\mathcal{M}$ in minimum time. The last requirement may be ensured by *branching the motion* into two non-intersecting sub-tubes that join each other when reaching M at the final optimal time, which is now properly defined.

Presented in this paper is a rigorous approach to solving such problems. Here, firstly, we state the main mathematical problem of *tube target control* under obstacles for the virtual tubes that describe the two types of containers. Then, given the motion of described containers, we secondly indicate the feedback control strategies for each team member within its container. Such solution would steer the team to the target set avoiding external obstacles and internal collisions within the team. We thus separate the original problem into two sub-problems: the one of describing the ellipsoidal container tubes responsible for avoiding external obstacles while reaching the target set and the other—of obtaining the feedback controls for team members inside the related containers under internal collision avoidance. The latter also satisfy the joint external state constraints enforced by related container tube.

The described new types of feedback control problems are formulated on finite time-intervals and their solutions are a response to new mathematical challenges, that should also include design of appropriate numerical techniques. The present work is based on results obtained in papers [7, 8, 12]. The discussed methods rely on Hamiltonian techniques for systems with matrix-valued state variable treating which would require tensor-valued relations. Here such matrix may either describe the configuration of an ellipsoid tube or constructed of state vectors of the team. Solvability conditions for the target team control problem are obtained in the form of equations of the modified Hamilton–Jacobi–Bellman type. The final control solution in terms of feedback strategies is then proposed for the team.

Obtained relations thus indicate how to solve considered problems to the end. This is done by applying the theory of trajectory tubes (see [14]).

2 Initial Positions

We consider a team of m members $x^j \in \mathbb{R}^n$, each of which is described as a ball of radius r,—a safety zone around x^j, $\mathcal{B}_r\left(x^j\right) = x^j + \mathcal{B}_r(0)$, where

$$\mathcal{B}_r(c) = \left\{x : \langle x - c, x - c\rangle \leqslant r^2\right\}.$$

The team consists of two sub-teams with two different safety radii: m_1 members with safety radius r and m_2—with safety radius R, $0 < r < R$, and $m_1 + m_2 = m$.

Assume that members $1, 2, \ldots, m_1$ have safety radius r, and members $m_1 + 1, m_1 + 2, \ldots, m$ have safety radius R. We will further construct two branching virtual ellipsoidal motions for each sub-team. These are $\mathbf{E}_r[t] = \mathcal{E}(q_r(t), Q_r(t))$, $\mathbf{E}_R[t] = \mathcal{E}(q_R(t), Q_R(t))$—(the *virtual containers* for each sub-team), such that

$$\mathbf{E}_r[t] \bigcap \mathbf{E}_R[t] = \emptyset, \quad t \in [t_0, \theta],$$

In order to do so, we first construct the initial ellipsoids for each sub-team.

Let $x_i = \{x_1^i, x_2^i, \ldots, x_n^i\} \in \mathbb{R}^n$ denote the position of the ith team member. Consider two axis-aligned bounding boxes:

$$B_r(\{x_i\}_{i=1}^{m_1}) = \Big\{ x \in \mathbb{R}^n : \min_{i=1,\ldots,m_1} x_k^i - r \leqslant x_k \leqslant \leqslant \max_{i=1,\ldots,m_1} x_k^i + r,\ k = 1, \ldots, n \Big\},$$

$$B_R(\{x^i\}_{i=m_1+1}^{m}) = \Big\{ x \in \mathbb{R}^n : \min_{i=m_1+1,\ldots,m} x_k^i - R \leqslant x_k \leqslant \leqslant \max_{i=m_1+1,\ldots,m} x_k^i + R,\ k = 1, \ldots, n \Big\}.$$

Denote an ellipsoid in $\mathbb{R}^n$ with center q and configuration matrix $Q = Q^\top > 0$ as

$$\mathcal{E}(q, Q) = \{x \in \mathbb{R}^n : \langle x - q, Q^{-1}(x - q)\rangle \leqslant 1\}.$$

Now the initial ellipsoids can be obtained as those of minimal volume that contain the appropriate bounding box:

$$\begin{gathered} \det Q_r \to \min, \\ \langle p_i - q_r, Q_r^{-1}(p_i - q_r)\rangle \leqslant 1,\ i = 1, \ldots, 2^n, \\ \det Q_R \to \min, \\ \langle w_i - q_R, Q_R^{-1}(w_i - q_R)\rangle \leqslant 1,\ i = 1, \ldots, 2^n, \end{gathered}$$

where p_i and w_i are, respectively, the corner points of boxes $B_r(\{x^i\}_{i=1}^{m_1})$ and $B_R(\{x^i\}_{i=m_1+1}^{m})$.

These problems can be efficiently solved using methods of convex analysis, see [1, 17].

3 The Main Problem

We now proceed with the main control problem of this paper. Assume that each team member has the following Newtonian dynamics:

$$\ddot{x}^i = f(t, x^i, \dot{x}^i, u^i),\ t \in [t_0, \theta].$$

Here u^i are the individual controls bounded by geometrical restrictions $u^i \in \mathcal{P} \subset \mathbb{R}^p$, where $\mathcal{P}$ is a convex compact set. We further use a matrix notation. Namely, set

$$x = \{x^1, x^2, \ldots, x^m\},\ x^j \in \mathbb{R}^n,$$

and

$$\mathbf{x}^j = [x^{j'}; \dot{x}^{j'}]' \in \mathbb{R}^{2n},\ \mathbf{X} = [\mathbf{x}^1, \ldots, \mathbf{x}^m] \in \mathbb{R}^{2n \times m}.$$

For each sub-team we introduce related boxes $\mathbf{X}_r, \mathbf{X}_R$ as

$$\begin{aligned} \mathbf{X}_r &= [\mathbf{x}^1, \ldots, \mathbf{x}_1^m] \in \mathbb{R}^{2n \times m_1}, \\ \mathbf{X}_R &= [\mathbf{x}^{m_1+1}, \ldots, u^m] \in \mathbb{R}^n \\ U_r &= \{u^1, \ldots, u_1^m\} \in \mathbb{R}_1^m, \\ U_R &= \{u^{m_1+1}, \ldots, u^m\} \in \mathbb{R}_2^m, \end{aligned}$$

Hence, we have

$$\begin{aligned} \dot{\mathbf{X}}_r &= \mathbf{F}(t, \mathbf{X}_r, U_r), \\ \dot{\mathbf{X}}_R &= \mathbf{F}(t, \mathbf{X}_R, U_R). \end{aligned}$$

We further introduce *distance matrices*

$$\mathbf{D}_r[t] = \{D_{ij}^r[t]\}_{i,j=1}^m,\ \mathbf{D}_R[t] = \{D_{ij}^R[t]\}_{i,j=1}^m$$

with elements

$$D_{ij}[t] = D(\mathcal{B}_r\left(x^i(t)\right), \mathcal{B}_r\left(x^j(t)\right)),$$

where $D(X, Y)$ is the Euclidian distance between two convex compact sets:

$$D(X, Y) = \min\{\|x - y\| \mid x \in X,\ y \in Y\}.$$

We thus have

$$D_{ij}[t] = \max\{0, \left\|x^i(t) - x^j(t)\right\| - 2r\}.$$

Also given are target sets $\mathcal{M}^r = \mathcal{E}(m_r, M_r)$, $\mathcal{M}^R = \mathcal{E}(m_R, M_R)$ and an array of disjoint static or moving ellipsoids $E_i = \mathcal{E}(z^{(i)}, Z^{(i)}),\ i = 1, \ldots, i_z$, that serve as obstacles to be avoided.

The aim of the constrained controls U_r, U_r would be to transfer the tubes $\mathbf{E}_r$, $\mathbf{E}_R$ from their given starting states of type $\mathbf{E}_r[t_0]$, $\mathbf{E}_R[t_0]$ to related subsets $\mathbf{E}_r[\vartheta] \subset \mathcal{M}^r$ (and similar for $\mathbf{E}_R$) in times $\vartheta_r - t_0,\ \vartheta_R - t_0$, accordingly, under following type of constraints (when placing lower lower indexes r and R):

$$k_-^2 \leqslant \sum_{j=1}^{n} \lambda_j(Q(t)) = \mathrm{Trace}(Q'(t)Q(t)) \leqslant k_+^2 \tag{1}$$

on the sum of squared semiaxes of $\mathbf{E}[t]$. Here Trace Q stands for the trace of matrix Q and $\lambda_j(Q)$ for its eigenvalues. The lower bound k_-^2 in (1) must be large enough to ensure inclusion of the respective sub-system, while the upper k_+^2 is to ensure availability of communications within the team inside.

The mentioned restrictions should also be complemented by external state constraints $\mathbf{E}[t] \cap E_i = \varnothing$. The state restrictions on $\mathbf{E}[t]$ are therefore *a combination of external an internal constraints*.

Now the subproblems of target team control for each of our sub-systems may be formulated as follows.

The Time-Optimal Problem for Tubes $\mathbf{E}_r[t]$, $\mathbf{E}_R[t]$.

Design a class of closed-loop control strategies U_r, U_R with set-valued maps $\mathbf{E}_r[t] = \mathcal{E}(q_r(t), Q_r(t))$, and $\mathbf{E}_R[t] = \mathcal{E}(q_R(t), Q_R(t))$ (the *virtual containers*), such that they do not collide

$$\mathbf{E}_r[t] \bigcap \mathbf{E}_R[t] = \emptyset, \quad t \in [t_0, \theta], \tag{2}$$

while the following conditions are fulfilled:

1. The pair of sub-teams is steered towards the pair of target sets $\mathcal{M}^r$, $\mathcal{M}^R$ so that:

$$\mathbf{X}_r[t_0] \underset{u(\cdot)}{\longrightarrow} \mathbf{X}_r[\theta] \subseteq \mathcal{M}^r + \varepsilon\mathcal{B}_1(0) = \mathcal{M}^r_\varepsilon,$$

$$\mathbf{X}_R[t_0] \underset{u(\cdot)}{\longrightarrow} \mathbf{X}_R[\theta] \subseteq \mathcal{M}^R + \varepsilon\mathcal{B}_1(0) = \mathcal{M}^R_\varepsilon,$$

   and

$$\mathcal{M}^r \bigcup \mathcal{M}^R \subset \mathcal{M}, \; \mathcal{M}^r \bigcap \mathcal{M}^R = \emptyset,$$

$$\mathcal{M}^r_\varepsilon \bigcup \mathcal{M}^R_\varepsilon \supset \mathcal{M},$$

   jointly, within the minimally possible time $\theta^0 - t_0$.
2. Collisions between the team members are avoided due to existence of safety radii r, R. This requires

$$D_{ij}[t] \geqslant \sigma > 0, \; i \neq j.$$

3. The *team proximity conditions* are true, namely,

$$\mathcal{B}_r\left(x^j(t)\right) \subset \mathbf{E}_r[t], \; j = 1, \ldots, m_1, \; t \in [t_0, \theta],$$
$$\mathcal{B}_R\left(x^j(t)\right) \subset \mathbf{E}_R[t], \; j = m_1 + 1, \ldots, m, \; t \in [t_0, \theta].$$

4. Collisions with external obstacles $\mathbf{E}_k[t]$ are avoided:

$$D(\mathbf{E}_r[t], \mathbf{E}_k[t]) > 0, \; k = 1, \ldots, k_0, \; t \in [t_0, \theta],$$
$$D(\mathbf{E}_R[t], \mathbf{E}_k[t]) > 0, \; k = 1, \ldots, k_0, \; t \in [t_0, \theta].$$

5. The volume of the container must ensure *capacity of the team*, allowing enough space for placing its members within, under conditions (2), (3). This requires relations of type

$$\lambda_-^r \leqslant \langle Q_r(t), Q_r(t) \rangle \leqslant \lambda_+^r, \; t \in [t_0, \theta], \; 0 < \lambda_-^r < \lambda_+^r,$$

$$\lambda_-^R \leqslant \langle Q_R(t), Q_R(t) \rangle \leqslant \lambda_+^R, \; t \in [t_0, \theta], \; 0 < \lambda_-^R < \lambda_+^R.$$

where values $\lambda_-^r, \lambda_+^r, \lambda_-^R, \lambda_+^R$ are given.

Here the external obstacles are described by an array of ellipsoids—fixed $\mathbf{E}_k = \mathcal{E}(q_k, Q_k)$, or moving $\mathbf{E}_k[t]$.

The containers $\mathbf{E}[t]$ and external obstacles $\mathbf{E}_k[t]$ are selected as ellipsoidal-valued functions that produce *ellipsoidal tubes*.

The equations of motion for such container tubes $\mathbf{E}[t] = \mathcal{E}(q(t), Q(t))$ are taken for—the centers $q^{(c)}(t)$ as

$$\dot{q}^{(c)} = H(t)q^{(c)} + C(t)v, \; q^{(c)}(t_0) = q_0^{(c)}, \; \langle v, v \rangle \leqslant \mu^2.$$

and for the configuration matrices $Q^{(c)}(t)$ as

$$\dot{Q}^{(c)} = T(t)Q^{(c)} + Q^{(c)}T(t) + V(t), \; Q^{(c)}(t_0) = Q_0^{(c)}, \; \text{Trace}\, V \leqslant \nu^2.$$

Note that for the ellipsoidal containers, the parameters $q(t)$, $Q(t)$ are controlled through inputs $\{v, V\}$.

The above equations are now applied to design of two related types of Hamilton–Jacobi–Bellman (HJB) equations, namely,

$$\partial V^{(c)}/\partial t + \min_v \{ \langle \partial V^{(c)}/\partial q^{(c)}, dq^{(c)}(t)/dt \rangle \} = 0, \tag{3}$$

and

$$\partial V^{(c)}/\partial t + \min_V \{ \langle \partial V^{(c)}/\partial Q^{(c)}, dQ^{(c)}(t)/dt \rangle \} = 0. \tag{4}$$

where the matrix derivative of type $\partial V^{(c)}/\partial Q^{(c)}$ is treated in the sense of Frechet.[1]

Applying these equations to our case, the lower indexes for $V^{(c)}(t, q^{(c)})$ should be $V_r^{(c)}(t, q_r^{(c)})$ and $V_R^{(c)}(t, q_R^{(c)})$, while for $V^{(c)}(t, Q^{(c)})$ they should be $V_R^{(c)}(t, q_R^{(c)})$ and $V_R^{(c)}(t, Q_R^{(c)})$.

[1]See R. Coleman, Calculus on normed vector spaces, Universitext, Springer, 2012.

4 The Problem Solution

We will design the solution using dynamic programming techniques.

Introduce the next two system positions:

$$Z_r[t] = \{t, \mathbf{X}_r, \mathbf{E}_r\} \in \mathbb{R} \times \mathbb{R}^{2n\times m_1} \times \operatorname{conv}\mathbb{R}^n),$$

$$Z_R[t] = \{t, \mathbf{X}_R, \mathbf{E}_R\} \in \mathbb{R} \times \mathbb{R}^{2n\times m_2} \times \operatorname{conv}\mathbb{R}^n).$$

Here $\operatorname{conv}\mathbb{R}^n$ is the space of all compact convex sets in $\mathbb{R}^n$.

The *total position* of the system will then be

$$\mathbf{Z}[t] = \{t, \mathbf{X}_r, \mathbf{X}_R, \mathbf{E}_r, \mathbf{E}_R\} \in \mathbb{R} \times \mathbb{R}^{2n\times m} \times \mathbb{R}^{2n\times m} \times \operatorname{conv}\mathbb{R}^n \times \operatorname{conv}\mathbb{R}^n.$$

We now first indicate the optimal time functions for steering individually each of the two tubes $\mathbf{E}_r[t]$, $\mathbf{E}_R[t]$ to their target sets, avoiding obstacles, within $(t_0 \leqslant t \leqslant \vartheta)$.

Let $W_r[t]$ be the set of positions $\{t, \mathbf{E}_r[t]\}$, from which the problem of steering tube $\mathbf{E}_r[t]$ towards target subset $\mathcal{M}^r$, at time $t = \vartheta$, while avoiding the obstacles $\mathbf{E}_k$, is solvable. Then it generates the *solvability tube* $W_r[\tau]$, $\tau \in [t, \vartheta]$. A similar definition applies to solvability tube $W_R[t]$. The definition of such sets are also given in [14], Chaps. 3, 10.

Further introduce functions

$$T_r(t, \mathbf{E}_r) = \max_{\mathbf{E_r}(\cdot)\in W_r[\cdot]} \min_{U_r(\cdot)} \in \mathcal{P}_1^m\{\theta - t : \mathbf{E_r}(\cdot),\ \mathbf{E_r}[\theta] \in \mathcal{M}^{\mathbf{r}} \mid \mathbf{E_r} = \mathbf{E_r[t]}\},$$

$$T_R(t, \mathbf{E}_R) = \max_{\mathbf{E_R}\in W_R[\cdot]} \min_{U_R(\cdot)} \in \mathcal{P}_2^m\{\theta - t : E_R[\cdot],\ \mathbf{E}_R(\theta) \in \mathcal{M}^R \mid \mathbf{E}_R = \mathbf{E}_R[t]\},$$

with optimizers $\{\vartheta_r^0(t), \vartheta_R^0(t)\}$.

Each of the indicated functions represents a problem of controlling an ellipsoidal-valued tube. The techniques for solving such problems are described in [14], Chaps. 2.6 and 8.4.

Then *the optimal group solution time* $T^0 = \vartheta^0 - t_0$ could be found as

$$T^0 = T(t_0, \mathbf{E}_r[t_0], \mathbf{E}_R[t_0]), \tag{5}$$

where

$$\mathbf{X}_r[\vartheta^0] \subset \mathcal{M}_\varepsilon^r,\ \mathbf{X}_R[\vartheta^0] \subset \mathcal{M}_\varepsilon^R,\ \ \vartheta^0 = \max\{\vartheta_r^0, \vartheta_R^0\}.$$

and the given target sets are

$$\mathcal{M}^r \subset \operatorname{conv}\{\mathcal{M}\}, \mathcal{M}^R \subset \operatorname{conv}\{\mathcal{M}\}.$$

Since the previous HJB equations for the containers' dynamics do not depend on team members, there should be a *time synchronization of the containers and the team.* It must be checked that the motion of the containers would be consistent with the internal motions of the team. Such synchronization is done by tuning the container parameters since these are virtual.

Assume

$$\mathbf{T}^i_r(t,\mathbf{x}^0) = \min_{u_i \in \mathcal{P}}\{t : x^i(t,\mathbf{x}^0,u_i) \in \mathcal{M} \text{ under condition (5)}\},$$
$$\mathbf{T}^j_R(t,\mathbf{x}^0) = \min_{u_j \in \mathcal{P}}\{t : \mathbf{x}^j(t,\mathbf{x}^0,u_j) \in \mathcal{M} \text{ under condition (5)}\},$$

where $i = 1,2,\ldots,m_1$ and $j = m_1+1,\ldots,m$, are the *individual* optimal times for each team member. Obtaining them is a standard problem of optimal control.

Next, define

$$\mathbf{T}_r(t_0) = \max_{i=1,2,\ldots,m_1} \mathbf{T}^i_r(t_0,\mathbf{x}^0_i), \ \mathbf{T}_R = \max_{j=m_1+1,\ldots,m} \mathbf{T}^j_R(t_0,\mathbf{x}^0_j). \tag{6}$$

The *initial time synchronization* will then consist in tuning the container parameters, ensuring:

$$\theta^0 - t_0 \geqslant \max\{\mathbf{T}_r, \mathbf{T}_R\}. \tag{7}$$

5 The Motions of the Teams

We now describe the motions of the teams within the containers $\mathbf{E}_r[t]$, $\mathbf{E}_R[t]$ in terms of the Hamiltonian formalism. Since motions within both containers are similar, we do that for only the first r-container.

Here each team member must satisfy both internal and external state constraints. Hence we firstly consider the external convex state constraints that keeps each team member within its container, having

$$\mathcal{B}_r\left(x^j(t)\right) \in \mathbf{E}_r(t)\forall t \in [t_0,\theta], \ \forall j.$$

We further use the Hausdorff semi-distance $h_+(A,B)$ for sets A, B:

$$h_+(A,B) = \min\{\varepsilon > 0 : \ B + \varepsilon\mathcal{B}_\varepsilon(0) \supseteq A\}$$

Then, for an individual motion $x^j(t)$, the control $U^j(t,\mathbf{X})$ that keeps it within container $\mathbf{E}[t]$ must produce

$$h^+_{jq}(t) = (h^2_+(x^j(t),\mathcal{E}_r(t)))_+ \geqslant 0,$$

which is ensured with

$$U^{jq}(t,\mathbf{X}) = \begin{cases} u \in \mathcal{P}: \dfrac{d\{h^{+}_{jq}(t)\}}{dt} \leqslant 0,\ d_r^2(t,\mathbf{x}^j) > 0, \\ \mathcal{P}, & \text{otherwise.} \end{cases}$$

The corresponding additional terms in the related HJB equation given below will be

$$H^{+}_{jq}(t) = (h^2_{+}(x^j(t), \mathcal{E}_r(t)))_{+}.$$

Next we proceed to internal state constraints that provide collision avoidance between team members while moving inside the corresponding container. Recall that the distance matrix $\mathbf{D}_r[t] = \{D_{ij}[t]\}_{i,j=1}^{m}$ is defined as

$$D_{ij}[t] = D\left(\mathcal{B}_r\left(x^{(i)}(t)\right), \mathcal{B}_r\left(x^{(j)}(t)\right)\right).$$

The controls that provide $D_{ij}[t] \geqslant \sigma$ should satisfy condition:

$$U^{ij}(t,\mathbf{X}) = \begin{cases} u \in \mathcal{P}: \dfrac{d\{D_{ij}(t,\mathbf{x})\}}{dt} \geqslant 0,\ D_{ij}(t,\mathbf{x}) = 0, \\ \mathcal{P}, & \text{otherwise.} \end{cases}$$

The corresponding additional terms in the related HJB equation below will be

$$H_{ij}[t] = (h^{+}(r^2 - \left\|x^i(t) - x^j(t)\right\|^2))_{+},\ i > j,$$

hence the total term for internal non-collisions will be

$$H_{+}(t,\mathbf{x}_r,\mathcal{E}_r[t]) = \sum_{j=1}^{m_1} H^{+}_{jq}[t] + \sum_{1=i<j}^{m_1} H_{ij}[t].$$

Then we can introduce the value function

$$V^{\vartheta}(t,\mathbf{x}_r^0) =$$
$$= \min_u \left\{ \int_t^{\vartheta} H_{+}(\mathbf{x}_r, \mathcal{E}_r[t])dt + h_{+}(\mathbf{x}_r(\vartheta), \mathcal{M}^r) \mid \mathbf{x}_r(t) = \mathbf{x}_r^0,\ u_r \in \mathcal{P}_r \right\},$$

and the corresponding HJB equation for $V^{\vartheta}(t,\mathbf{x}_r)$ will be:

$$\frac{\partial V^{\vartheta}}{\partial t} + \min_{U_r}\left\{\left\langle \frac{\partial V^{\vartheta}}{\partial \mathbf{x}_r}, F(t,\mathbf{x}_r,U_r)\right\rangle + h_{+}(t,\mathbf{x}_r,\mathcal{E}_r[t])\right\} = 0,$$

with boundary condition

$$V^{\vartheta}(\vartheta, \mathbf{x}_r) = h_+(\mathbf{x}_r(\vartheta), \mathcal{M}^r).$$

Here $\partial V^{\vartheta}/\partial \mathbf{x}_r$ is again a *matrix derivative* understood in the sense of Frechet.

The solvability set for target control under present constrains will be

$$\mathcal{W}[t_0] = \{\mathbf{x}_r : V^{\vartheta}(t_0, \mathbf{x}_r) \leqslant 0\},$$

which consists of all individual starting positions $\mathbf{x}_r$ that allow an admissible control $U_r(\cdot)$ that solves the sub-problem "r" on $\mathcal{M}^r$ target control.

Theorem. Under condition

$$\mathcal{U}(t, \mathbf{x}_r) = \left(\bigcap_{i<j} U^{ij}(t, \mathbf{x}_r)\right) \bigcap \left(\bigcap_{i=1}^{n} U^{iq}(t, \mathbf{x}_r)\right) \neq \varnothing,$$

the group control problem "r" is solvable if and only if $\mathcal{W}[t_0] \neq \varnothing$. The situation for group control problem "R" is similar.

Remark It is possible that for solving group problems "r', "R" it would be necessary to have a *reconfiguration* of the team members within related containers.

6 The Overall Solution and Calculation Scheme

The solution with calculation scheme are realized on the basis of methods described in publications [2, 4, 6, 14].

Given pair of starting positions $\mathbf{X}_r[t_0]$, $\mathbf{X}_R[t_0]$,

1. Calculate optimal time $T^0 = \vartheta^* - t_0$ using formula (5).
2. Construct appropriate tube of *solvability sets* $W_r[t_0]$, $W_R[t_0]$ as level sets of $T(t), t = t_0$.The definitions of such sets are given in [14], Chaps. 3, 10.
3. Construct internal ellipsoidal approximations $W_r^-[t]$ of $W_r[t]$ and $W_R^-[t]$ of $W_R[t]$, using methods of ellipsoidal calculus [14], Chaps. 10.5, 10.6, [6].
4. Construct controls that aim the system towards the sets $W_r^-[t]$, $W_R^-[t]$ (see [2, 14], Chaps. 10.5, 10.6).

7 Conclusion

This paper describes a class of group target control problems for a system with Newtonian dynamics under joint optimal time-minimal performance. Its mostly open-loop solution is here achieved in terms of the Hamiltonian formalism and methods of convex analysis. Further on it would be useful to investigate obstacles that would require solutions by teams with a leader. Apart from that, the design of closed-loop controls would require exchange of on-line information in between the team members and also with a possible central control, center using related communication networks. Of interest are also new optimization problems for joint design of collective motion control with communication schemes.

References

1. Boyd, S., Vandenberghe, L.: Convex Optimization. Cambridge University Press, Boston (2004)
2. Krasovski, N.N., Subbotin, A.I.: Game-Theoretic Control Problems. Springer, Berlin (1988)
3. Kumar, V., Leonard, N., Morse, A.S. (eds.): Cooperative Control. Lecture Notes in Control and Information Sciences, vol. 309. Springer, Berlin (2004)
4. Kuntsevich, V.M.: Control Under Uncertainty: Guaranteed Results in Control and Identification Problems. Naukova Dumka, Kyiv (2006)
5. Kuntsevich, V.M., Gubarev, V.F., Kondratenko, Y.P., Lebedev, D.V., Lysenko, V.P. (eds.): Control Systems: Theory and Applications. Automation, Control and Robotics. River Publishers, Gistrup (2018)
6. Kurzhanski, A.B.: Comparison principle for equations of the Hamilton–Jacobi type in control theory. Proc. Steklov Inst. Math. **251**, 185–195 (2006)
7. Kurzhanski, A.B.: On a team control problem under obstacles. Proc. Steklov Inst. Math. **291**(1), S122–S142 (2015)
8. Kurzhanski, A.B.: Problem of collision avoidance for a group motion with obstacles. Trudy Inst. Mat. I Mekh. UrO RAN **21**(2), 134–149 (2015)
9. Kurzhanski, A.B., Kuntsevich, V.M.: Attainability domains for linear and some classes of nonlinear discrete systems and their control. J. Autom. Inf. Sci. **42**(1), 1–18 (2010)
10. Kurzhanski, A.B., Mesyats, A.I.: Ellipsoidal motions for applied control: from theory to computation. In: Proceedings of CDC 2013 (the 52nd IEEE Conference on Decision and Control), pp. 5816–5821 (2013)
11. Kurzhanski, A.B., Mesyats, A.I.: The mathematics of team control. In: Proceedings of MTNS 2014 (the 21st International Symposium on Mathematical Theory of Networks and Systems), pp. 1755–1758 (2014)
12. Kurzhanski, A.B., Mesyats, A.I.: The Hamiltonian formalism for problems of group control under obstacles. In: Proceedings of NOLCOS 2016 (the 10th IFAC Symposium on Nonlinear Control Systems) (2016)
13. Kurzhanski, A.B., Varaiya, P.: On synthesizing target controls under obstacle and collision avoidance. J. Frankl. Inst. **347**(1), 130–145 (2010)
14. Kurzhanski, A.B., Varaiya, P.: Dynamics and Control of Trajectory Tubes: Theory and Computation. Birkhäuser, Boston (2014)
15. Nijmeijer, H., Pettersen, K.Y., Gravdahl, J.T. (eds.): Group Coordination and Cooperative Control. Lecture Notes in Control and Information Sciences, vol. 336. Springer, Berlin (2006)
16. Olfati-Saber, R.: Flocking for multi-agent dynamic systems: algorithms and theory. IEEE Trans. Autom. Control **51**(3), 5816–5821 (2006)
17. Rockafellar, R.T.: Convex Analysis. Princeton University Press, Princeton (1972)

Model Predictive Control for Discrete MIMO Linear Systems

Vyacheslav F. Gubarev, M. D. Mishchenko and B. M. Snizhko

Abstract Mathematical model in the form of discrete linear system of equations with multiple input and multiple output (MIMO) describes different important for practice dynamic processes. Model predictive control (MPC) in recent years is a very intensively developing field of control theory and application. In given chapter, some of the possible realizations of MPC for discrete MIMO linear system are considered. Developed approach is based on the Cauchy formula for discrete linear system. Problem of control design is reduced to solving series of extremal problems including constrained optimization problems. It has succeeded to obtain final results in analytical form which is appropriate for solving different problems including control analysis and design.

Keywords Model predictive control · MIMO system · Discrete system · Optimization · Constraints · Stabilization

1 Introduction

In recent years, model predictive control (MPC), also referred to as moving horizon control or receding horizon control, became popular and widespread for control in the process industries and many other applications. It is worth mentioning that MPC problem may be treated as a finite horizon open-loop optimal control. Contemporary computer technology allows us to calculate control on-line using current

---

V. F. Gubarev
Space Research Institute of NAS and SSA of Ukraine, Kiev, Ukraine
e-mail: v.f.gubarev@gmail.com

M. D. Mishchenko (✉) · B. M. Snizhko
Institute for Applied System Analysis, National Technical University of Ukraine "Igor Sikorsky Kyiv Polytechnic Institute", Building No. 35, Prospect Peremohy, 37-A, Kyiv 03056, Ukraine
e-mail: mishenkomihailo@gmail.com

B. M. Snizhko
e-mail: dansnow@ukr.net

Y. P. Kondratenko et al. (eds.), *Advanced Control Techniques in Complex Engineering Systems: Theory and Applications*, Studies in Systems, Decision and Control 203, https://doi.org/10.1007/978-3-030-21927-7_4

measurements and, if necessary, moving horizon state estimation. There are many publications devoted to MPC, where different approaches to this problem have been developed for linear or nonlinear systems [1–5].

For linear systems with constraint, the work [6] represents a significant leap forward in the MPC theory, where a receding horizon control scheme with infinite prediction horizon and finite control horizon is proposed. For both stable and unstable systems nominal closed-loop stability is achieved by MPC control with constraints.

Optimal MPC control for a general class of constrained discrete time systems is considered in [7], including stability problem and moving horizon approximations.

In this chapter, an original approach to predictive control on finite moving horizon for discrete multiple input and multiple output (MIMO) linear systems is developed. Discrete analog of the Cauchy formula expressed through controllability matrix is used for this. Section 2 describes the mathematical formulation of the MPC problem which is based on linear discrete state-space model. Variational method of the MPC problem solving is developed in Sect. 3. Theoretical results developed in this section allow to write optimal control law in analytical form of standard linear feedback. There are two ways to solve optimal problem analytically. Both of them give the state set where the generated feedback conforms to the control constraints. Advanced optimal method of the MPC problem solving with constraints is considered in Sect. 4. Original method of consecutive approach to terminal state when it does not belong to permissible by control constraint set is proposed in Sect. 5. Since methods described in Sects. 4 and 5 involve exhaustive search, a specific search order is proposed in Sect. 6. Some practical results of MPC stabilization problem solving are described in Sect. 7.

2 Formulation of the Problem

Let there be a controlled discrete dynamic MIMO system. To implement a controlled process based on MPC in such a system, we consider a sliding interval $[k, k+s]$ where k determines the current process time, and s defines the size of the sliding interval or moving horizon. In this case, dependence $s = \mathrm{s}(k)$ is also assumed. In the linear case, the MIMO model has the form (2.1), where $\mathbf{x}_k$ is an $n \times 1$ state vector of system at the moment of time k; A is an $n \times n$ matrix; $\mathbf{u}_k$ is an $r \times 1$ vector of control; B is an $n \times r$ matrix.

$$\mathbf{x}_{k+1} = \mathrm{A}\mathbf{x}_k + \mathrm{B}\mathbf{u}_k \tag{2.1}$$

Prediction of the state vector at the moment $k+s$, i.e. at the end of the sliding interval, is determined by the Cauchy formula for a discrete process. For a sequence of controls $(\mathbf{u}_k, \mathbf{u}_{k+1}, \ldots, \mathbf{u}_{k+s-1})$ we get (2.2).

$$\mathbf{x}_{k+s} = \mathrm{A}^s\mathbf{x}_k + \mathrm{A}^{s-1}\mathrm{B}\mathbf{u}_k + \mathrm{A}^{s-2}\mathrm{B}\mathbf{u}_{k+1} + \cdots + \mathrm{B}\mathbf{u}_{k+s-1} \tag{2.2}$$

Formula (2.2) defines the state of the system at the end of the sliding interval by using the initial state $\mathbf{x}_k$ and a finite horizon control applied to the system. On its basis we set the task of synthesizing sequence $\{\mathbf{u}_j\}$, $j = k, k+1, \ldots, k+s-1$ that provides reaching a specified state $\mathbf{x}_{k+s}$ at the moment $k+s$. For this purpose, we write (2.2) in form (2.3).

$$\mathrm{B}_1\mathbf{u}_k + \mathrm{B}_2\mathbf{u}_{k+1} + \cdots + \mathrm{B}_s\mathbf{u}_{k+s-1} = \mathbf{x}_s^* \tag{2.3}$$

Here $\mathrm{B}_i = \mathrm{A}^{s-i}\mathrm{B}, i \in \overline{1..s}$ are $n \times r$ matrices, and $\mathbf{x}_s^* = \mathbf{x}_{k+s} - \mathrm{A}^s\mathbf{x}_k$ characterizes deviation between the target state at the moment $k+s$ and the state of the system when control values are equal to zero, i.e. with free motion on the sliding interval. Equation (2.3) allows a "soft" mode of reaching $\mathbf{x}_{k+s}$, when we have $\mathbf{u}_{k+s-1} = 0$ at the last step. Then the solution is determined for a subsequence $(\mathbf{u}_k, \mathbf{u}_{k+1}, \ldots, \mathbf{u}_{k+s-2})$. Obviously, for a fixed s increase of $\|\mathbf{x}_{k+s} - \mathrm{A}^s\mathbf{x}_k\|$ leads to increase of required control resource.

Let us denote the desired control on a moving horizon in the vector form by $\mathbf{u}(k,s) = \left(\mathbf{u}_{k+s-1}^{\mathrm{T}}, \mathbf{u}_{k+s-2}^{\mathrm{T}}, \ldots, \mathbf{u}_k^{\mathrm{T}}\right)^{\mathrm{T}}$. Then (2.3) can be written as (2.4), where $\Omega_s = (\mathrm{B}_s\mathrm{B}_{s-1}\ldots\mathrm{B}_1)$ is a controllability matrix composed of s blocks.

$$\Omega_s \cdot \mathbf{u}(k,s) = \mathbf{x}_s^* \tag{2.4}$$

From (2.4) we can deduce conditions under which problem (2.3) or (2.4) has a solution suitable for MPC. The length of moving horizon must be such that rank $\Omega_s = n$. If the system (2.1) is controllable, then such s exists. It is easy to determine the minimum value of $s = s_0$, starting from which this condition will be satisfied. To find it, one must take initial s such that the number of columns Ω_s will be greater than or equal to n, and strictly less than n for $s-1$. If for this s rank of Ω_s will be less than n then it is necessary to increment s by one. Equality rank $\Omega_s = n$ for this s means that we have found s_0. Choosing the length of the sliding interval $s \geq s_0$ ensures the solvability of the MPC synthesis problem. It is recommended to take s close (or equal) to s_0, while ensuring that (2.4) is an underdetermined system, i.e. the number of equations is less than the number of unknown variables. In this case (2.4) specifies the set of controls that implement MPC. On this set we should find the one that is most effective in terms of practical application. Some possible ways to determine it are presented in the following sections.

3 Variational Method of the Problem Solving

By applying the variational principle to this problem, it is possible to find a solution that, on the one hand, satisfies the controls' constraints, and, on the other hand, would have properties which are useful for practical implementation. But firstly, we must consider the constraints on the control variables. The simplest variant of such constraints is component constraints on the allowable vector $\mathbf{u}(k, s)$. Let $u_i(k, s)$ be an i-th component of $\mathbf{u}(k, s)$. We set up an interval within which it can change as described in (3.1).

$$-u_{0i} \leq u_i(k, s) \leq u_{0i}, i = \overline{1..sr} \quad (3.1)$$

For the sake of simplicity, we assume that $u_{0i} = b_{\mathrm{u}}$ for all i. Such a simplification is not fundamental for the method under consideration and in many cases is acceptable from a practical point of view.

The most suitable case for an MPC with constraints is the case when (2.4) is an underdetermined system of linear equations. In this case, the solution can be obtained on the basis of the variational principle, for example, from the extremal problem (3.2), where $\mathrm{U}_{\mathrm{per}}$ is represented by linear constraints in the form of equalities (2.4).

$$\min_{\mathbf{u}(k,s)\in \mathrm{U}_{\mathrm{per}}} \sum_{j=1}^{sr} u_j^2(k, s) \quad (3.2)$$

As a result, we have a quadratic programming problem with constraints in the form of linear equations. It can be reduced to a system of linear equations by using Lagrange multipliers. Expression (3.2) is equivalent to the unconditional extremum problem which is described as (3.3), where $\boldsymbol{\lambda}$ is a vector of Lagrange multipliers which should be taken such that the Eq. (2.4) is satisfied.

$$\sum_{j=1}^{sr} u_j^2(k, s) + \left\langle \boldsymbol{\lambda}, \Omega_s \cdot \mathbf{u}(k, s) - \mathrm{x}_s^* \right\rangle \to \min \quad (3.3)$$

We will write the controllability matrix Ω_s in the form of (3.4), where $\boldsymbol{\Omega}_i^{(s)}$, $i \in \overline{1..sr}$ are corresponding columns of Ω_s.

$$\Omega_s = \left(\boldsymbol{\Omega}_1^{(s)} \boldsymbol{\Omega}_2^{(s)} \ldots \boldsymbol{\Omega}_{sr}^{(s)} \right) \quad (3.4)$$

The necessary optimality condition for (3.3), i.e. the Euler equation which takes into account (3.4), will have the form of (3.5).

$$2u_j(k, s) + \left\langle \boldsymbol{\lambda}, \boldsymbol{\Omega}_j^{(s)} \right\rangle = 0, \ j = \overline{1..sr} \quad (3.5)$$

If the matrix Ω_s has full rank, i.e. the system (2.1) is controllable, then condition (3.5) will be sufficient. It is necessary to add the system of equations (2.4) which will help us to determine the Lagrange multipliers. As a result, we get system of linear equations (3.6).

$$\begin{cases} 2u_j(k,s) + \left\langle \boldsymbol{\lambda}, \boldsymbol{\Omega}_j^{(s)} \right\rangle = 0,\ j = \overline{1..sr} \\ \Omega_s \cdot \mathbf{u}(k,s) = \mathbf{x}_s^* \end{cases} \tag{3.6}$$

It is easy to see that (3.6) gives the linear system of dimension $n + sr$ for finding an extended vector $\begin{pmatrix} \mathbf{u}(k,s) \\ \boldsymbol{\lambda} \end{pmatrix}$. The peculiarity of the system (3.6) allows us to write its solution in a simple analytical form. We represent the matrix Ω_s in the block form (3.7), where Ω_{s1} is composed of n $\boldsymbol{\Omega}_j^{(s)}$ vectors and Ω_{s2} contains all remaining.

$$\Omega_s = (\Omega_{s1}\Omega_{s2}) \tag{3.7}$$

Matrix Ω_{s1} is square and its constituent vectors $\boldsymbol{\Omega}_j^{(s)}$ must be chosen in a way so that matrix Ω_{s1} is not singular, i.e. det $\Omega_{s1} \neq 0$. This condition is feasible because the rank of the matrix Ω_s is equal to n. After this choice, we represent the vector $\mathbf{u}(k,s)$ in the form of two compound vectors $\mathbf{u}_1^{(s)}$ and $\mathbf{u}_2^{(s)}$. Vector $\mathbf{u}_1^{(s)}$ has dimension n and consists of those vector components $\mathbf{u}(k,s)$ that match the matrix Ω_{s1}. Then the second equation in (3.6) can be written as (3.8).

$$(\Omega_{s1}\Omega_{s2})\begin{pmatrix} \mathbf{u}_1^{(s)} \\ \mathbf{u}_2^{(s)} \end{pmatrix} = \mathbf{x}_s^* \tag{3.8}$$

As a result, it is possible to get (3.9) from the first equation in (3.6).

$$\mathbf{u}_1^{(s)} = -\frac{1}{2}\Omega_{s1}^{\mathrm{T}} \cdot \boldsymbol{\lambda},\ \mathbf{u}_2^{(s)} = -\frac{1}{2}\Omega_{s2}^{\mathrm{T}} \cdot \boldsymbol{\lambda} \tag{3.9}$$

Taking into account that det$\Omega_{s1} \neq 0$, the first equation in (3.9) can be used to find $\boldsymbol{\lambda}$, as it is described in (3.10).

$$\boldsymbol{\lambda} = -2\left(\Omega_{s1}^{\mathrm{T}}\right)^{-1}\mathbf{u}_1^{(s)} \tag{3.10}$$

By substituting $\boldsymbol{\lambda}$ in the second part of (3.9), we get (3.11).

$$\mathbf{u}_2^{(s)} = \Omega_{s2}^{\mathrm{T}} \cdot \left(\Omega_{s1}^{\mathrm{T}}\right)^{-1} \cdot \mathbf{u}_1^{(s)} \tag{3.11}$$

If the system (3.8) is written in the form (3.12), then we get (3.13) by substituting $\mathbf{u}_2^{(s)}$ with (3.11).

$$\Omega_{s1}\mathbf{u}_1^{(s)} + \Omega_{s2}\mathbf{u}_2^{(s)} = \mathbf{x}_s^* \tag{3.12}$$

$$\left[\Omega_{s1} + \Omega_{s2}\Omega_{s2}^{\mathrm{T}} \cdot \left(\Omega_{s1}^{\mathrm{T}}\right)^{-1}\right] \cdot \mathbf{u}_1^{(s)} = \mathbf{x}_s^* \tag{3.13}$$

Equation (3.13) allows us to express $\mathbf{u}_1^{(s)}$ through $\mathbf{x}_s^*$, and we can also express $\mathbf{u}_2^{(s)}$ in analytical form through $\mathbf{x}_s^*$ by using (3.11). As a result, we will have (3.14).

$$\begin{aligned}\mathbf{u}_1^{(s)} &= \left[\Omega_{s1} + \Omega_{s2}\Omega_{s2}^{\mathrm{T}} \cdot \left(\Omega_{s1}^{\mathrm{T}}\right)^{-1}\right]^{-1} \cdot \mathbf{x}_s^* \\ \mathbf{u}_2^{(s)} &= \Omega_{s2}^{\mathrm{T}} \cdot \left(\Omega_{s1}^{\mathrm{T}}\right)^{-1}\left[\Omega_{s1} + \Omega_{s2}\Omega_{s2}^{\mathrm{T}} \cdot \left(\Omega_{s1}^{\mathrm{T}}\right)^{-1}\right]^{-1} \cdot \mathbf{x}_s^*\end{aligned} \tag{3.14}$$

Thus, the resulting solution (3.14) allows to express each component of control vector $\mathbf{u}(k, s)$ on a moving interval through $\mathbf{x}_s^*$. They can be represented as linear functions of $\mathbf{x}_s^*$, as it is described in (3.15), where $\mathbf{k}_q^{(s)}$ depends only on the parameters of the model (2.1) and the length of the sliding interval.

$$u_q(k, s) = \left\langle \mathbf{k}_q^{(s)}, \mathbf{x}_s^* \right\rangle \tag{3.15}$$

By using the representation (3.15), we can take into account the constraints and establish the admissible membership set for $\mathbf{x}_s^*$, for each element of which controls (3.15) are realizable, i.e. satisfy the given constraints (3.1). Such a set can be built for any s, beginning with s_0.

Each of the $u_q(s, r)$ components is contained in the bounds, as it is described in (3.16), where s is fixed and greater than or equal to s_0. Intersection of sets (3.16) for all $q \in \overline{1..sr}$ will give a polyhedron in the state space of those $\mathbf{x}_s^*$ which are reachable for a given s.

$$-b_{\mathrm{u}} \leq \left\langle \mathbf{k}_q^{(s)}, \mathbf{x}_s^* \right\rangle \leq b_{\mathrm{u}},\ q = \overline{1..sr} \tag{3.16}$$

By constructing such sets in series for $s = s_0$, $s = s_0 + 1, \ldots$ and so on, we get the achievable sets of $\mathbf{x}_s^*$ that are consistent with the control capabilities with different control horizon on a sliding interval. It is expected that these sets expand with increasing s, i.e. the more current system's state deviates from the desired one, the larger horizon length is required for the implementation of the MPC with constraints (3.16).

All bounds defined by (3.16) contain the point $\mathbf{x}_s^* = 0$. It is easy to calculate the distance from the point $\mathbf{x}_s^* = 0$ to each surface of the polyhedron. If these distances are not very different from each other, then the admissible set can be approximated by a hypersphere with radius equal to the smallest distance to the hyperplanes. If there is a large difference between the specified distances, an attempt can be made to improve the result by changing the formulation of the extremal problem, considering a weighted functional instead of (3.2). The reason for this approach is that the coefficients of $\mathbf{k}_q^{(s)}$ in (3.15) depend only on matrices A, B and used variant of functional (3.2). We cannot change the matrices of the system model, but the optimization criterion may be different. Therefore, it is proposed to include weight coefficients into the criterion (3.2), as a result, instead of (3.3) we will have (3.17), where γ_i, $i = \overline{1..sr}$ are weights.

$$\sum_{i=1}^{sr} \gamma_i u_i^2(k,s) + \langle \boldsymbol{\lambda}, \Omega_s \cdot \mathbf{u}(k,s) - \mathbf{x}_s^* \rangle \to \min \tag{3.17}$$

Accordingly, (3.9) will transform into (3.18), where Γ_1 and Γ_2 are square diagonal matrices of dimensions $n \times n$ and $(sr - n) \times (sr - n)$ respectively.

$$\Omega_{s1}^{\mathrm{T}} \boldsymbol{\lambda} = -2\Gamma_1 \mathbf{u}_1^{(s)}, \Omega_{s2}^{\mathrm{T}} \boldsymbol{\lambda} = -2\Gamma_2 \mathbf{u}_2^{(s)} \tag{3.18}$$

The matrix Γ_1 has first n of γ_i values on its diagonal, and Γ_2 has the rest $sr - n$ of γ_i values on its diagonal. The first system in (3.18) allows us to find $\boldsymbol{\lambda}$, as it is described in (3.19), and the second allows to express $\mathbf{u}_2^{(s)}$ through $\mathbf{u}_1^{(s)}$, as in (3.20).

$$\boldsymbol{\lambda} = -2(\Omega_{s1}^{\mathrm{T}})^{-1} \cdot \Gamma_1 \mathbf{u}_1^{(s)} \tag{3.19}$$

$$\mathbf{u}_2^{(s)} = \Gamma_2^{-1} \Omega_{s2}^{\mathrm{T}} \cdot (\Omega_{s1}^{\mathrm{T}})^{-1} \cdot \Gamma_1 \mathbf{u}_1^{(s)} \tag{3.20}$$

The relation (3.20) after substituting it into (3.12) gives an equation (3.21) for finding $\mathbf{u}_1^{(s)}$.

$$\left[\Omega_{s1} + \Omega_{s2}\Gamma_2^{-1} \cdot \Omega_{s2}^{\mathrm{T}} \cdot (\Omega_{s1}^{\mathrm{T}})^{-1} \cdot \Gamma_1\right] \cdot \mathbf{u}_1^{(s)} = \mathbf{x}_s^* \tag{3.21}$$

We will show that the appropriate choice of γ_i can change the distance to the surfaces of the polyhedron in the desired direction. Let the diagonal matrix Γ_1 has the same γ_j equal to γ_1. Similarly, the diagonal matrix Γ_2 consists of the same numbers γ_2. Then (3.21) and (3.20) take the form of (3.22).

$$\begin{gathered} \left[\Omega_{s1} + \tfrac{\gamma_1}{\gamma_2} \Omega_{s2} \cdot \Omega_{s2}^{\mathrm{T}} \cdot (\Omega_{s1}^{\mathrm{T}})^{-1}\right] \cdot \mathbf{u}_1^{(s)} = \mathbf{x}_s^* \\ \mathbf{u}_2^{(s)} = \tfrac{\gamma_1}{\gamma_2} \Omega_{s2}^{\mathrm{T}} \cdot (\Omega_{s1}^{\mathrm{T}})^{-1} \cdot \mathbf{u}_1^{(s)} \end{gathered} \tag{3.22}$$

As follows from the first equation in (3.22), if the hyperplane with the smallest distance to zero is among $\mathbf{u}_1^{(s)}$, then it is possible to increase this distance by choosing $\gamma_1 > \gamma_2$, but at the same time, according to the second equation, it will bring to decrease the distance from zero to hyperplanes defined by $\mathbf{u}_2^{(s)}$. This circumstance can be utilized in the form of iterative scheme for finding weight coefficients that approximate distances to hyperplanes, i.e. increase small and increase large ones.

In addition to the iterative procedure, other methods for calculating suitable weights can be used, especially when the iterative schemes are ineffective. For example, they may be constructed on the base of (3.20), (3.21) to find weights which maximize the minimum distances to the surfaces of the polyhedron. The methods of stochastic optimization, in particular, differential evolution algorithms, including sorting, can be used for this purpose.

By using the described methods, it is possible to develop a control strategy that is most suitable for the task in each specific case of solving an applied problem.

In particular, it is first necessary to find the smallest s, permissible membership set of which contains $\mathbf{x}_s^*$ which is determined from the measured or estimated values. Then an appropriate control is selected from the database containing (3.15) for different s. The database of control parameters is calculated in advance for given system model s.

4 Exhaustive Search Method

In the previous section, the MPC task was considered at first as a task without control restrictions. After obtaining its solution in analytical form, it became possible to take into account the restrictions on control by considering that the affiliation area for $\mathbf{x}_{k+s}$ is defined by these restrictions. In cases where it is known a priori that the terminal state belongs to the reachability region for the initial state $\mathbf{x}_k$, the MPC problem should be solved as an optimization problem with regard to the control constraints. It can be written, for example, as a quadratic programming problem with constraints in the form of equalities (2.4) and inequalities (3.1). Necessary and sufficient conditions for optimality follow directly from the Kuhn-Tucker theorem [8]. On the basis of these conditions, a system of equations is written, from which an iterative scheme for finding a solution can be proposed.

In contradiction to the previous method, we will not shrink the admissible membership set of $\mathbf{x}_s^*$ for our convenience. Instead, we will perform an exhaustive search of the optimal control (in terms of (4.2)) over multiple subproblems derived from the main problem by adding some artificial constraints.

4.1 Definition of Admissible Control Set

Firstly, we need a formal definition of set of all controls $\mathbf{u}(k, s)$ which conform to constraints and will lead system from initial state $\mathbf{x}_k$ to final state $\mathbf{x}_{k+s}$ during s steps. Obviously, this definition is conjunction of (2.4) and (3.1). It can be written as (4.1), where $\mathbf{u}_0 = (u_{01}, \ldots, u_{0sr})^{\mathrm{T}}$.

$$\mathrm{U}_{\text{per}}\left(\mathbf{x}_s^*, \mathbf{u}_0\right) = \left\{ \mathbf{u}(k, s) : \begin{cases} \Omega_s \cdot \mathbf{u}(k, s) = \mathbf{x}_s^* \\ u_j(k, s) \leq u_{0j},\ j = \overline{1..sr} \\ -u_j(k, s) \leq u_{0j},\ j = \overline{1..sr} \end{cases} \right\} \tag{4.1}$$

4.2 Finding the Optimal Control

Let us assume that $\mathrm{U_{per}}\left(\mathbf{x}_s^*, \mathbf{u}_0\right)$ is not empty. In this case we need to choose an optimal control from this set. We will call "the optimal" a control, described as (4.2).

$$\mathbf{u}^*(k, s) \overset{\text{def}}{=} \underset{\mathbf{u}(k,s)\in \mathrm{U_{per}}\left(\mathbf{x}_s^*, \mathbf{u}_0\right)}{\arg\min} \frac{1}{2}\sum_{j=1}^{sr} u_j^2(k, s) \tag{4.2}$$

Thus, we have a quadratic programming problem with constraints in the form of linear equalities and weak inequalities. The Lagrange function for this problem is (4.3) where $\boldsymbol{\lambda}$, $\mathbf{v}^{(-)}$ and $\mathbf{v}^{(+)}$ are vectors of Lagrange multipliers.

$$\begin{aligned} \mathrm{L}\left(\mathbf{u}(k, s), \boldsymbol{\lambda}, \mathbf{v}^{(-)}, \mathbf{v}^{(+)}\right) = \frac{1}{2}\sum_{j=1}^{sr} u_j^2(k, s) + \left\langle \boldsymbol{\lambda}, \Omega_s \cdot \mathbf{u}(k, s) - \mathbf{x}_s^* \right\rangle + \\ + \left\langle \mathbf{v}^{(-)}, -\mathbf{u}(k, s) - \mathbf{u}_0 \right\rangle + \left\langle \mathbf{v}^{(+)}, \mathbf{u}(k, s) - \mathbf{u}_0 \right\rangle \end{aligned} \tag{4.3}$$

According to the Kuhn–Tucker theorem, as described in [8], the solution $\mathbf{u}^*(k, s)$ of the problem (4.1), (4.2) and its corresponding Lagrange multipliers $\boldsymbol{\lambda}^*$, $\mathbf{v}^{(-)*}$ and $\mathbf{v}^{(+)*}$ conform system (4.4).

$$\begin{cases} \frac{\partial \mathrm{L}}{\partial \mathrm{u}(k,s)}\left(\mathbf{u}^*(k, s), \boldsymbol{\lambda}^*, \mathbf{v}^{(-)*}, \mathbf{v}^{(+)*}\right) = 0 \\ \Omega_s \cdot \mathbf{u}^*(k, s) = \mathbf{x}_s^* \\ v_j^{(-)*}\left(-u_j^*(k, s) - u_{0j}\right) = 0, \quad j = \overline{1..sr} \\ v_j^{(+)*}\left(u_j^*(k, s) - u_{0j}\right) = 0, \quad j = \overline{1..sr} \\ \mathbf{v}^{(-)*} \geq 0 \\ \mathbf{v}^{(+)*} \geq 0 \end{cases} \tag{4.4}$$

The first equation of (4.4) can be rewritten as (4.5) or (4.6).

$$u_j^*(k, s) + \left\langle \boldsymbol{\lambda}^*, \boldsymbol{\Omega}_j^{(s)} \right\rangle = v_j^{(-)*} - v_j^{(+)*}, j = \overline{1..sr} \tag{4.5}$$

$$\mathbf{u}^*(k, s) + \Omega_s^{\mathrm{T}} \boldsymbol{\lambda}^* = \mathbf{v}^{(-)*} - \mathbf{v}^{(+)*} \tag{4.6}$$

4.3 Control Vector Index Subsets

Let us define subsets (4.7), (4.8) and (4.9) of $\overline{1..sr}$ index set:

$$\mathrm{I_{min}} = \left\{ j \in \overline{1..sr} : v_j^{(-)*} > 0 \right\} \tag{4.7}$$

$$\mathrm{I}_{\max} = \left\{ j \in \overline{1..sr} : v_j^{(+)*} > 0 \right\} \quad (4.8)$$

$$\mathrm{I}_0 = \left\{ j \in \overline{1..sr} : v_j^{(-)*} = 0, v_j^{(+)*} = 0 \right\} \quad (4.9)$$

From the system (4.4), we can derive (4.10) and (4.11).

$$\forall j \in \mathrm{I}_{\min} u_j^*(k,s) = -u_{0j}, \mathrm{v}_j^{(+)*} = 0 \quad (4.10)$$

$$\forall j \in \mathrm{I}_{\max} u_j^*(k,s) = u_{0j}, \mathrm{v}_j^{(-)*} = 0 \quad (4.11)$$

Also, we have no such information about whether $u_j^*(k,s)$ resides inside or at one of the borders of $\left[-u_{0j}, u_{0j}\right]$ for $j \in \mathrm{I}_0$.

So, it is obvious that $\mathrm{I}_{\min} \cap \mathrm{I}_{\max} = \emptyset$, $\mathrm{I}_0 \cap \mathrm{I}_{\min} = \emptyset$, $\mathrm{I}_0 \cap \mathrm{I}_{\max} = \emptyset$, as well as that $\mathrm{I}_0 \cup \mathrm{I}_{\min} \cup \mathrm{I}_{\max} = \overline{1..sr}$.

Thus, we can split (4.6) using $\mathrm{I}_{\min}$, $\mathrm{I}_{\max}$ and I_0. But to describe such division in laconic way we need to introduce a specific notation.

From now on to describe a matrix or a vector, which consists of rows or elements corresponding to indices in a specific set, we will write this set in a subscript to the initial matrix or vector.

For example, $\mathbf{u}_{\mathrm{I}_0}^*(k,s)$ is a vector which consists of elements from $\mathbf{u}^*(k,s)$ corresponding to indices in I_0, as well as $\left(\Omega_s^{\mathrm{T}}\right)_{\mathrm{I}_0}$ is a matrix which consists of rows from Ω_s^{T} corresponding to the same indices.

Also, to describe a matrix which consists of columns corresponding to indices in a specific set, we will write this set in a superscript.

For example, $\Omega_s^{\mathrm{I}_0}$ is a matrix which consists of columns from Ω_s corresponding to indices in I_0.

By the way, it is obvious that $\left(\Omega_s^{\mathrm{T}}\right)_{\mathrm{I}_0} = \left(\Omega_s^{\mathrm{I}_0}\right)^{\mathrm{T}}$.

So, by splitting the system of Eqs. (4.6) we get (4.12).

$$\begin{cases} \mathbf{u}_{\mathrm{I}_0}^*(k,s) + \left(\Omega_s^{\mathrm{T}}\right)_{\mathrm{I}_0} \boldsymbol{\lambda}^* = 0 \\ \mathbf{v}_{\mathrm{I}_{\min}}^{(+)*} = 0; \quad \mathbf{u}_{\mathrm{I}_{\min}}^*(k,s) + \left(\Omega_s^{\mathrm{T}}\right)_{\mathrm{I}_{\min}} \boldsymbol{\lambda}^* = \mathbf{v}_{\mathrm{I}_{\min}}^{(-)*} > 0 \\ \mathbf{v}_{\mathrm{I}_{\max}}^{(-)*} = 0; \quad \mathbf{u}_{\mathrm{I}_{\max}}^*(k,s) + \left(\Omega_s^{\mathrm{T}}\right)_{\mathrm{I}_{\max}} \boldsymbol{\lambda}^* = -\mathbf{v}_{\mathrm{I}_{\max}}^{(+)*} < 0 \end{cases} \quad (4.12)$$

4.4 Applying Artificial Constraints

To continue solving this problem, we need to make an assumption regarding contents of $\mathrm{I}_{\min}$, $\mathrm{I}_{\max}$ and I_0. There are 3^{sr} variants of their contents, and to find a solution, we need to check them sequentially until we find the right one.

There are three main groups of variants with fundamentally different ways of finding a solution:

(1) those, where number of indices in I_0 is equal to dimension of $\boldsymbol{\lambda}^*$, i.e. card $I_0 = \dim \boldsymbol{\lambda}^* = n$, which means that the first equation in (4.12) is a determined system.
(2) those, where number of indices in I_0 is strictly greater than dimension of $\boldsymbol{\lambda}^*$, i.e. card $I_0 > \dim \boldsymbol{\lambda}^* = n$, which means that the first equation in (4.12) is an overdetermined system.
(3) those, where number of indices in I_0 is strictly less than dimension of $\boldsymbol{\lambda}^*$, i.e. card $I_0 < \dim \boldsymbol{\lambda}^* = n$, which means that the first equation in (4.12) is an underdetermined system.

To find the optimal control we need to search over all possible variants of contents of I_{min}, I_{max} and I_0 and try to solve this problem for each case separately. This requires to perform large amount of computations, but with modern computing technologies it is still possible.

4.5 The Case of Determined System

In the case of determined system we can get Eq. (4.13) for $\boldsymbol{\lambda}^*$ from (4.12) if $\left(\Omega_s^{\mathrm{T}}\right)_{I_0}$ is invertible. Otherwise we conclude that this particular variant of I_0 does not give us the problem's solution.

$$\boldsymbol{\lambda}^* = -\left(\left(\Omega_s^{\mathrm{T}}\right)_{I_0}\right)^{-1} \mathbf{u}_{I_0}^*(k,s) \tag{4.13}$$

By applying (4.13) to the inequalities in (4.12) we get (4.14).

$$\begin{cases} \mathbf{u}_{I_{min}}^*(k,s) - \left(\Omega_s^{\mathrm{T}}\right)_{I_{min}} \left(\left(\Omega_s^{\mathrm{T}}\right)_{I_0}\right)^{-1} \mathbf{u}_{I_0}^*(k,s) > 0 \\ \mathbf{u}_{I_{max}}^*(k,s) - \left(\Omega_s^{\mathrm{T}}\right)_{I_{max}} \left(\left(\Omega_s^{\mathrm{T}}\right)_{I_0}\right)^{-1} \mathbf{u}_{I_0}^*(k,s) < 0 \end{cases} \tag{4.14}$$

We already know that $\mathbf{u}_{I_{min}}^*(k,s) = -\mathbf{u}_{0I_{min}}$ and $\mathbf{u}_{I_{max}}^*(k,s) = \mathbf{u}_{0I_{max}}$. So, (4.14) can be further transformed for practical usefulness into (4.15).

$$\begin{cases} \left(\Omega_s^{\mathrm{T}}\right)_{I_{min}} \left(\left(\Omega_s^{\mathrm{T}}\right)_{I_0}\right)^{-1} \mathbf{u}_{I_0}^*(k,s) < -\mathbf{u}_{0I_{min}} \\ \left(\Omega_s^{\mathrm{T}}\right)_{I_{max}} \left(\left(\Omega_s^{\mathrm{T}}\right)_{I_0}\right)^{-1} \mathbf{u}_{I_0}^*(k,s) > \mathbf{u}_{0I_{max}} \end{cases} \tag{4.15}$$

If we split the second equation from (4.4) using I_{min}, I_{max} and I_0, we get (4.16).

$$\Omega_s^{I_0} \cdot \mathbf{u}_{I_0}^*(k,s) = \mathbf{x}_s^* - \Omega_s^{I_{min}} \cdot \mathbf{u}_{I_{min}}^*(k,s) - \Omega_s^{I_{max}} \cdot \mathbf{u}_{I_{max}}^*(k,s) \tag{4.16}$$

Thus, we can directly calculate $\mathbf{u}_{I_0}^*(k,s)$ in a way described in (4.17).

$$\mathbf{u}_{I_0}^*(k,s) = \left(\Omega_s^{I_0}\right)^{-1}\left(\mathbf{x}_s^* + \Omega_s^{I_{min}} \cdot \mathbf{u}_{0I_{min}} - \Omega_s^{I_{max}} \cdot \mathbf{u}_{0I_{max}}\right) \tag{4.17}$$

If this result conforms (4.15) and (3.1), then it is the optimal control. Otherwise, we need to try another variant of I_{min}, I_{max} and I_0.

4.6 The Case of Overdetermined System

In this case we need to split I_0 into its subsets $I_{0,1}$ and $I_{0,2}$ so that card $I_{0,1} = \dim \boldsymbol{\lambda}^*$, $I_{0,1} \cup I_{0,2} = I_0$, $I_{0,1} \cap I_{0,2} = \emptyset$ and $\Omega_s^{I_{0,1}}$ is invertible. We also should choose that variant of $I_{0,1}$'s contents which gives us $\Omega_s^{I_{0,1}}$ with the smallest condition number, so that computations will be more precise.

With this additional splitting we get (4.18) from the first equation of (4.12).

$$\begin{cases} \mathbf{u}_{I_{0,1}}^*(k,s) + \left(\Omega_s^T\right)_{I_{0,1}} \boldsymbol{\lambda}^* = 0 \\ \mathbf{u}_{I_{0,2}}^*(k,s) + \left(\Omega_s^T\right)_{I_{0,2}} \boldsymbol{\lambda}^* = 0 \end{cases} \tag{4.18}$$

Thus, similarly to the previous case, we get Eq. (4.19) for $\boldsymbol{\lambda}^*$.

$$\boldsymbol{\lambda}^* = -\left(\left(\Omega_s^T\right)_{I_{0,1}}\right)^{-1} \mathbf{u}_{I_{0,1}}^*(k,s) \tag{4.19}$$

By applying (4.19) to the second equation of (4.18) we get (4.20).

$$\mathbf{u}_{I_{0,2}}^*(k,s) = \left(\Omega_s^T\right)_{I_{0,2}}\left(\left(\Omega_s^T\right)_{I_{0,1}}\right)^{-1} \mathbf{u}_{I_{0,1}}^*(k,s) \tag{4.20}$$

And, as in previous case, we get (4.21), (4.22), (4.23).

$$\begin{cases} \mathbf{u}_{I_{min}}^*(k,s) - \left(\Omega_s^T\right)_{I_{min}}\left(\left(\Omega_s^T\right)_{I_{0,1}}\right)^{-1} \mathbf{u}_{I_{0,1}}^*(k,s) > 0 \\ \mathbf{u}_{I_{max}}^*(k,s) - \left(\Omega_s^T\right)_{I_{max}}\left(\left(\Omega_s^T\right)_{I_{0,1}}\right)^{-1} \mathbf{u}_{I_{0,1}}^*(k,s) < 0 \end{cases} \tag{4.21}$$

$$\begin{cases} \left(\Omega_s^T\right)_{I_{min}}\left(\left(\Omega_s^T\right)_{I_{0,1}}\right)^{-1} \mathbf{u}_{I_{0,1}}^*(k,s) < -\mathbf{u}_{0I_{min}} \\ \left(\Omega_s^T\right)_{I_{max}}\left(\left(\Omega_s^T\right)_{I_{0,1}}\right)^{-1} \mathbf{u}_{I_{0,1}}^*(k,s) > \mathbf{u}_{0I_{max}} \end{cases} \tag{4.22}$$

$$\begin{aligned} &\Omega_s^{I_{0,1}} \cdot \mathbf{u}_{I_{0,1}}^*(k,s) + \Omega_s^{I_{0,2}} \cdot \mathbf{u}_{I_{0,2}}^*(k,s) = \\ &= \mathbf{x}_s^* - \Omega_s^{I_{min}} \cdot \mathbf{u}_{I_{min}}^*(k,s) - \Omega_s^{I_{max}} \cdot \mathbf{u}_{I_{max}}^*(k,s) \end{aligned} \tag{4.23}$$

By substituting $\mathbf{u}_{I_{0,2}}^*(k,s)$ in (4.23) with (4.20), we get (4.24) which can be rewritten as (4.25).

$$\begin{gathered}\Omega_s^{\mathrm{I}_{0,1}} \cdot \mathbf{u}_{\mathrm{I}_{0,1}}^*(k,s) + \Omega_s^{\mathrm{I}_{0,2}} \left(\Omega_s^{\mathrm{T}}\right)_{\mathrm{I}_{0,2}} \left(\left(\Omega_s^{\mathrm{T}}\right)_{\mathrm{I}_{0,1}}\right)^{-1} \cdot \mathbf{u}_{\mathrm{I}_{0,1}}^*(k,s) \\ = \mathbf{x}_s^* - \Omega_s^{\mathrm{I}_{\min}} \cdot \mathbf{u}_{\mathrm{I}_{\min}}^*(k,s) - \Omega_s^{\mathrm{I}_{\max}} \cdot \mathbf{u}_{\mathrm{I}_{\max}}^*(k,s)\end{gathered} \tag{4.24}$$

$$\begin{gathered}\left(\Omega_s^{\mathrm{I}_{0,1}} + \Omega_s^{\mathrm{I}_{0,2}} \left(\Omega_s^{\mathrm{T}}\right)_{\mathrm{I}_{0,2}} \left(\left(\Omega_s^{\mathrm{T}}\right)_{\mathrm{I}_{0,1}}\right)^{-1}\right) \cdot \mathbf{u}_{\mathrm{I}_{0,1}}^*(k,s) = \\ = \mathbf{x}_s^* - \Omega_s^{\mathrm{I}_{\min}} \cdot \mathbf{u}_{\mathrm{I}_{\min}}^*(k,s) - \Omega_s^{\mathrm{I}_{\max}} \cdot \mathbf{u}_{\mathrm{I}_{\max}}^*(k,s)\end{gathered} \tag{4.25}$$

So, we can write down a computational formula (4.26) for $\mathbf{u}_{\mathrm{I}_{0,1}}^*(k,s)$.

$$\begin{gathered}\mathbf{u}_{\mathrm{I}_{0,1}}^*(k,s) = \left(\Omega_s^{\mathrm{I}_{0,1}} + \Omega_s^{\mathrm{I}_{0,2}} \left(\Omega_s^{\mathrm{T}}\right)_{\mathrm{I}_{0,2}} \left(\left(\Omega_s^{\mathrm{T}}\right)_{\mathrm{I}_{0,1}}\right)^{-1}\right)^{-1} \cdot \\ \cdot \left(\mathbf{x}_s^* + \Omega_s^{\mathrm{I}_{\min}} \cdot \mathbf{u}_{0\mathrm{I}_{\min}} - \Omega_s^{\mathrm{I}_{\max}} \cdot \mathbf{u}_{0\mathrm{I}_{\max}}\right)\end{gathered} \tag{4.26}$$

If this result conforms (4.22) and (3.1), then it is the optimal control. Otherwise, we need to try another variant of $\mathrm{I}_{\min}$, $\mathrm{I}_{\max}$ and I_0.

4.7 *The Case of Underdetermined System*

If the first equation from (4.12) is underdetermined, we can't express $\boldsymbol{\lambda}^*$ by using $\mathbf{u}_{\mathrm{I}_0}^*(k,s)$, as it was done in (4.13) and (4.19). So, the best we can do with this equation is (4.28).

$$\mathbf{u}_{\mathrm{I}_0}^*(k,s) = -\left(\Omega_s^{\mathrm{T}}\right)_{\mathrm{I}_0} \boldsymbol{\lambda}^* \tag{4.28}$$

So, to find $\mathbf{u}_{\mathrm{I}_0}^*(k,s)$ we first need to find $\boldsymbol{\lambda}^*$. As in previous two cases, we can split the second equation from (4.4) using $\mathrm{I}_{\min}$, $\mathrm{I}_{\max}$ and I_0 and get (4.16). By applying (4.28) to it, we get (4.29) which can be also written as (4.30).

$$-\Omega_s^{\mathrm{I}_0}\left(\Omega_s^{\mathrm{T}}\right)_{\mathrm{I}_0} \boldsymbol{\lambda}^* = \mathbf{x}_s^* - \Omega_s^{\mathrm{I}_{\min}} \cdot \mathbf{u}_{\mathrm{I}_{\min}}^*(k,s) - \Omega_s^{\mathrm{I}_{\max}} \cdot \mathbf{u}_{\mathrm{I}_{\max}}^*(k,s) \tag{4.29}$$

$$-\Omega_s^{\mathrm{I}_0}\left(\Omega_s^{\mathrm{T}}\right)_{\mathrm{I}_0} \boldsymbol{\lambda}^* = \mathbf{x}_s^* + \Omega_s^{\mathrm{I}_{\min}} \cdot \mathbf{u}_{0\mathrm{I}_{\min}} - \Omega_s^{\mathrm{I}_{\max}} \cdot \mathbf{u}_{0\mathrm{I}_{\max}} \tag{4.30}$$

It would be great, if $\Omega_s^{\mathrm{I}_0}\left(\Omega_s^{\mathrm{T}}\right)_{\mathrm{I}_0}$ were invertible, but its rank is not full. So, in general case, the last system of equations has no solution.

5 Optimal Aim Approach

The methods of solving the MPC problem described above can be used only in the case when $\mathbf{x}_s^*$ belongs to an admissible region. This section describes the method by which we can make a sequential approach to desired point, e. g. reach the most close reachable point to the desirable during s-long time interval. In other words, this method produces an MPC for time interval which is insufficient to reach the desired point, while ensuring that the system will eventually reach the desired point, even if we have insufficient computational power to calculate all corresponding controls for required time interval at the same time.

To do this, we need to solve a distinct MPC problem for every next s-long time interval, until we reach the desired point. The solution of each of those problems transfers the system to the projection of the desired terminal state onto the reachable set. At the last iteration, when the endpoint becomes reachable, we can use a control obtained by applying one of the methods presented in the previous sections.

Let us describe a method for calculating the MPC for such iterative process. Let there be a criterion function (5.1).

$$\frac{1}{2}\left(\left\langle \Omega_s \mathbf{u}(k,s) - \mathbf{x}_s^*, \Omega_s \mathbf{u}(k,s) - \mathbf{x}_s^* \right\rangle + \alpha \langle \mathbf{u}(k,s), \mathbf{u}(k,s)\rangle\right) \tag{5.1}$$

The element $\mathbf{u}(k, s)$ at which (5.1) has a minimum on the set defined by control constraints (3.1) will be considered the sought MPC for current s-long time interval. To ensure that there exists only one such control, a stabilizer $\langle \mathbf{u}(k, s), \mathbf{u}(k, s)\rangle$ is used as a part of (5.1). This stabilizer highlights the normal solution on the set of all $\mathbf{u}(k, s)$ that minimizes the discrepancy in the implementation of Eq. (2.4). The solution of this minimization problem converges to one with the smallest value of $\langle \mathbf{u}(k, s), \mathbf{u}(k, s)\rangle$ among the normal ones, as the scalar parameter α approaches zero. If the discrepancy minimizing element $\mathbf{u}^*(k, s)$ is unique, then we can use $\alpha = 0$. In other cases, α should be chosen small enough, while assuring that corresponding solution retains its stability in regard to the errors in measurement of the right-hand side of (2.4), i.e. $\mathbf{x}_s^*$. As a result, the control obtained in this way leads the system into a sufficiently small neighborhood of a point which is a projection of the terminal state onto the reachable state's set. In this case, the parameter α must be consistent with the measurement or estimation errors of $\mathbf{x}_s^*$. This ensures that sensitivity of the solution of the MPC problem to possible realizations of the error is small enough.

Thus, finding of the MPC reduces to the extremal problem (5.2), where $\mathrm{U}_{\mathrm{per}}$ is as it is described in (4.1).

$$\min_{\mathbf{u}(k,s)\in \mathrm{U}_{\mathrm{per}}} \frac{1}{2}\left(\left\langle \Omega_s \mathbf{u}(k,s) - \mathbf{x}_s^*, \Omega_s \mathbf{u}(k,s) - \mathbf{x}_s^* \right\rangle + \alpha \langle \mathbf{u}(k,s), \mathbf{u}(k,s)\rangle\right) \tag{5.2}$$

As provided by the Lagrange method, the problem (5.2) can be reduced to finding the saddle point of Lagrange function (5.3), where Lagrange multipliers $v_j^{(+)}$ and $v_j^{(-)}$ are components of the vectors $\mathbf{v}^{(+)}$ and $\mathbf{v}^{(-)}$.

$$\begin{aligned} &\mathrm{L}\big(\mathbf{u}(k,s),\mathbf{v}^{(-)},\mathbf{v}^{(+)}\big)= \\ &=\frac{1}{2}\big(\langle\Omega_s\mathbf{u}(k,s)-\mathbf{x}_s^*,\Omega_s\mathbf{u}(k,s)-\mathbf{x}_s^*\rangle+\alpha\langle\mathbf{u}(k,s),\mathbf{u}(k,s)\rangle\big)+ \\ &+\langle\mathbf{v}^{(-)},-\mathbf{u}(k,s)-\mathbf{u}_0\rangle+\langle\mathbf{v}^{(+)},\mathbf{u}(k,s)-\mathbf{u}_0\rangle \end{aligned} \tag{5.3}$$

According to the Kuhn-Tucker theorem which establishes the necessary and sufficient conditions for the optimality of the criterion (5.2), we come to system (5.4), where Ω_s^{T} is the transposed matrix Ω_s, E is the unit matrix, $\mathbf{u}_0$ is a vector whose components are equal to b_{u}.

$$\begin{cases} \big(\Omega_s^{\mathrm{T}}\Omega_s+\alpha\mathrm{E}\big)\mathbf{u}(k,s)=\mathbf{x}_s^{**}+\mathbf{v}^{(-)*}-\mathbf{v}^{(+)*} \\ v_j^{(-)*}\Big(-u_j^*(k,s)-u_{0j}\Big)=0,\quad j=\overline{1..sr} \\ v_j^{(+)*}\Big(u_j^*(k,s)-u_{0j}\Big)=0,\quad j=\overline{1..sr} \\ \mathbf{v}^{(-)*}\geq 0 \\ \mathbf{v}^{(+)*}\geq 0 \end{cases} \tag{5.4}$$

We can see that the last four statements of (5.4) are the same as in (4.4), so we can define $\mathrm{I_{min}}$, $\mathrm{I_{max}}$ and $\mathrm{I_0}$ in the same way, as it was done in Subsect. 4.3. In this approach we will also make exhaustive search over every possible variant of contents of $\mathrm{I_{min}}$, $\mathrm{I_{max}}$ and $\mathrm{I_0}$.

Let there be $\mathbf{w}=\mathbf{v}^{(-)}-\mathbf{v}^{(+)}$, $\Omega=\Omega_s^{\mathrm{T}}\Omega_s+\alpha\mathrm{E}$, $\mathbf{x}_s^{**}=\Omega_s^{\mathrm{T}}\mathbf{x}_s^*$. Then the first equation in (5.4) can be written as (5.5).

$$\Omega\cdot\mathbf{u}(k,s)=\mathbf{x}_s^{**}+\mathbf{w} \tag{5.5}$$

Since we can vary α, let us assume that Ω is invertible. So, we can conclude (5.6) from (5.5).

$$\mathbf{u}(k,s)=\Omega^{-1}\cdot\mathbf{x}_s^{**}+\Omega^{-1}\cdot\mathbf{w} \tag{5.6}$$

If we split (5.6) using $\mathrm{I_{min}}\cup\mathrm{I_{max}}$, and $\mathrm{I_0}$, we get (5.7).

$$\begin{cases} \mathbf{u}_{\mathrm{I_0}}(k,s)=\big(\Omega^{-1}\big)_{\mathrm{I_0}}\cdot\mathbf{x}_s^{**}+\big(\Omega^{-1}\big)_{\mathrm{I_0}}\cdot\mathbf{w} \\ \mathbf{u}_{\mathrm{I_{min}}\cup\mathrm{I_{max}}}(k,s)=\big(\Omega^{-1}\big)_{\mathrm{I_{min}}\cup\mathrm{I_{max}}}\cdot\mathbf{x}_s^{**}+\big(\Omega^{-1}\big)_{\mathrm{I_{min}}\cup\mathrm{I_{max}}}\cdot\mathbf{w} \end{cases} \tag{5.7}$$

We can also conclude that $\mathbf{w}_{\mathrm{I_0}}=0$, $\mathbf{w}_{\mathrm{I_{min}}}>0$ and $\mathbf{w}_{\mathrm{I_{max}}}<0$, so (5.7) transforms into (5.8).

$$\begin{cases} \mathbf{u}_{I_0}(k,s) = \left(\Omega^{-1}\right)_{I_0} \cdot \mathbf{x}_s^{**} + \left(\Omega^{-1}\right)_{I_0}^{I_{\min}\cup I_{\max}} \cdot \mathbf{w}_{I_{\min}\cup I_{\max}} \\ \mathbf{u}_{I_{\min}\cup I_{\max}}(k,s) = \left(\Omega^{-1}\right)_{I_{\min}\cup I_{\max}} \cdot \mathbf{x}_s^{**} + \left(\Omega^{-1}\right)_{I_{\min}\cup I_{\max}}^{I_{\min}\cup I_{\max}} \cdot \mathbf{w}_{I_{\min}\cup I_{\max}} \end{cases} \tag{5.8}$$

Since we know actual values of $\mathbf{u}_{I_{\min}\cup I_{\max}}(k,s)$ for every variant of $I_{\min}$ and $I_{\max}$, we can directly calculate $\mathbf{w}_{I_{\min}\cup I_{\max}}$ and $\mathbf{u}_{I_0}(k,s)$, as described in (5.9) (5.10).

$$\mathbf{w}_{I_{\min}\cup I_{\max}} = \left(\left(\Omega^{-1}\right)_{I_{\min}\cup I_{\max}}^{I_{\min}\cup I_{\max}}\right)^{-1} \left(\mathbf{u}_{I_{\min}\cup I_{\max}}(k,s) - \left(\Omega^{-1}\right)_{I_{\min}\cup I_{\max}} \cdot \mathbf{x}_s^{**}\right) \tag{5.9}$$

$$\mathbf{u}_{I_0}(k,s) = \left(\Omega^{-1}\right)_{I_0} \cdot \mathbf{x}_s^{**} + \left(\Omega^{-1}\right)_{I_0}^{I_{\min}\cup I_{\max}} \cdot \left(\left(\Omega^{-1}\right)_{I_{\min}\cup I_{\max}}^{I_{\min}\cup I_{\max}}\right)^{-1} \left(\mathbf{u}_{I_{\min}\cup I_{\max}}(k,s) - \left(\Omega^{-1}\right)_{I_{\min}\cup I_{\max}} \cdot \mathbf{x}_s^{**}\right) \tag{5.10}$$

If $\mathbf{u}_{I_0}(k,s)$ satisfies the condition $\left|u_j(k,s)\right| < b_u$, $j \in I_0$, then we have found the solution of the problem. Otherwise, we need to try another variant of $I_{\min}$, $I_{\max}$ and I_0.

Comment When we calculate $\mathbf{w}_{I_{\min}\cup I_{\max}}$ from (5.9), some of w_i may turn out to be zero. This is permissible, since in this case there are no contradictions with (5.4). Condition (5.4) admits equality to zero $v_j^{(+)}$ or $v_j^{(-)}$ and $u_j(k,s) = \pm b_u$ simultaneously.

It is advisable to include next scheme into the procedure for determining of the solution regularization parameter α.

Let us choose a decreasing sequence $\{\alpha_q\}$, $q = 0, 1, 2, \ldots$ and solve the optimization problem for values α_q starting from α_0. The obtained values w_i for α_0 are used as initial fitting for determination of $u_j(k,s)$ with α_1. In formula (5.6), the matrix Ω is calculated for $\alpha = \alpha_1$. After that, we find the solution according to the described iterative scheme. The last obtained solution $u_j(k,s)$ for α_1 is then used as the initial approximation (value w_i) in the next transition step the value α_2. This process continues to fairly small values α_q at which the solution remains stable. Stability is determined by the variation of the obtained values w_i. If stability is preserved at any α_q, then we can get a solution as close as possible to normal. Otherwise, we will have an approximated normal solution with minimum α_q permitted by stability condition. Such a solution can be regarded as quasioptimal.

The control which transfer the system from $\mathbf{x}_k$ to the point closest to $\mathbf{x}_{k+s}$ can be calculated according to formula (2.4). Starting from this point, it is necessary to continue approaching the shortest distance from the permissible region to $\mathbf{x}_{k+s}$. As mentioned above, at the last stage, when $\mathbf{x}_{k+s}$ will be in the permissible set restricted by the control constraints, the MPC can be found using the method described in Sect. 3.

6 The Order of the Exhaustive Search

The methods described in Sects. 4 and 5 both involves examination of every possible variant of I_{min}, I_{max} and I_0. With this approach actual number of variants we will examine before we find the right one, as well as required computation time, depends only on our luck. Thus, there is a question, which order of the search is better. In this section we will propose an order which, in our opinion, allows us to find the solution faster in most cases.

Let us pick a variant with $I_0 = \overline{1..sr}$ as the first we consider. For every rejected variant we generate next variant to consider in the following way. We move from I_0 to I_{min} those $i \in I_0$ which violate the left-hand inequality of (3.1). We move from I_0 to I_{max} those $i \in I_0$, which violate the right-hand inequality of (3.1). If we are solving the problem with the approach described in Sect. 4, we also move from I_{min} and I_{max} to I_0 those $i \in I_{min} \cup I_{max}$ which violate corresponding inequalities in (4.22). In an unfortunate case when we generate an already rejected variant, we can consider instead another random generated variant among those still not rejected.

7 Stabilization Problem

If we substitute $\mathbf{x}_{k+s} = 0$ into the expression $\mathbf{x}_s^* = \mathbf{x}_{k+s} - A^s\mathbf{x}_k$, we will have the stabilization problem. In this kind of problem the system should be at rest at time $t = k + s$. If we use the soft approach with $\mathbf{u}_{k+s} = 0$, the system will remain in it after that indefinitely in case of absence of disturbing influences. A system with stable eigenvalues of the matrix A will be in the equilibrium neighborhood for sufficiently small perturbations. However, when stable eigenvalues are close to the unit circle, there exists possibility that system may leave this neighborhood. Therefore, the stabilizing control has form of negative feedback $u_q(k, s) = -\langle \mathbf{k}_q^{(s)}, \mathbf{x}_k \rangle$, where $\mathbf{k}_q^{(s)} = (A^s)^T \cdot \mathbf{k}_q^{(s)}$. This should constantly ensure presence of the system in the neighborhood of the equilibrium point. Moreover, for unstable systems with significant increments it will require huge control resources. Measurement or estimation errors of current state will not allow us to keep the system in zero equilibrium with high precision. It will always reside in its neighborhood, size of which is determined by characteristics of data errors. Sometimes this neighborhood is called an invariant set, and membership to it is supported by a stabilization system [9]. The set of reachable states defined by the control constraints (3.1) has an important role in this, because such stabilization is ensured for sufficiently arbitrary small disturbances. For strictly stable systems the matrix A^s is contractive, while the matrix A^{-s} on the contrary expands membership sets of $\mathbf{x}_{k+s}$. The degree of compression or expansion depends on s. In turn, the choice of the appropriate s depends on what kind of system we are dealing with, namely: well-stable, close to neutral or unstable.

8 Simulation Results

Feasibility of the proposed approach to MPC problem solving has been tested by numerical simulation for the method described in Sect. 3. A stable system (2.1) of dimension five was considered. The vector of input dimension three provided full controllability for each of inputs. Admissible control was set defined by the condition (3.1), in which $u_{0i} = 1, i = 1, 2, 3$. Thus, the minimum value $s = s_0$ was equal to two which corresponded to the diagonal matrices Γ_1 of dimension five and Γ_2 of dimension one. The polyhedron (3.1) which defined the permissible area $\mathbf{x}_s^*$, with good controllability gave an acceptable result even in the case $\gamma_i = 1, i = 1..3s$. This is confirmed by computational experiments in which Γ_1 and Γ_2 were identity diagonal matrices, but s changed from $s_0 = 2$ to $s = 15$. With the increase of S, the admissible membership set for $\mathbf{x}_s^*$ also widened. In this case, the surfaces of the polyhedron closest and most distant behaved differently. The closest edge has slightly increased its distance to zero. While the outermost edge has increased its distance to zero by more than 25 times.

Studies of the dependence of the admissible set on the weight coefficients of the matrices Γ_1 and Γ_2 were carried out in two ways. In the first case, an optimization problem should be first solved for $\gamma_i = 1$. The distances from zero to all surfaces are calculated. Their inverse values were chosen as weights for problem (3.17). After its solution polyhedron may be estimated from (3.1). As a result a more substantial distance from zero to the closest of hyperplanes and a significant decrease in distance to zero of the most distant hyperplane was obtained. Therefore, the confirmation of the result follows from relations (3.22). By alteration of weighs it is possible to achieve an increase of the minimum distances from zero to the hyperplane at cost of reducing the maximum distances.

In the second case, a random search of the weights in the criterion (3.17) was implemented. Obtained results of such search allowed us to carry out a comparative analysis of the obtained polyhedrons. A polyhedron with the smallest difference between the maximum and minimum distances from zero to its hyperplanes was found. The obtained result is fully consistent with the method without weights alteration, e. g. the increase in the smallest distance, achieved at cost of a decrease in the largest distance, was not large. It was not possible to obtain any significant increase of the admissible set through the variation of the weight coefficients. Therefore, their role in MPC is not essential.

9 Conclusion

Theoretical results developed in the chapter allows us to construct and implement in practice different control strategies based on MPC for discrete linear MIMO systems. Using finite horizon leads to nonlocal analysis and synthesis of control processes in contrary to the conventional control theory where local consideration with feedback is applied. Due to this new possibilities for control strategy design and other understanding feed forward and feedback arise. In the proposed approach there is no need to reduce this problem to the fundamental concept of stability according to Liapunov.

MPC is well coordinated with nonlocal state estimation using the data on moving interval. Such state estimation method is described, for example, in [10, 11]. Another MPC dignity consists in possibility to implement nonstochastic treatment for additive input perturbation and measurement errors. Nonstochastic uncertainty interpretation is associated with researches made by V. M. Kuntsevich during past years. His name is associated with so-called method of ellipsoidal estimation which gave rise to the guaranteed approach to control problem under unknown but bounded uncertainty, as well as fundamental results on attainability and invariant sets for linear and some classics of nonlinear systems [12]. So next step for development of the theoretical results obtained in this chapter should be dealing with taking in account input perturbation and measurement errors described as unknown but bounded uncertainty.

References

1. Garcia, C.E., Prett, D.M., Morari, M.: Model predictive control: theory and practice—a survey. Automatica **25**, 335–347 (1989)
2. Rawlings, J.B., Muske, K.R.: The stability of constrained receding horizon control. IEEE Trans. Autom. Control AC **38**(10), 1512–1516 (1993)
3. Mayne, D.Q.: Optimization in model based control. In: IFAC Symposium, Helsingor, Denmark, pp. 229–242 (1995)
4. van den Boom, T.J.J.: Model based predictive control: status and perspective. In: Symposium on Control, Optimization and Supervision, CESA'96 IMACS Multiconference, Lille, pp. 1–12 (1996)
5. Richalet, J., Rault, A., Testud, J.L., Papon, J.: Model predictive heuristic control: application to industrial processes. Automatica **14**, 413–428 (1978)
6. Rawlings, J.B., Mayne, D.Q.: Model Predictive Control: Theory and Design. Nob Hill Publishing, Madison, WI (2009)
7. Keerthi, S.S., Gilbert, E.G.: Optimal infinite-horizon feedback laws for a general class of constrained discrete-time systems: stability and moving-horizon approximations. J. Optim. Theory Appl. **57**(2), 265–293 (1988)
8. Karmanov, V.G.: Mathematical Programming. Nauka, Moscow
9. Kuntsevich, V.M.: Control under Uncertainty: Guaranteed Results in Control and Identification Problems. Naukova Dumka, Kyiv (2006)
10. Gubarev, V.F., Shevchenko, V.M., Zhykov, A.O., Gummel, A.V.: State estimation for systems subjected to bounded uncertainty using moving horizon approach. In: Preprints of the 15th IFAC Symposium on System Identification, Saint-Malo, France, pp. 910–915 (2009)
11. Kuntsevich, V.M., Gubarev, V.F., Kondratenko, Y.P., Lebedev, D.V., Lysenko, V.P.: Control Systems: Theory and Applications. Series in Automation, Control and Robotics. River Publishers (2018)
12. Kuntsevich, V.M., Kurzhanski, A.B.: Attainability domains for linear and some classes of nonlinear discrete systems and their control. J. Autom. Inf. Sci. **42**(1), 1–18 (2010). https://doi.org/10.1615/JAutomatInfScien.v42.i1

Krasovskii's Unification Method and the Stability Defect of Sets in a Game Problem of Approach on a Finite Time Interval

Vladimir N. Ushakov

Abstract A nonlinear conflict-controlled system in a finite-dimensional Euclidean space on a finite time interval is considered, in which the controls of the players are constrained by geometric restrictions. We study a game problem of the approach of the system to a compact target set in the phase space of the system at a fixed instant of time. The problem is investigated in the frame of the positional formalization proposed by N.N. Krasovskii. We study the central property of stability in the theory of positional differential games and, in particular, the generalization of this property such as the stability defect of sets in the space of game positions.

Keywords Control · Stability · Unification · Stable bridge · Derived set · Stability defect · Approach problem · Hamiltonian of a system

1 Introduction

In the paper, a nonlinear conflict-controlled system on a finite time interval with geometric constraints on controls of the players is considered. A game problem of approach to a compact target set at a fixed (terminal) instant of time is studied. This problem is formalized as a positional differential game [2]. In the framework of this problem, the stability property introduced in [3, 6–8], which is the central notion in positional differential games, is investigated.

In fact, the stability property is the property of weak invariance of a set in the position space of the control system with respect to a number of differential inclusions whose structure is determined by the conflict-controlled system. This property identifies stable bridges [8] in the position space of a differential game, which are the main elements of resolving constructions in the approach problem. At present, there are several different definitions of stability, which are equivalent to each other;

V. N. Ushakov (✉)
Krasovskii Institute of Mathematics and Mechanics, Ural Branch of the Russian Academy of Sciences, 16, S. Kovalevskaya Str., 620990 Yekaterinburg, Russia
e-mail: ushak@imm.uran.ru

Y. P. Kondratenko et al. (eds.), *Advanced Control Techniques in Complex Engineering Systems: Theory and Applications*, Studies in Systems, Decision and Control 203, https://doi.org/10.1007/978-3-030-21927-7_5

i.e. they identify the same stable bridges. However, it is important which formulation is taken as a basis for studying the game-theoretic problem of approach. In our opinion, the formulation of the stability property proposed in [4, 5] and based on unification appeared to be effective both from the theoretical point of view and in terms of applications.

The present paper supplements investigations of the stability property and is based on unification constructions. It extends the stability concept by considering not only sets with the stability property but also a few other sets in the space of positions of the game approach problem.

The key idea of the extension of the stability notion is as follows: for a closed set in the space of game positions, a certain nonnegative function is defined on the boundary of this set. This function estimates the degree of inconsistence between the set and the dynamics of the conflict-controlled system from the stability standpoint. The paper is related to the studies [1–18] dealing with the invariance and weak invariance of dynamical systems.

2 Game Problem of Approach on a Finite Time Interval

Let a control system on the time interval $[t_0, \vartheta]$ $(t_0 < \vartheta < \infty)$ be given as follows

$$\frac{dx}{dt} = f(t, x, u, v), \qquad x(t_0) = x^{(0)}, \\ u \in P, \quad v \in Q; \tag{2.1}$$

here, x is the m-dimensional phase vector of the system, u is the control of the first player, v is the control of the second player; and P and Q are compact sets in Euclidean spaces $\mathbb{R}^p$ and $\mathbb{R}^q$, respectively.

The following conditions are imposed on system (2.1).

Condition A The function $f(t, x, u, v)$ is defined and continuous in all variables t, x, u, v and, for any bounded and closed domain $\Omega \subset [t_0, \vartheta] \times \mathbb{R}^m$, there exists a constant $L = L(\Omega) \in (0, \infty)$ such that

$$\left\| f(t, x^{(1)}, u, v) - f(t, x^{(2)}, u, v) \right\| \le L \left\| x^{(1)} - x^{(2)} \right\|, \\ (t, x^{(i)}, u, v) \in \Omega \times P \times Q, \quad i = 1, 2;$$

here $\|f\|$ is the norm of the vector f in the Euclidean space.

Condition B There exists a constant $\gamma \in (0, \infty)$ such that

$$\|f(t, x, u, v)\| \le \gamma(1 + \|x\|), \quad (t, x, u, v) \in [t_0, \vartheta] \times \mathbb{R}^m \times P \times Q.$$

Approach Problem. Let $(t_0, x^{(0)}) \in [t_0, \vartheta] \times \mathbb{R}^m$. The first player has to choose the control $u = u(t, x)$ to ensure that the phase vector $x = x(t)$ of system (2.1) hits a given compact set $M \subset \mathbb{R}^m$ at time $\vartheta: \; x(\vartheta) \in M$.

It is obvious that the game problem of approach is solvable not for any initial position $(t_0, x^{(0)})$. Hence, we are faced with an important question or even with the problem of finding a maximal closed set W^0 in the space of positions (t, x) such that the approach problem is solvable for any starting position $(t_*, x_*) \in W^0$. Note that Conditions A and B imposed on the conflict-controlled system (2.1) and the definition of the set W^0 imply that W^0 is a compact set in $[t_0, \vartheta] \times \mathbb{R}^m$.

The issue of finding the set W^0 in $[t_0, \vartheta] \times \mathbb{R}^m$ is a serious mathematical problem, which is generally unsolvable due to the complexity of the approach problem. An effective analytical description of the set W^0 is possible only in rather simple cases. This fact places the problem of finding approximately the set W^0 in $[t_0, \vartheta] \times \mathbb{R}^m$ at the forefront of research for some fairly wide classes of conflict-controlled systems.

The set W^0 has the following very important property of weak invariance by nature: W^0 is the maximal u-stable bridge [7, 8].

Sometimes, trying to identify W^0, one finds some compact set W^* in $[t_0, \vartheta] \times \mathbb{R}^m$. This raises the question to which extent the set W^* is stable (u-stable). The question, however, requires correction. To correctly formulate this question, we give a definition of the stability defect of the set W^* (see [17, 18]) in Sect. 4 of the present paper. The stability defect of the set W^* is a nonnegative number. It is shown in Sect. 4 how the notion of the stability defect of the set W^* can be used to construct for the set W^* a set $\mathcal{W}^*$ that contains W^* and possesses the stability property.

3 Stability Property in the Approach Problem

The stability property of a compact set $W \in [t_0, \vartheta] \times \mathbb{R}^m$ can be characterized as the property of weak invariance with respect to some family of differential inclusions whose structure is induced by system (2.1). We describe these inclusions in more detail.

To do this, we introduce the function (Hamiltonian of system (2.1))

$$H(t, x, l) = \max_{u \in P} \min_{v \in Q} \langle l, f(t, x, u, v) \rangle, \quad (t, x, l) \in [t_0, \vartheta] \times \mathbb{R}^m \times \mathbb{R}^m,$$

where $\langle l, f \rangle$ is the inner product of the vectors l and f from $\mathbb{R}^m$.

Suppose that some compact set $W^* \in [t_0, \vartheta] \times \mathbb{R}^m$ is given. We consider this set together with the set W^0.

Taking into account Conditions A and B, the definition of the set W^0 and applying the Gronwall–Bellman lemma, we can show that some closed domain

$$\Omega = \big\{(t, x) \colon \; t \in [t_0, \vartheta], \;\; x \in \mathbb{U}(\mathbb{O}; r(t))\big\},$$
$$r(t) = (r_0 + \gamma(t - t_0))\, e^{\gamma(t - t_0)}, \quad r_0 \in (0, \infty),$$

contains $W^0 \cup W^*$. In addition, all solutions $x(t)$ $((t_*, x(t_*)) = (t_*, x_*) \in \Omega,\ t \in [t_*, \vartheta])$, of the differential inclusion

$$\frac{dx}{dt} \in F(t, x) = \text{co}\{f(t, x, u, v):\ u \in P,\ v \in Q\}$$

satisfy the inclusion $(t, x(t)) \in \Omega$.

Here we denote $\mathbb{U}(\mathbb{O}; r) = \{h \in \mathbb{R}^m : ||h|| \le r\}$, where $r \in (0, \infty)$.

We assume $G = \mathbb{U}(\mathbb{O}; \hat{r})$ $(\hat{r} \in (0, \infty))$ to be a ball in $\mathbb{R}^m$ containing all $F(t, x)$, $(t, x) \in \Omega$.

We also assume for $(t, x, l) \in \Omega \times S,\ \ S = \{l \in \mathbb{R}^m : ||l|| = 1\}$ that

$$F_l(t, x) = \Pi_l(t, x) \cap F(t, x),$$
$$\Pi_l(t, x) = \Big\{f \in \mathbb{R}^m : \langle l, f\rangle \le H(t, x, l)\Big\}.$$

It follows from the definition of the sets $F_l(t, x)$ that these sets are ball segments in the space $\mathbb{R}^m$ satisfying the inclusion

$$F_l(t, x) \subset G, \quad (t, x, l) \in \Omega \times S.$$

Let us define the family $\mathcal{L}$ as a set of set-valued mappings $(t, x) \in \Omega$ that corresponds to the vectors $l \in S$.

We define the stability property using the family $\mathcal{L}$ and pre-assuming that

$X_l(t^*, t_*, x_*)$ is the reachable set of the differential inclusion
$\dot{x} \in F_l(t, x),\ \ x(t_*) = x_*,\ \ $ at time $t^* \in [t_*, \vartheta]$;

$$X_l^{-1}(t_*, t^*, X^*) = \Big\{x_* \in \mathbb{R}^m :\ X_l(t^*, t_*, x_*) \cap X^* \ne \varnothing\Big\}, \quad X^* \subset \mathbb{R}^m.$$

Definition 1 The stable absorption operator π in the approach problem is the mapping

$$(t_*, t^*, X^*) \longmapsto \pi(t_*, t^*, X^*) = \bigcap_{l \in S} X_l^{-1}(t_*, t^*, X^*) \subset \mathbb{R}^m,$$
$$(t_*, t^*, X^*) \in \Delta \times 2^{\mathbb{R}^m};$$

here $\Delta = \{(t_*, t^*):\ t_0 \le t_* < t^* \le \vartheta\} \subset [t_0, \vartheta] \times [t_0, \vartheta]$.

Definition 2 A compact set $W \subset \Omega$ is called a stable (u-stable) bridge in the approach problem if

$$W(\vartheta) \subset M \ \text{ and } \ \mathrm{W(t_*) \subset ß(t_*, t^*, W(t^*)), \ \ (t_*, t^*) \in \Delta};$$

here, we denote $W(t_*) = \{x \in \mathbb{R}^m :\ (t_*, x_*) \in W\},\ t_* \in [t_0, \vartheta]$.

We now cite another formulation of the stability property expressed also in terms of the family $\mathcal{L}$ but having an infinitesimal character (see [17]). Let us give this formulation in the form of a theorem.

Theorem 1 *A nonempty compact set $W \subset \Omega$ is a stable bridge in the approach problem if and only if*

(1) $W(\vartheta) \subset M$;

(2) $\overrightarrow{D}\, W(t,x) \cap F_l(t,x) \neq \varnothing, \quad t \in [t_0, \vartheta], \quad (t,x,l) \in \partial W \times S.$

Here, $\overrightarrow{D}\, W(t,x) = \Big\{ d \in \mathbb{R}^m: \quad d = \lim\limits_{k\to\infty} (t_k - t)^{-1}(w_k - x), \quad \{(t_k, w_k)\}$ *is a sequence in* W, *where* $t_k \downarrow t$ *as* $k \to \infty$ *and* $\lim\limits_{k\to\infty} w_k = x \Big\}$.

Remark 1 The stable bridge W^0 satisfies the equality $W^0(\vartheta) = M$. In addition, the bridge W^0, as any other stable bridge, possesses the property of continuity: if $(t_*, t^*) \in \Delta$ and $W^0(t_*) \neq \varnothing$, then $W^0(t_*) \neq \varnothing$.

The definition of W^0 implies the relation

$$X_l(t^*, t_*, x_*) \cap W^0(t^*) \neq \varnothing \quad \text{for } (t_*, x_*, l) \in W^0 \times S,$$

from which, in view of the inclusion $X_l(t^*, t_*, x_*) \subset \mathbb{U}(x_*; (t^* - t_*)\hat{r})$, we deduce

$$\mathbb{U}(x_*; (t^* - t_*)\hat{r}) \cap W^0(t^*) \neq \varnothing, \quad (t_*, x_*) \in W^0.$$

It follows from the latter relation that

$$h\left(W^0(t_*), W^0(t^*)\right) \leq (t^* - t_*)\hat{r}, \quad (t_*, t^*) \in \Delta. \tag{3.1}$$

Here the notation $h\left(W^{(1)}, W^{(2)}\right) = \max\limits_{w^{(1)} \in W^{(1)}} \min\limits_{w^{(2)} \in W^{(2)}} ||w^{(1)} - w^{(2)}||$ is the Hausdorff deviation of the compact set $W^{(1)} \subset \mathbb{R}^m$ from the compact set $W^{(2)} \subset \mathbb{R}^m$. It follows from (3.1) that

$$G \cap \overrightarrow{D}\, W^0(t_*, x_*) \neq \varnothing, \quad (t_*, x_*) \in W^0, \quad t_* \in [t_0, \vartheta).$$

4 The Stability Defect of the Set in the Approach Problem

We give a definition of the *stability defect* of the set $W^* \subset \Omega$ (see [18]). As a preliminary, we constraint the set W^* by certain conditions.

We assume that $W^* \subset \Omega$ from Sect. 3 is continuous, as well as W^0, and $W^*(\vartheta) = M$.

We slightly strengthen the assumption of the continuity property.

Condition C.1 For some $R \in [\hat{r}, \infty)$, we have

$$h\left(W^*(t_*), W^*(t^*)\right) \leq R\,(t^* - t_*), \quad (t_*, t^*) \in \Delta.$$

Condition C.1 is an analog of condition (3.1), which holds for the set W^0. In Condition C.1, the weakening is made for the constant R: this constant obeys the inclusion $R \in [\hat{r}, \infty)$. The constraint on R is not a matter of principle.

The following condition symmetric to Condition C.1 is also assumed.

Condition C.2 $h(W^*(t^*), W^*(t_*)) \leq R(t^* - t_*)$, $(t_*, t^*) \in \Delta$, where R is from Condition C.1.

Let us combine Conditions C.1 and C.2 into one condition.

Condition C For some $R \in [\hat{r}, \infty)$, we have

$$d\left(W^*(t_*), W^*(t^*)\right) \leq R(t^* - t_*), \quad (t_*, t^*) \in \Delta;$$

here, $d\left(W^{(1)}, W^{(2)}\right)$ is the Hausdorff distance between the compact sets $W^{(1)}$ and $W^{(2)}$ from $\mathbb{R}^m$.

Condition C means that the set-valued mapping $t \mapsto W^*(t)$, $t \in [t_0, \vartheta]$, is Lipschitz with Lipschitz constant $R \in [\hat{r}, \infty)$. Condition C is not a too strict constraint on the compact set W^*. It implies that

$$\overrightarrow{D}\, W^*(t_*, x_*) \cap \mathbb{U}(\mathbb{O}; R) \neq \varnothing, \quad (t_*, x_*) \in \partial W^*, \quad t_* \in [t_0, \vartheta).$$

The latter condition is similar to the condition

$$\overrightarrow{D}\, W^0(t_*, x_*) \cap G \neq \varnothing, \quad (t_*, x_*) \in \partial W^0, \quad t_* \in [t_0, \vartheta),$$

which holds for the maximal stable bridge W^0.

We now proceed directly to the definition of the *stability defect* of the set W^*.

To every point $(t_*, x_*) \in \partial W^*$, $t_* \in [t_0, \vartheta)$, we associate the number

$$\varepsilon(t_*, x_*) = \sup_{l \in S} \rho\left(\overrightarrow{D}\, W^*(t_*, x_*),\ F_l(t_*, x_*)\right) \geq 0;$$

here, $\rho\left(W^{(1)}, W^{(2)}\right) = \inf\left\{||w^{(1)} - w^{(2)}||\colon \left(w^{(1)}, w^{(2)}\right) \in W^{(1)} \times W^{(2)}\right\}$ is the "distance" between the sets $W^{(1)}$ and $W^{(2)}$ from $\mathbb{R}^m$.

The number $\varepsilon(t_*, x_*)$ is called the *stability defect* of the set W^* at the point $(t_*, x_*) \in \partial W^*$, $t_* \in [t_0, \vartheta)$.

In what follows, we need some additional propositions concerning the sets $F_l(t, x)$, $l \in S$. Thus, we assume that the following conditions are satisfied.

Condition D.1 The following inequality holds:

$$H_*(t, x, l) < H(t, x, l) < H^*(t, x, l), \quad (t, x, l) \in \Omega \times S.$$

Condition D.2 There exists $\lambda = \lambda(L) \in (0, \infty)$ such that

$$d\big(F_l(t, x^{(1)}),\ F_l(t, x^{(2)})\big) \leq \lambda\, ||x^{(1)} - x^{(2)}||, \quad l \in S, \quad (t, x^{(i)}) \in \Omega, \quad i = 1, 2.$$

Note that, under Condition D.1, the set-valued mapping $(t, x, l) \mapsto F_l(t, x)$ is continuous in the Hausdorff metric on the compact set $\Omega \times S$. Hence, the mapping $(t, x, l) \mapsto F_l(t, x)$ is uniformly continuous on $\Omega \times S$. This uniform continuity implies that there exists such a scalar function $\omega^*(\delta) \downarrow 0$ as $\delta \downarrow 0$ for which the following inequality is satisfied:

$$d\big(F_l(t_*, x_*),\ F_l(t^*, x^*)\big) \leq \omega^*\left(|t_* - t^*| + ||x_* - x^*||\right),$$
$$(t_*, x_*) \text{ and } (t^*, x^*) \in \Omega, \quad l \in S.$$

We introduce the notation $\omega(\delta) = \delta\,\omega^*((1 + \hat{r})\delta)$, $\delta > 0$, where $\hat{r}$ is the radius of the ball G from Sect. 3. The function $\omega(\delta)$ will be used in the proof of the main statement of the paper, that is, Theorem 2 below.

Note that the set $\overrightarrow{D}\,W^*(t_*, x_*)$ involved in the expression for $\varepsilon(t_*, x_*)$ is not necessarily compact in $\mathbb{R}^m$. However, in further considerations, it is convenient to deal with a compact set in $\mathbb{R}^m$. In this connection, we replace the set $\overrightarrow{D}\,W^*(t_*, x_*)$ by some its subset, which is compact in $\mathbb{R}^m$, so that the value $\varepsilon(t_*, x_*)$ is preserved. Namely, we introduce the set

$$\overrightarrow{D}^{\nabla} W^*(t_*, x_*) = \overrightarrow{D}\,W^*(t_*, x_*) \cap \mathbb{U}(\mathbb{O}; 3R).$$

According to the definition of the number R, the inclusion $G \subset \mathbb{U}(\mathbb{O}; R)$ is true and, hence, $F_l(t_*, x_*) \subset \mathbb{U}(\mathbb{O}; R)$ for $(t_*, x_*) \in \partial W^*$, $t_* \in [t_0, \vartheta)$, $l \in S$. In addition, it follows from Condition C that $\overrightarrow{D}\,W^*(t_*, x_*) \cap \mathbb{U}(\mathbb{O}; R) \neq \varnothing$. Hence, we deduce the equality

$$\rho\big(\overrightarrow{D}^{\nabla} W^*(t_*, x_*),\ F_l(t_*, x_*)\big) = \rho\big(\overrightarrow{D}\,W^*(t_*, x_*),\ F_l(t_*, x_*)\big),$$
$$(t_*, x_*) \in \partial W^*, \quad t_* \in [t_0, \vartheta), \quad l \in S.$$

Taking this equality into account, we obtain

$$\varepsilon(t_*, x_*) = \sup_{l \in S} \rho\big(\overrightarrow{D}^{\nabla} W^*(t_*, x_*),\ F_l(t_*, x_*)\big), \quad (t_*, x_*) \in \partial W^*, \quad t_* \in [t_0, \vartheta).$$

Suppose for $t_* \in [t_0, \vartheta)$ that

$$\varepsilon(t_*) = \sup_{(t_*, x_*) \in \Lambda(t_*)} \varepsilon(t_*, x_*);$$

here, we define $\Lambda(t_*) = \partial W^* \cap \Gamma_{t_*}$, $\Gamma_{t_*} = \{(t, x)\colon\ t = t_*\}$.

The value $\varepsilon(t_*)$ is called the *stability defect of the set* W^* *at time* $t_* \in [t_0, \vartheta)$.

Thus, we have defined the scalar function $\varepsilon(t)$, $t \in [t_0, \vartheta)$, which can be extended to the entire interval $[t_0, \vartheta]$ by setting $\varepsilon(\vartheta) = 0$. The function $\varepsilon(t)$ for $t \in [t_0, \vartheta]$ is some characteristic of the instability of the set $W^* \subset \Omega$.

From the definitions of the sets $\overrightarrow{D}^{\nabla} W^*(t_*, x_*)$ and $F_l(t_*, x_*)$ it follows that $\varepsilon(t_*, x_*) \leq 2R$ for any $(t_*, x_*) \in \partial W^*$ and, hence, $\varepsilon(t_*) \leq 2R$ for any $t_* \in [t_0, \vartheta)$.

Remark 2 It is possible to define the same scalar function $\varepsilon(t)$ for $t \in [t_0, \vartheta)$ in a different way. Namely, we associate every triple $(t_*, x_*, l) \in \partial W^* \times S$, $t_* \in [t_0, \vartheta)$, with the number

$$\varepsilon_l(t_*, x_*) = \rho\big(\overrightarrow{D}^{\nabla} W^*(t_*, x_*),\ F_l(t_*, x_*)\big).$$

We set

$$\varepsilon_l(t_*) = \sup_{(t_*, x_*) \in \Lambda(t_*)} \varepsilon_l(t_*, x_*), \quad t_* \in [t_0, \vartheta), \quad l \in S.$$

At the same time, the function $\varepsilon_l(t) \geq 0$ on $[t_0, \vartheta)$ arises, which represents some characteristic of the degree of (weak) noninvariance of the set W^* with respect to the differential inclusion $\dot{x} \in F_l(t, x)$ corresponding to a chosen $l \in S$. The relation

$$\varepsilon(t) = \sup_{l \in S} \varepsilon_l(t) \tag{4.1}$$

is valid. In our opinion, representation (4.1) is convenient to study properties of the function $\varepsilon(t)$. So, if we succeeded in showing for a certain game problem that the functions $\varepsilon_l(t)$, $t \in [t_0, \vartheta]$, are Lipschitz for all $l \in S$ with one and the same Lipschitz constant, then the function $\varepsilon(t)$, $t \in [t_0, \vartheta]$, is also Lipschitz with the same Lipschitz constant.

In some game problems of approach, the family $\mathcal{L} = \{(t, x) \mapsto F_l(t, x) : l \in S\}$ can be narrowed down to an equivalent finite or countable family $\mathcal{L}_* = \{(t, x) \mapsto F_l(t_*, x_*) : l \in S_*\}$ $(S_* \in S)$, which distinguishes the same u-stable bridges W in the domain Ω, as the family $\mathcal{L}$ does. In such problems, the function $\varepsilon(t)$ from (4.1) can be replaced by another function $\varepsilon(t)$, which is defined by the equality

$$\varepsilon(t) = \sup_{l \in S_*} \varepsilon_l(t), \quad t \in [t_0, \vartheta). \tag{4.2}$$

In this case, when the set S_* is finite, we obtain

$$\varepsilon(t) = \max_{l \in S_*} \varepsilon_l(t), \quad t \in [t_0, \vartheta). \tag{4.3}$$

But if the functions $\varepsilon_l(t)$, $t \in [t_0, \vartheta)$, corresponding to vectors $l \in S_*$ proved to be Lebesgue measurable in these game problems, then the function (4.3) is also Lebesgue measurable on $[t_0, \vartheta)$. Thereby, for the considered game problems, we state that the function $\varepsilon(t)$ satisfies Condition E.1 introduced below (see "Condition E.1"); this condition is important for our further reasoning.

Our brief explanations show that representation (4.2) or (4.3) turns out to be useful in the study of various properties of the function $\varepsilon(t)$, $t \in [t_0, \vartheta]$.

If some set $W^* \subset \Omega$ is a u-stable bridge in the approach game problem under consideration, then, by Theorem 1, we have

$$\overrightarrow{D}\, W^*(t_*, x_*) \cap F_l(t_*, x_*) \neq \varnothing, \quad (t_*, x_*) \in \partial W^*, \ \ t_* \in [t_0, \vartheta), \ \ l \in S,$$

and, consequently,

$$\overrightarrow{D}^{\nabla} W^*(t_*, x_*) \cap F_l(t_*, x_*) \neq \varnothing, \quad (t_*, x_*) \in \partial W^*, \ \ t_* \in [t_0, \vartheta), \ \ l \in S.$$

This implies the equality $\varepsilon(t_*, x_*) = 0$, $(t_*, x_*) \in \partial W^*$, $t_* \in [t_0, \vartheta)$, and hence $\varepsilon(t) = 0$, $t \in [t_0, \vartheta)$.

On the other hand, if the equality $\varepsilon(t) = 0$, $t \in [t_0, \vartheta)$, holds for some (nonempty) closed set $W^* \subset \Omega$ $(W^*(\vartheta) = M)$, then

$$\overrightarrow{D}^{\nabla} W^*(t_*, x_*) \cap F_l(t_*, x_*) \neq \varnothing, \quad (t_*, x_*) \in \partial W^*, \ \ t_* \in [t_0, \vartheta), \ \ l \in S,$$

which entails the u-stability of the set W^* and (due to the equality $W^*(\vartheta) = M$) the fact that W^* is a u-stable bridge.

This reasoning leads us to the proposition that, in the case where a small function $\varepsilon(t)$ for $t \in [t_0, \vartheta)$ corresponds to the set $W^* \subset \Omega$, the rule of extremal aiming at stable bridges (known in the theory of differential games and proposed by Krasovskii [7, 8]) applied to the set W^* must ensure that the motion $x(t)$ of system (2.1) hits a small ε-neighborhood $M_\varepsilon(t)$ of the target set M. The second proposition in the case if the function $\varepsilon(t)$ on $[t_0, \vartheta)$ is small consists in the fact that the set W^* can be replaced by some closed set $\mathcal{W}^* \subset \Omega$ containing W^*, not greatly differing from W^* in the Hausdorff metric, and possessing the u-stability property.

To justify these assertions, we assume that the set W^* and the function $\varepsilon(t)$ satisfy two additional conditions.

Condition E.1 There exists a function $\varphi^*(\delta) \geq 0$, $\delta \in (0, \vartheta - t_0)$, $\varphi^*(\delta) \downarrow 0$ as $\delta \downarrow 0$, such that

$$h\left(x_* + \delta \overrightarrow{D}^{\nabla} W^*(t_*, x_*), W^*(t_* + \delta)\right) \leq \delta\, \varphi^*(\delta),$$
$$(t_*, x_*) \in \partial W^*, \ \ t_* \in [t_0, \vartheta), \ \ \delta \in (0, \vartheta - t_*).$$

Condition E.2 The function $\varepsilon(t)$ is Lebesgue measurable on $[t_0, \vartheta]$.

Let us introduce a closed set $\mathcal{W}^* \subset [t_0, \vartheta] \times \mathbb{R}^m$, where $\mathcal{W}^*(t) = W^*(t) + \mathbb{U}(\mathbb{O}; \varkappa(t))$, $\varkappa(t) = \int_{t_0}^{t} e^{\lambda(t-\tau)} \varepsilon(\tau) d\tau$, $t \in [t_0, \vartheta]$, and $\varkappa(t)$ is the Lebesgue integral of the measurable function $e^{\lambda(t-\tau)}\varepsilon(\tau)$, $\tau \in [t_0, t]$.

The following main statement is true.

Theorem 2 *$\mathcal{W}^*$ is a u-stable bridge in the problem of approach of system (2.1) at time ϑ to the target set $M_{\varkappa(\vartheta)}$.*

Proof We prove the theorem by contradiction. Assume that, for some $(\tau_*, t^*) \in \Delta$, $(\tau_*, w_*) \in \partial\mathcal{W}^*$, and $l \in S$,

$$\mathcal{W}^*(t^*) \cap X_l(t^*; \tau_*, w_*) \neq \varnothing. \tag{4.4}$$

The assumption implies that $\tau_* < t^*$; i.e. the interval $[\tau_*, t^*]$ is nondegenerate. Setting $X^* = \{(t, x) : \ t \in [\tau_*, t^*], x \in X_l(t; \tau_*, w_*)\}$, we obtain that the set $\mathcal{W}^* \cap X^*$ is a nonempty compact set in $[\tau_*, t^*] \times \mathbb{R}^m$. Then the set $T^* = pr_t(\mathcal{W}^* \cap X^*)$, which is the orthogonal projection of the set $\mathcal{W}^* \cap X^*$ onto the axis t, is a compact set contained in $[\tau_*, t^*]$. In view of (4.4), we obtain $t_* = \max\limits_{t \in T^*} t < t^*$.

Let us consider the interval $[t_*, t^*]$. We have the relations

$$\begin{aligned} \mathcal{W}^*(t_*) \cap X_l(t_*; \tau_*, w_*) &\neq \varnothing, \\ \mathcal{W}^*(t) \cap X_l(t; \tau_*, w_*) &= \varnothing, \quad t \in (t_*, t^*]. \end{aligned} \tag{4.5}$$

Consider an arbitrary point $x_* \in \mathcal{W}^*(t_*) \cap X_l(t_*; \tau_*, w_*)$. It follows from (4.5) that $(t_*, x_*) \in \partial\mathcal{W}^*$. It follows from $x_* \in X_l(t_*; \tau_*, w_*)$ that $X_l(t; t_*, x_*) \subset X_l(t; \tau_*, w_*)$, $t \in (t_*, t^*]$ and, hence, $\mathcal{W}^*(t) \cap X_l(t; t_*, x_*) = \varnothing$, $t \in (t_*, t^*]$.

As a result, assuming the contrary to the statement of Theorem 2, we conclude that there are such $(t_*, t^*) \in \Delta$, $(t_*, x_*) \in \partial\mathcal{W}^*$, $l \in S$, for which

$$\mathcal{W}^*(t) \cap X_l(t; t_*, x_*) = \varnothing, \quad t \in (t_*, t^*].$$

We introduce the scalar function

$$\rho(t) = \rho\big(X_l(t; t_*, x_*), \ W^*(t)\big), \quad t \in [t_*, t^*].$$

The function $\rho(t)$ satisfies the relations

$$\rho(t_*) = \varkappa(t_*), \quad \rho(t) > \varkappa(t) \ \text{ for } \ t \in (t_*, t^*].$$

Since the set-valued mappings $t \mapsto X_l(t) = X_l(t; t_*, x_*)$ and $t \mapsto W^*(t)$ are Lipschitz on $[t_*, t^*]$ in the Hausdorff metric, it follows that $\rho(t)$ on $[t_*, t^*]$ is a Lipschitz function and, hence, the derivative $\dot{\rho}(t)$ of the function $\rho(t)$ exists almost everywhere on $[t_*, t^*]$.

Let ξ_* be an arbitrary point from (t_*, t^*) at which the derivative $\dot{\rho}(\xi_*)$ exists. We consider the reachable set $X_l(\xi_*; t_*, x_*)$ and a point $z_* \in X_l(\xi_*; t_*, x_*)$ such that $\rho(\xi_*) = \rho(z_*, W^*(\xi_*))$; here $\rho(z, W)$ is the distance from the point z to the compact set W in $\mathbb{R}^m$.

Let y_* be the closest to z_* point on the set $W^*(\xi_*)$. Since $\xi_* \in (t_*, t^*)$ and $z_* \notin W^*(\xi_*)$, it follows that $y_* \in \partial W^*(\xi_*)$. We take the point $(\xi_*, y_*) \in \partial W^*$. By the definition of the value $\varepsilon(\xi_*, y_*)$, we have

$$\rho\big(F_l(\xi_*, y_*), \ \overrightarrow{D}^{\nabla} W^*(\xi_*, y_*)\big) \leq \varepsilon(\xi_*, y_*). \tag{4.6}$$

Let $f \in F_l(\xi_*, y_*)$ and $d \in \overrightarrow{D}^{\nabla} W^*(\xi_*, y_*)$ be vectors satisfying the equality

$$\|f - d\| = \rho\big(F_l(\xi_*, y_*),\ \overrightarrow{D}^{\nabla} W^*(\xi_*, y_*)\big). \tag{4.7}$$

Since $d \in \overrightarrow{D}^{\nabla} W^*(\xi_*, y_*)$, there exists sequences $\{\delta_k\}$ ($\delta_k \downarrow 0$ as $k \to \infty$) and $\{y^*(\xi_* + \delta_k)\}$ $\big(y^*(\xi_* + \delta_k) \in W^*(\xi_* + \delta_k), k = 1, 2, \ldots\big)$ that satisfy the equalities

$$y^*(\xi_* + \delta_k) = y_* + \delta_k d + \delta_k \varphi^*(\delta_k), \quad k = 1, 2, \ldots, \tag{4.8}$$

where the function $\varphi^*(\delta)$ is introduced in Condition E.1.

For the vector $f \in F_l(\xi_*, y_*)$, there exists a vector $f^* \in F_l(\xi_*, z_*)$ satisfying the inequality

$$\|f - f^*\| \le d\big(F_l(\xi_*, y_*),\ F_l(\xi_*, z_*)\big) \le \lambda \|y_* - z_*\|, \tag{4.9}$$

where $\lambda = \lambda(\Omega)$ (see Condition D.2).

The sequence $\{\delta_k\}$ generates the sequence $\{z_* + \delta_k f^*\}$. For each point $z_* + \delta_k f^*$ of this sequence, there exists a point $z(\xi_* + \delta_k)$ in the set $X_l(\xi_* + \delta_k; \xi_*, z_*)$ such that satisfies the inequality

$$\|z(\xi_* + \delta_k) - (z_* + \delta_k f^*)\| \le\ d\big(X_l(\xi_* + \delta_k; \xi_*, z_*),\ z_* + \delta_k F_l(\xi_*, z_*)\big) \le \\ \le \omega(\delta_k). \tag{4.10}$$

Since $X_l(\xi_* + \delta_k; \xi_*, z_*) \subset X_l(\xi_* + \delta_k; t_*, x_*)$, it follows that $z(\xi_* + \delta_k) \subset X_l(\xi_* + \delta_k; t_*, x_*)$.

Relations (4.6)–(4.10) imply the inequalities

$$\begin{aligned} &\|(z_* + \delta_k f^*) - (y_* + \delta_k f)\| \le (1 + \delta_k \lambda)\|z_* - y_*\|, \\ &\|(y_* + \delta_k f) - (y_* + \delta_k d)\| \le \delta_k \varepsilon(\xi_*, y_*) \le \delta_k \varepsilon(\xi_*), \\ &\|(y_* + \delta_k d) - y^*(\xi_* + \delta_k)\| \le \delta_k \varphi^*(\delta_k), \quad k = 1, 2, \ldots. \end{aligned} \tag{4.11}$$

Taking into account the inclusions $z(\xi_* + \delta_k) \subset X_l(\xi_* + \delta_k; t_*, x_*)$, $y^*(\xi_* + \delta_k) \in W^*(\xi_* + \delta_k)$, and inequalities (4.10), (4.11), we find the estimate

$$\rho(\xi_* + \delta_k) \le \|z(\xi_* + \delta_k) - y^*(\xi_* + \delta_k)\| \le \omega(\delta_k) + \\ + (1 + \delta_k \lambda)\|z_* - y_*\| + \delta_k \varepsilon(\xi_*) + \delta_k \varphi^*(\delta_k), \quad k = 1, 2, \ldots. \tag{4.12}$$

From inequality (4.12), in view of $\rho(\xi_*) = \|z_* - y_*\|$, we obtain

$$\delta_k^{-1}\big(\rho(\xi_* + \delta_k) - \rho(\xi_*)\big) \le \lambda \rho(\xi_*) + \varepsilon(\xi_*) + \delta_k^{-1}\omega(\delta_k) + \varphi^*(\delta_k), \ \ k = 1, 2, \ldots.$$

Hence, for any point $\xi_* \in (t_*, t^*)$ at which the function $\rho(\xi)$ is differentiable, we have the estimate

$$\dot{\rho}(\xi_*) \le \lambda \rho(\xi_*) + \varepsilon(\xi_*). \tag{4.13}$$

Inequality (4.13), which holds for almost all ξ_* from $[t_*, t^*]$, implies

$$\rho(t) \leq e^{\lambda(t-t_*)}\rho(t_*) + \int_{t_*}^{t} e^{\lambda(t-\tau)}\varepsilon(\tau)d\tau, \quad t \in [t_*, t^*].$$

Taking into account the equality $\rho(t_*) = 0$, we find that

$$\rho(t) \leq \int_{t_*}^{t} e^{\lambda(t-\tau)}\varepsilon(\tau)d\tau, \quad t \in [t_*, t^*].$$

Then, keeping in mind the definition of the function $\varkappa(t)$, $t \in [t_*, t^*]$, we obtain

$$\rho(t) \leq \varkappa(t), \quad t \in (t_*, t^*]. \tag{4.14}$$

Inequality (4.14) contradicts the inequality written on Section "The Stability Defect of the Set in the Approach Problem":

$$\rho(t) > \varkappa(t), \quad t \in (t_*, t^*].$$

Consequently, the contrary assumption is false. Theorem 2 is proved. □

Since, by Theorem 2, the set $\mathcal{W}^*$ is a u-stable bridge in the problem of approach of conflict-controlled system (2.1) to the target set $M_{\varkappa(\vartheta)}$, there exists a positional control strategy $u^*(t, x)$ of the first player guaranteeing that the phase vector $x(\vartheta)$ of system (2.1) hits $M_{\varkappa(\vartheta)}$ in the case where the initial point $x(t_0) = x^{(0)}$ of the motion $x(t)$, $t \in [t_0, \vartheta]$, satisfies the inclusion $(t_0, x^{(0)}) \in \mathcal{W}^*$.

5 An Example of the Set W^* Satisfying Conditions C, E.1 and E.2

Suppose that a given set $W^* \subset [t_0, \vartheta] \times \mathbb{R}^m$ can be represented in the form

$$W^* = \{(t, x) \in \Omega : \quad \varphi(t, x) \leq 0\}, \tag{5.1}$$

where the function $\varphi(t, x)$ is defined and continuous on $\mathbb{R}^1 \times \mathbb{R}^m$ together with its partial derivatives $\dfrac{\partial\varphi(t, x)}{\partial t}$, $\dfrac{\partial\varphi(t, x)}{\partial x_i}$, $i = 1, 2, \ldots, m$.

The following condition is assumed to be satisfied:

$$h(t_*, x_*) = \text{grad}\,\varphi(t_*, x_*) \neq 0, \quad (t_*, x_*) \in \partial W^*, \quad t_* \in [t_0, \vartheta];$$

here $\partial W^* = \{(t_*, x_*) \in [t_0, \vartheta] \times \mathbb{R}^m : \quad \varphi(t_*, x_*) = 0\}$.

Let $(t_*, x_*) \in [t_0, \vartheta] \times \mathbb{R}^m$. Then

$$\varphi(t, x) = \varphi(t_*, x_*) + \langle h(t_*, x_*), x - x_* \rangle + \\ + \frac{\partial \varphi(t_*, x_*)}{\partial t}(t - t_*) + \omega_{(t_*,x_*)}(t - t_*, x - x_*), \quad (5.2)$$

where

$$\lim_{(|t-t_*|+||x-x_*||)\downarrow 0} \big(|t - t_*| + ||x - x_*||\big)^{-1} \omega_{(t_*,x_*)}(t - t_*, x - x_*) = 0. \quad (5.3)$$

Let us fix some $\varepsilon > 0$ and introduce the set $\Omega^\varepsilon = \Omega_\varepsilon \cap \big([t_0, \vartheta] \times \mathbb{R}^m\big)$; here, Ω_ε is the closed ε-neighborhood of the set Ω in the space $\mathbb{R}^1 \times \mathbb{R}^m$.

The function $\omega_{(t_*,x_*)}(t - t_*, x - x_*)$ involved in representation (5.2) of the function $\varphi(t, x)$ can be represented as

$$\omega_{(t_*,x_*)}(t - t_*, x - x_*) = \big(|t - t_*| + ||x - x_*||\big)\, \xi_{(t_*,x_*)}(t - t_*, x - x_*), \\ (t, x) \in [t_0, \vartheta] \times \mathbb{R}^m,$$

where the function $\xi_{(t_*,x_*)}(t - t_*, x - x_*)$ satisfies the inequality

$$\max_{\substack{(t_*,x_*),\ (t,x) \in \Omega^\varepsilon, \\ |t-t_*|+||x-x_*|| \le \rho}} \big|\xi_{(t_*,x_*)}(t - t_*, x - x_*)\big| \le \xi^*(\rho), \quad \rho > 0,$$

in which the function $\xi^*(\rho)$ is defined on $(0, \infty)$ and $\xi^*(\rho) \downarrow 0$ as $\rho \downarrow 0$.

Hence, we have

$$\varphi(t^*, x^*) \le \varphi(t_*, x_*) + \langle h(t_*, x_*), x - x_* \rangle + \tfrac{\partial \varphi(t_*,x_*)}{\partial t}(t - t_*) + \\ + \big(|t^* - t_*| + ||x^* - x_*||\big)\, \xi^*(|t^* - t_*| + ||x^* - x_*||), \quad (5.4)$$

where (t_*, x_*) and (t^*, x^*) belong to Ω^ε.

Taking into account estimate (5.4), we show that the set W^* satisfies Condition C.

Let a point (t_*, x_*) obey the inclusion $(t_*, x_*) \in [t_0, \vartheta] \times \partial W(t_*)$ and suppose that $r_* \in (0, \infty)$, $R \in (0, \infty)$, and Δ_t and Δ_x are specified by the relations

$$\max_{(t_*,x_*)\in \partial W^*} \left| \frac{\partial \varphi(t_*, x_*)}{\partial t} \right| \le r_*, \quad h_* < r_*, \quad R = \frac{2r_*}{h_*}$$

with $h_* = \min\limits_{(t,x)\in \partial W^*} ||h(t, x)|| > 0$, $\Delta_t = t^* - t_*$, and

$$\Delta_x = x^* - x_* = -R\, ||h(t_*, x_*)||^{-1} h(t_*, x_*) |\Delta_t|;$$

here, (t^*, x^*) is some point from $[t_0, \vartheta] \times \mathbb{R}^m$.

Then the estimate

$$\varphi(t^*, x^*) \leq -R\,||h(t_*, x_*)||\,|\Delta_t| + r_*|\Delta_t| + \\ +(1+R)\,|\Delta_t|\,\xi^*\big((1+R)|\Delta_t|\big), \tag{5.5}$$

is valid; i.e.

$$\varphi(t^*, x^*) \leq \Big\{ -r_* + \xi^*\big((1+R)|\Delta_t|\big)\,(1+R)\Big\}\,|\Delta_t|. \tag{5.6}$$

It should be noted that estimates (5.5) and (5.6) are correct only if the point (t^*, x^*) is contained in Ω^ε along with the point $(t_*, x_*) \in \partial W^* \subset \Omega^\varepsilon$. In connection with this remark, we assume that estimates (5.5) and (5.6) are considered for sufficiently small $|\Delta_t|$: $|\Delta_t| \leq \delta_2 = \dfrac{1}{2}\min\big((\vartheta - t_*), \delta_1\big)$, where $\delta_1 = (h_*/2r_*)\varepsilon$. For these Δ_t, the inequality $|\Delta_t| \leq \dfrac{\varepsilon}{2}$ and $|\Delta_x| = ||x^* - x_*|| \leq \dfrac{\varepsilon}{2}$ holds and, consequently, $(t^*, x^*) \in \Omega^\varepsilon$ along with $(t_*, x_*) \in \Omega^\varepsilon$.

Let us choose $\delta_* \in (0, \delta_2)$ so small that

$$\xi^*\big((1+R)\delta_*\big) \leq (1+R)^{-1} r_*.$$

Taking account of the limit relation $\xi^*(\rho) \downarrow 0$ as $\rho \downarrow 0$, we obtain

$$-r_* + \xi^*\big((1+R)|\Delta_t|\big)\,(1+R) \leq 0, \quad |\Delta_t| \leq \delta_*. \tag{5.7}$$

It follows from inequalities (5.6) and (5.7) that

$$\varphi(t^*, x^*) \leq 0. \tag{5.8}$$

Inequality (5.8) means that $x^* \in W^*(t^*)$ for $|\Delta_t| \leq \delta_*$. This implies that the following inequality is valid for any point $x_* \in \partial W^*(t_*)$, $t_* \in [t_0, \vartheta]$:

$$\rho(x_*, W^*(t^*)) \leq ||x_* - x^*|| = R|\Delta_t|. \tag{5.9}$$

It follows from (5.9) that the Hausdorff deviation of the set $W^*(t_*)$ from $W^*(t^*)$ satisfies the estimate

$$h\big(W^*(t_*), W^*(t^*)\big) \leq R|t_* - t^*|, \quad |\Delta_t| \leq \delta_*. \tag{5.10}$$

For the same reasons, we have the estimate

$$h\big(W^*(t^*), W^*(t_*)\big) \leq R|t^* - t_*|, \quad |\Delta_t| \leq \delta_*. \tag{5.11}$$

Estimates (5.10) and (5.11) imply Condition C:

$$d\big(W^*(t_*), W^*(t^*)\big) \le R(t^* - t_*), \quad (t_*, t^*) \in \Delta.$$

We now prove that the set W^* satisfies Condition E.1. Condition E.1 seems to be more sophisticated than Condition C, since its formulation involves elements of infinitesimal constructions, namely, the set $\overrightarrow{D}^{\nabla} W^*(t_*, x_*) = \overrightarrow{D}\, W^*(t_*, x_*) \cap \mathbb{U}(\mathbb{O}; 3R)$ at the point $(t_*, x_*) \in \partial W^*$, $t_* \in [t_0, \vartheta)$.

To this end, we consider the derived set $\overrightarrow{D}\, W^*(t_*, x_*)$ at the point $(t_*, x_*) \in \partial W^*$, $t_* \in [t_0, \vartheta)$ of the set-valued mapping $t \mapsto W^*(t)$. This set can be represented in the form (see, e.g. [1])

$$\overrightarrow{D}\, W^*(t_*, x_*) = \left\{ d \in \mathbb{R}^m :\ \langle h(t_*, x_*), d\rangle + \frac{\partial \varphi(t_*, x_*)}{\partial t} \le 0 \right\},$$

and, hence, $\overrightarrow{D}\, W^*(t_*, x_*)$ is a closed half-space in $\mathbb{R}^m$.

In this case, the compact set $\overrightarrow{D}^{\nabla} W^*(t_*, x_*)$ in $\mathbb{R}^m$ is a ball segment in the space $\mathbb{R}^m$. Then the set $x_* + \delta\, \overrightarrow{D}^{\nabla} W^*(t_*, x_*)$ with $\delta > 0$ from Condition E.1 can be represented in the form

$$\begin{aligned} x_* + \delta\, \overrightarrow{D}^{\nabla} W^*(t_*, x_*) = \Big\{ x_\delta^d = x_* + \delta d :& \\ \langle h(t_*, x_*),\ x_\delta^d - x_*\rangle + \frac{\partial \varphi(t_*, x_*)}{\partial t}\,\delta \le 0, &\quad (x_\delta^d - x_*) \in \mathbb{U}(\mathbb{O}; 3R\delta) \Big\}. \end{aligned}$$

Let $\delta > 0$ and $d \in \overrightarrow{D}^{\nabla} W^*(t_*, x_*)$. These relations imply the inclusion

$$x_\delta^d = x_* + \delta d \in x_* + \delta \overrightarrow{D}^{\nabla} W^*(t_*, x_*).$$

Together with the point x_δ^d, we consider the point $x_{\gamma(\delta)} = x_\delta^d - \gamma(\delta) s$, $s = ||h(t_*, x_*)||^{-1} h(t_*, x_*)$; here $\gamma(\delta)$ is some number from $(0, \infty)$.

The dependence of γ on δ (i.e. the function $\gamma(\delta) > 0$, $\delta > 0$) will be found later. The following equality is true:

$$\begin{aligned} \varphi(t_* + \delta, x_{\gamma(\delta)}) &= \varphi(t_* + \delta, x_\delta^d) + \\ &\quad + \langle h(t_* + \delta, x_\delta^d),\ x_{\gamma(\delta)} - x_\delta^d\rangle + \omega_{(t_*+\delta, x_\delta^d)}(0, x_{\gamma(\delta)} - x_\delta^d) = \\ &= \varphi(t_*, x_*) + \langle h(t_*, x_*),\ x_\delta^d - x_*\rangle + \frac{\partial \varphi(t_*, x_*)}{\partial t}\,\delta + \omega_{(t_*, x_*)}(\delta, x_\delta^d - x_*) + \\ &\quad + \langle h(t_* + \delta, x_\delta^d),\ x_{\gamma(\delta)} - x_\delta^d\rangle + \omega_{(t_*+\delta, x_\delta^d)}(0, x_{\gamma(\delta)} - x_\delta^d). \end{aligned}$$

Since $\varphi(t_*, x_*) = 0$ and $\langle h(t_*, x_*), d\rangle + \dfrac{\partial \varphi(t_*, x_*)}{\partial t} \le 0$, we have

$$\varphi(t_* + \delta, x_{\gamma(\delta)}) \leq -\gamma(\delta)||h(t_*, x_*)|| - \gamma(\delta)\langle h(t_* + \delta, x^d_\delta) - h(t_*, x_*), s\rangle + \\ + \omega_{(t_*,x_*)}(\delta, \delta d) + \omega_{(t_*+\delta,\delta d)}(0, -\gamma(\delta)s). \quad (5.12)$$

The second term on the right-hand side of inequality (5.12) is constrained by the estimate

$$-\gamma(\delta)\ \langle h(t_* + \delta, x^d_\delta) - h(t_*, x_*), s\rangle \leq \gamma(\delta)\ ||h(t_* + \delta, x^d_\delta) - h(t_*, x_*)||.$$

Setting $\delta_0 = \min\{(\vartheta - t_*),\ (1 + 3R)^{-1}\varepsilon\}$, we obtain that $(t_* + \delta, x^d_\delta) \in \Omega^\varepsilon$ holds for $\delta \in (0, \delta_0)$ and, hence, the estimate

$$\|h(t_* + \delta, x^d_\delta) - h(t_*, x_*)\| \leq \chi((1 + 3R)\delta), \quad \delta \in (0, \delta_0),$$

is valid; here, the function $\chi(\rho)$, $\rho \in (0, \infty)$, is defined by the equality

$$\chi(\rho) = \max\Big\{\ \|h(t_*, x_*) - h(t^*, x^*)\|:\ \ (t_*, x_*) \text{ and } (t^*, x^*)\ \in\ \Omega^\varepsilon, \\ |t_* - t^*| + ||x_* - x^*|| \leq \rho\Big\}.$$

This implies the estimate

$$-\ \gamma(\delta)\ \langle h(t_* + \delta, x^d_\delta) - h(t_*, x_*), s\rangle \leq \gamma(\delta)\ \chi((1 + 3R)\delta), \quad \delta \in (0, \delta_0). \quad (5.13)$$

The last two summands on the right-hand side of inequality (5.12) are constrained for $\delta \in (0, \delta_0)$ by the estimates

$$\omega_{(t_*,x_*)}(\delta, \delta d) \leq (1 + ||d||)\ \delta\ \xi^*((1 + ||d||)\delta) \leq \\ \leq (1 + 3R)\ \delta\ \xi^*((1 + 3R)\delta), \quad (5.14)$$

$$\omega_{(t_*+\delta,\delta d)}(0, -\gamma(\delta)s) \leq \gamma(\delta)\ \xi^*(\gamma(\delta)). \quad (5.15)$$

Using (5.12) and taking into account (5.13)–(5.15), we obtain the following estimate for $\delta \in (0, \delta_0)$:

$$\varphi(t_* + \delta,\ x_{\gamma(\delta)}) \leq -\gamma(\delta)||h(t_*, x_*)|| + \gamma(\delta)\chi((1 + 3R)\delta) + \\ + (1 + 3R)\ \delta\ \xi^*((1 + 3R)\delta) + \gamma(\delta)\ \xi^*(\gamma(\delta)) \leq \\ \leq \gamma(\delta)\big(-h_* + \chi((1 + 3R)\delta)\big) + \gamma(\delta)\ \xi^*(\gamma(\delta)) + \\ + (1 + 3R)\ \delta\ \xi^*((1 + 3R)\delta).$$

Let $\delta^0 \in (0, \delta_0)$ satisfy the inequality $\chi((1 + 3R)\delta^0) \leq \dfrac{h_*}{2}$. Then, keeping in mind the monotone decrease of the function $\chi(\rho)$ as $\rho \downarrow 0$, we find that

$$\varphi(t_* + \delta,\ x_{\gamma(\delta)}) \leq \gamma(\delta)\left(-\frac{h_*}{2} + \xi^*(\gamma(\delta))\right) +$$
$$+(1+3R)\,\delta\,\xi^*((1+3R)\delta), \quad \delta \in (0,\delta^0). \tag{5.16}$$

We write inequality (5.16) in the form

$$\varphi(t_* + \delta,\ x_{\gamma(\delta)}) \leq \gamma(\delta)\left(-\frac{h_*}{4} + \xi^*(\gamma(\delta))\right) +$$
$$+\left(-\gamma(\delta)\tfrac{h_*}{4} + (1+3R)\,\delta\,\xi^*((1+3R)\delta)\right), \quad \delta \in (0,\delta^0). \tag{5.17}$$

Setting $\gamma(\delta) = \left(\frac{4}{h_*}\right)(1+3R)\,\delta\,\xi^*((1+3R)\delta)$, $\delta \in (0,\delta^0)$, we observe that the second term on the right-hand side of inequality (5.17) vanishes. Consequently, the following estimate holds for this function $\gamma(\delta)$ and for $\delta \in (0,\delta^0)$:

$$\varphi(t_* + \delta,\ x_{\gamma(\delta)}) \leq \gamma(\delta)\left(-\frac{h_*}{4} + \xi^*(\gamma(\delta))\right).$$

Since $\gamma(\delta) \downarrow 0$ as $\delta \downarrow 0$ and $\xi^*(\rho) \downarrow 0$ as $\rho \downarrow 0$, we have the inequality

$$-\frac{h_*}{4} + \xi^*(\gamma(\delta)) \leq 0$$

for sufficiently small $\delta^1 \in (0,\delta^0)$ and $\delta \in (0,\delta^1]$. The latter inequality implies

$$\varphi(t_* + \delta,\ x_{\gamma(\delta)}) \leq 0 \quad \text{for } \delta \in (0,\delta^1].$$

It follows that we can take the function

$$\varphi^*(\delta) = \frac{\gamma(\delta)}{\delta} = \frac{4}{h_*}\,(1+3R)\,\xi^*((1+3R)\delta) \tag{5.18}$$

as the "part" of the function $\varphi^*(\delta)$ in Condition E.1 corresponding to the domain of definition $(0,\delta^1]$.

There are two possibilities for the number δ^1:

(1) $\delta^1 \geq \vartheta - t_0$,

(2) $\delta^1 < \vartheta - t_0$.

In the case when possibility 1 is realized, we take the function $\varphi^*(\delta)$ on $(0, \vartheta - t_0]$ as in (5.18). Then, for any point $(t_*, x_*) \in \partial W^*$, $t_* \in [t_0, \vartheta)$, there is a function $\varphi^*(\delta)$ defined on $(0, \vartheta - t_*]$ and satisfying Condition E.1.

In the case when possibility 2 is realized, the function $\varphi^*(\delta)$ in Condition E.1 is defined on $(0,\delta^1]$ by equality (5.18). Then we extend the definition of the function $\varphi^*(\delta)$ to the half-open interval $(\delta^1, \vartheta - t_0]$. To do this, we consider an arbitrary point $(t_*, x_*) \in \partial W^*$, $t_* \in [t_0, \vartheta)$ and $\delta \in (\delta^1, \vartheta - t_*]$.

Let us estimate the value $\delta^{-1}h(x_* + \delta\overrightarrow{D}^{\nabla}W^*(t_*, x_*),\ W^*(t_* + \delta))$ from above. Condition C implies the relation

$$W^*(t_* + \delta) \cap \mathbb{U}(x_*;\ \delta R) \neq \varnothing, \quad \delta \in (0, \vartheta - t_*].$$

In addition, the inclusion

$$x_* + \delta\overrightarrow{D}^{\nabla}W^*(t_*, x_*) \subset \mathbb{U}(x_*;\ 3R\,\delta), \quad \delta \in (0, \vartheta - t_*].$$

holds. These two relations imply

$$h(x_* + \delta\overrightarrow{D}^{\nabla}W^*(t_*, x_*),\ W^*(t_* + \delta)) \leq 4R\,\delta, \quad \delta \in (0, \vartheta - t_*].$$

Therefore, the inequality

$$\delta^{-1}\,h(x_* + \delta\overrightarrow{D}^{\nabla}W^*(t_*, x_*),\ W^*(t_* + \delta)) \leq 4R$$

is true for $t_* \in [t_0, \vartheta)$, $(t_*, x_*) \in \partial W^*$, and $\delta \in (\delta^1, \vartheta - t_*]$.

By specifying the function

$$\varphi^*(\delta) = \varphi^*(\delta^1) + 4R + (\delta - \delta^1), \tag{5.19}$$

on the half-open interval $(\delta^1, \vartheta - t_*]$, we obtain the estimate

$$\delta^{-1}\,h(x_* + \delta\overrightarrow{D}^{\nabla}W^*(t_*, x_*),\ W^*(t_* + \delta)) \leq \varphi^*(\delta)$$

for $\delta \in (\delta^1, \vartheta - t_*]$.

The function $\varphi^*(\delta)$ for $\delta \in (\delta^1, \vartheta - t_*]$ decreases monotonically as δ decreases; in addition,

$$\varphi^*(\delta^1) < \varphi^*(\delta), \quad \delta \in (\delta^1, \vartheta - t_*]. \tag{5.20}$$

Thus, in the case when possibility 2 is realized, we have defined the function $\varphi^*(\delta)$, $\delta \in (0, \vartheta - t_*]$, by equalities (5.18), (5.19). This function decreases monotonically as δ decreases and satisfies the relation $\varphi^*(\delta) \downarrow 0$ as $\delta \downarrow 0$.

Let us extend the function $\varphi^*(\delta)$ from the half-open interval $(0, \vartheta - t_*]$ to the half-open interval $(0, \vartheta - t_0]$ according to formula (5.19). This results in the function $\varphi^*(\delta)$ for $\delta \in (0, \vartheta - t_0]$ that satisfies the relation $\varphi^*(\delta) \downarrow 0$ as $\delta \downarrow 0$. The function $\varphi^*(\delta)$ satisfies Condition E.1.

Further, we prove that Condition E.2 holds in the example under consideration. To do this, we calculate the function $\varepsilon(t_*, x_*)$, $t_* \in [t_0, \vartheta)$, $(t_*, x_*) \in \partial W^*$.

Recall that the function $\varepsilon(t_*, x_*)$ is defined by the formula

$$\varepsilon(t_*, x_*) = \sup_{l \in S} \rho\big(\overrightarrow{D}^{\nabla}W^*(t_*, x_*),\ F_l(t_*, x_*)\big).$$

For every point $(t_*, x_*) \in \partial W^*$, $t_* \in [t_0, \vartheta)$, two cases are possible:
1. $\varepsilon(t_*, x_*) > 0$; 2. $\varepsilon(t_*, x_*) = 0$.

Suppose that *case* 1 is realized. We take an arbitrary set $F_l(t_*, x_*)$ from the family of sets $\{F_l(t_*, x_*): \ l \in S\}$. It follows from the definition of the number R (see Section "The Stability Defect of the Set in the Approach Problem") that the inclusion $F_l(t_*, x_*) \subset \mathbb{U}(\mathbb{O};\ R)$ holds. Since the sets $F_l(t_*, x_*)$ and $\overrightarrow{D}^{\nabla} W^*(t_*, x_*)$ are compact in $\mathbb{R}^m$, there exists a pair (f_l, d_l) of points $f_l \in F_l(t_*, x_*)$ and $d_l \in \overrightarrow{D}^{\nabla} W^*(t_*, x_*)$ at which the value $\rho\big(\overrightarrow{D}^{\nabla} W^*(t_*, x_*),\ F_l(t_*, x_*)\big)$ is attained:

$$\| f_l - d_l \| = \rho\big(\overrightarrow{D}^{\nabla} W^*(t_*, x_*),\ F_l(t_*, x_*)\big).$$

Let it happen that $\| f_l - d_l \| > 0$ holds for the chosen set $F_l(t_*, x_*)$. The set $\overrightarrow{D}^{\nabla} W^*(t_*, x_*)$ is a ball segment in the space $\mathbb{R}^m$, that is, the intersection of the ball $\mathbb{U}(\mathbb{O};\ 3R)$ and the closed half-space $\overrightarrow{D}\, W^*(t_*, x_*) = \left\{d \in \mathbb{R}^m:\ \langle h(t_*, x_*),\ d\rangle \le -\dfrac{\partial \varphi(t_*, x_*)}{\partial t}\right\}$.

We assume that $\Gamma_* = \left\{d \in \mathbb{R}^m:\ \langle h(t_*, x_*),\ d\rangle = -\dfrac{\partial \varphi(t_*, x_*)}{\partial t}\right\}$ is the hyperplane in $\mathbb{R}^m$ that is the boundary of the set $\overrightarrow{D}\, W^*(t_*, x_*)$: $\Gamma_* = \partial \overrightarrow{D}\, W^*(t_*, x_*)$.

It follows from the relations $f_l \in \mathbb{U}(\mathbb{O};\ R)$ and $f_l \notin \overrightarrow{D}^{\nabla} W^*(t_*, x_*)$ that $\mathbb{U}(\mathbb{O};\ R) \setminus \overrightarrow{D}^{\nabla} W^*(t_*, x_*) \neq \varnothing$. Along with this relation, we have $\mathbb{U}(\mathbb{O};\ R) \cap \overrightarrow{D}\, W^*(t_*, x_*) \neq \varnothing$. Consequently, $\mathbb{U}(\mathbb{O};\ R) \cap \Gamma_* \neq \varnothing$.

Taking into account the geometry of the mutual location of the sets $\mathbb{U}(\mathbb{O};\ R)$, $\mathbb{U}(\mathbb{O};\ 3R)$, and Γ_*, we conclude that, for any point $g \in \mathbb{U}(\mathbb{O};\ R) \setminus \overrightarrow{D}^{\nabla} W^*(t_*, x_*)$, the nearest point in the set $\overrightarrow{D}^{\nabla} W^*(t_*, x_*)$ is such a point $g_* \in \Gamma_*$ for which the vector $g_* - g$ is the normal vector to the hyperplane Γ_*.

Taking into account this fact and the inequality $h_{F_{l_*}(t_*, x_*)}(l_*) \le h_{F_l(t_*, x_*)}(l_*)$, which holds for the unification family of sets $\{F_s(t_*, x_*):\ s \in S\}$, we find that the value $\sup\limits_{l \in S} \rho\big(\overrightarrow{D}^{\nabla} W^*(t_*, x_*),\ F_l(t_*, x_*)\big) = \sup\limits_{l \in S} \| f_l - d_l \|$ is attained at the vector $l_* = -||h(t_*, x_*)||^{-1} h(t_*, x_*) \in S$.

According to the definition of the point f_{l_*}, the equality $\langle l_*, f_{l_*}\rangle = H(t_*, x_*, l_*)$ holds and hence, in the case 1 under consideration, we have

$$\begin{aligned}
\varepsilon(t_*, x_*) &= \langle l_*, d_{l_*}\rangle - \langle l_*, f_{l_*}\rangle = \\
&= ||h(t_*, x_*)||^{-1} \Big(\langle -h(t_*, x_*), d_{l_*}\rangle - H(t_*, x_*, -h(t_*, x_*)) \Big) = \\
&= ||h(t_*, x_*)||^{-1} \left(\frac{\partial \varphi(t_*, x_*)}{\partial t} - H(t_*, x_*, -h(t_*, x_*)) \right).
\end{aligned}$$

Setting $H_*(t_*, x_*, l) = \min\limits_{u \in P} \max\limits_{v \in Q} \langle l,\ f(t_*, x_*, u, v)\rangle$, we obtain the following formula for the stability defect $\varepsilon(t_*, x_*)$ of the set W^* at the point (t_*, x_*) in case 1:

$$\varepsilon(t_*, x_*) = ||h(t_*, x_*)||^{-1} \left(\frac{\partial\varphi(t_*, x_*)}{\partial t} + H_*(t_*, x_*, h(t_*, x_*)) \right) > 0.$$

Let us consider *case* 2: $\varepsilon(t_*, x_*) = 0$.

We show that the equality $\varepsilon(t_*, x_*) = 0$ holds in the example under consideration if and only if

$$||h(t_*, x_*)||^{-1} \left(\frac{\partial\varphi(t_*, x_*)}{\partial t} + H(t_*, x_*, h(t_*, x_*)) \right) \leq 0.$$

Indeed, let $\varepsilon(t_*, x_*) = 0$. This equality means that $F_l(t_*, x_*) \cap \overrightarrow{D} W^*(t_*, x_*) \neq \varnothing$, $l \in S$. In particular, we have the relation $F_{l_*}(t_*, x_*) \cap \overrightarrow{D} W^*(t_*, x_*) \neq \varnothing$.

We denote by d_* an arbitrary point from Γ_* for which the relation

$$\langle l_*, d_* \rangle \leq h_{F_{l_*}(t_*, x_*)}(l_*) = H(t_*, x_*, l_*) \tag{5.21}$$

is satisfied; i.e.

$$||h(t_*, x_*)||^{-1} \Big(\langle -h(t_*, x_*), d_* \rangle - H(t_*, x_*, -h(t_*, x_*)) \Big) \leq 0. \tag{5.22}$$

It follows from the inclusion $d_* \in \Gamma_*$ that $\langle -h(t_*, x_*),\ d_* \rangle = -\dfrac{\partial\varphi(t_*, x_*)}{\partial t}$. Hence, inequality (5.22) transforms into the inequality

$$||h(t_*, x_*)||^{-1} \left(\frac{\partial\varphi(t_*, x_*)}{\partial t} + H(t_*, x_*, h(t_*, x_*)) \right) \leq 0. \tag{5.23}$$

We have shown that the equality $\varepsilon(t_*, x_*) = 0$ implies (5.23).

We suppose now that (5.23) holds. Inequality (5.21) can be derived from this equality. Then, given the relation

$$H(t_*, x_*, l_*) = h_{F_{l_*}(t_*, x_*)}(l_*) \leq h_{F_l(t_*, x_*)}(l), \quad l \in S,$$

we obtain that $\langle l_*, d_* \rangle \leq h_{F_l(t_*, x_*)}(l_*)$, $l \in S$. This inequality means that

$$F_l(t_*, x_*) \cap \overrightarrow{D} W^*(t_*, x_*) \neq \varnothing, \quad l \in S;$$

consequently, $\varepsilon(t_*, x_*) = 0$.

Thus, the equivalence of the equality $\varepsilon(t_*, x_*) = 0$ and (5.23) is established.

The above arguments show that, in the example under consideration, the stability defect $\varepsilon(t_*, x_*)$ of the set W^* at the point $(t_*, x_*) \in \partial W^*$, $t_* \in [t_0, \vartheta)$, is defined by the relations

$$\varepsilon(t_*, x_*) = \max \big\{0, k(t_*, x_*)\big\}, \tag{5.24}$$

$$k(t_*, x_*) = ||h(t_*, x_*)||^{-1} \left(\frac{\partial \varphi(t_*, x_*)}{\partial t} + H(t_*, x_*, h(t_*, x_*)) \right).$$

Hence, the following representation is valid for $t \in [t_*, \vartheta)$ in the example under consideration:

$$\varepsilon(t_*) = \max_{(t_*, x_*) \in \Lambda(t_*)} \varepsilon(t_*, x_*) = \max\{0, \zeta(t_*)\},$$

$$\zeta(t_*) = \max_{(t_*, x_*) \in \Lambda(t_*)} k(t_*, x_*).$$

In view of the conditions imposed on the function $\varphi(t_*, x_*)$, we obtain that the function $k(t_*, x_*)$ is continuous on $\partial W*$. In addition, the set $\Lambda(t_*)$ depends continuously on t_* in the Hausdorff metric on $[t_0, \vartheta)$. Therefore, the function $\zeta(t)$ and the function $\varepsilon(t)$ are continuous on $[t_0, \vartheta)$. It follows that the function $\varepsilon(t)$ is measurable on $[t_0, \vartheta)$. Thus, it is proved that Condition E.2 holds.

In our opinion, one more notion is important for the theory of stable sets. This notion supplements the notion of stability defect of a set and relates to these points $(t_*, x_*) \in \partial W^*$, $t_* \in [t_0, \vartheta)$, at which the stability occurs, i.e. to the points (t_*, x_*) at which $\varepsilon(t_*, x_*) = 0$.

Definition 3 Let a point $(t_*, x_*) \in \partial W^*$, $t_* \in [t_0, \vartheta)$, be such that $\varepsilon(t_*, x_*) = 0$ (see (5.24) for the definition of the function $\varepsilon(t_*, x_*)$). The value $k(t_*, x_*)$ in (5.24) is called the stability index of the set W^* at the point $k(t_*, x_*)$.

6 Conclusion

In the paper, we have studied the notion of the stability defect of the set W^* in the space $[t_0, \vartheta) \times \mathbb{R}^m$ of positions (t, x) of the conflict-controlled system (2.1). The notion of the stability defect was introduced in [17, 18] and is studied here under certain conditions imposed on system (2.1) and the set W^*. An example of the set W^* with smooth boundary that satisfies these conditions together with system (2.1) is given. In the example, a formula for the local defect $\varepsilon(t_*, x_*)$ $\big((t_*, x_*) \in \partial W^*$, $t_* \in [t_0, \vartheta)\big)$ of the set W^* is derived. This formula can be used in the approximate calculation of the defect of the set W^* in a number of game approach problems.

Acknowledgements This work was supported by the Russian Foundation for Basic Research under Grant no. 18-01-00221. The author expresses the highest esteem for Academician Vsevolod Mikhailovich Kuntsevich and his scientific achievements and wishes him good health, wellness, and scientific longevity.

References

1. Guseinov, H.G., Subbotin, A.I., Ushakov, V.N.: Derivatives for multivalued mappings with applications to game-theoretical problems of control. Probl. Control. Inform. Theory **14**(3), 155–167 (1985)
2. Krasovskii, N.N.: Game problems in dynamics I. Eng. Cybern. **5**, 1–10 (1969)
3. Krasovskii, N.N.: Game problems on the encounter of motions. Nauka, Moscow (1970) [In Russian]
4. Krasovskii, N.N.: On the problem of the unification of differential games. Sov. Math. Dokl. **17**(1), 269–273 (1976)
5. Krasovskii, N.N.: Unification of differential games. Game Control Problems, Trydy Inst. Mat. Mekh. Ural. Nauchn. Tsentr Akad. Nauk SSSR, vol. 24, pp. 32–45. Sverdlovsk (1977) [In Russian]
6. Krasovskii, N.N., Subbotin, A.I.: Mixed control in a differential game. Sov. Math. Dokl. **10**, 1180–1183 (1969)
7. Krasovskii, N.N., Subbotin, A.I.: The structure of differential games. Sov. Math. Dokl. **11**, 143–147 (1970)
8. Krasovskii, N.N., Subbotin, A.I.: Positional Differential Games. Nauka, Moscow (1974) [In Russian]; English translation: Game-Theoretical Control Problems. Springer, New York (1988)
9. Kryazhimskii, A.V., Osipov, Yu.S.: Differential-difference game of encounter with a functional target set. J. Appl. Math. Mech. **37**(1), 1–10 (1973)
10. Kuntsevich, V.M.: Control under uncertainty: guaranteed results in control and identification problems. Naukova Dumka, Kyiv (2006) [In Russian]
11. Kuntsevich, V., Gubarev, V., Kondratenko, Y., Lebedev, D., Lysenko, V. (eds.): Control Systems: Theory and Applications. River Publishers Series in Automation, Control and Robotics. River Publishers, Gistrup, Delft (2018)
12. Kurzhanski, A.B., Varaiya, P.: Dynamic optimization for reachibility problems. J. Optim. Theory Appl. **108**(2), 227–251 (2001)
13. Pontryagin, L.S.: Linear differential games II. Sov. Math. Dokl. **8**, 910–912 (1967)
14. Subbotin, A.I., Chentsov, A.G.: Guaranteed Optimization in Control Problems. Nauka, Moscow (1981) [In Russian]
15. Subbotin, A.I., Subbotina, N.N.: The optimal result function in a control problem. Sov. Math. Dokl. **26**(2), 336–340 (1982)
16. Taras'yev, A.M., Ushakov, V.N., Khripunov, A.P.: On a computational algorithm for solving game control problems. J. Appl. Math. Mech. **51**(2), 167–172 (1988)
17. Ushakov, V.N., Latushkin, Ya.A.: The stability defect of sets in game control problems. Proc. Steklov Inst. Math. **255**(Suppl. 1), S198–S215 (2006)
18. Ushakov, V.N., Malev, A.G.: On the question of the stability defect of sets in an approach game problem. Proc. Steklov Inst. Math. **272**(Suppl. 1), S229–S254 (2011)

Control of Stochastic Systems Based on the Predictive Models of Random Sequences

Igor Atamanyuk, Janusz Kacprzyk, Yuriy P. Kondratenko and Marina Solesvik

Abstract This chapter is devoted to the development of the mathematical models of stochastic control systems based on the predictive models of random sequences. In particular, the linear algorithms of the forecast of a control object state for an arbitrary number of states of an investigated object and the values of a control parameter are obtained. In a mean-square sense, the algorithms allow one to get an optimal estimation of the future values of a forecasted parameter, in case true known values for an observation interval are used, provided that the measurements are made with errors. Using an arbitrary number of non-linear stochastic relations, a predictive model of a control system is obtained as well. The schemes that reflect the peculiarities of determining the parameters of a nonlinear algorithm and its functioning regularities are introduced in the work. Developed models allows one to take the peculiarities of the sequence of the change of control object parameters into full consideration and also to make full use of all known priori and posteriori information about the random sequence that was investigated. The algorithms obtained in this chapter can be used in different areas of human activity to solve a wide range of problems of the control of the objects of stochastic nature.

Keywords Control · Stochastic systems · Predictive models · Random sequences · Canonical expansion

---

I. Atamanyuk (✉)
Mykolayiv National Agrarian University, Mykolayiv, Ukraine
e-mail: atamanyuk@mnau.edu.ua

J. Kacprzyk
Systems Research Institute, Polish Academy of Sciences, Warsaw, Poland
e-mail: kacprzyk@ibspan.waw.pl

Y. P. Kondratenko
Petro Mohyla Black Sea National University, Mykolaiv, Ukraine
e-mail: yuriy.kondratenko@chmnu.edu.ua

M. Solesvik
Nord University Business School, Nord University, Bodø, Norway
e-mail: marina.solesvik@hvl.no

Y. P. Kondratenko et al. (eds.), *Advanced Control Techniques in Complex Engineering Systems: Theory and Applications*, Studies in Systems, Decision and Control 203, https://doi.org/10.1007/978-3-030-21927-7_6

1 Introduction

The control of dynamic objects in the conditions of absolute uncertainty regarding the characteristics of the environment disturbances and the object model remains one of the most important problems of the modern control theory [1–3]. In most cases, the functioning of the objects of the different nature is carried out in the conditions of interaction of a great number of random factors and therefore, a control object is not fully controlled [4].

Therefore, along with the mandatory requirement of ensuring the sustainability of the designed control systems, the primary task is to compensate as many uncontrolled disturbances acting on the control objects as possible [1, 3, 5, 6].

The tasks of control in the conditions of uncertainty are met in the different fields of science and engineering [7–12], particularly when controlling stationary and moving objects [13–16], economics [17, 18], medicine [19, 20], biology [21], etc. In these tasks, the evolution of the system is characterized by various conflict situations [22, 23], different parameters and external influences, the values of which are known imprecisely [3, 5, 24]. In particular, the movement of an aircraft in the atmosphere (including polar conditions [25]) occurs under the influence of a great number of different [26–28] poorly controlled or unpredictable factors, among which are external imprecisely known forces, dispersion of aerodynamic characteristics and structural parameters of an aircraft, bursts of wind, variations of atmospheric density, magnetic and gravitational fields of the Earth, etc.

Power plants and their elements (diesels, steam and gas turbines), the parameters of which parameters change during the operation [29] are also stochastic control systems. For example, the changes of temperature, air pressure and humidity essentially influence on the properties of a turbo-generator which in turn reduces or increases the effective capacity of a turbine with unchanged fuel consumption. Here, the operational characteristics of a gas turbine quality deteriorate over time because the contamination of the flow parts of a compressor and a turbine, residue formation and instability of fuel delivery cause the change of operating modes. The stochastic character of the change of controlled parameters can also be connected with the hindrances in the channel of a system observation.

When creating effective control systems for objects of this class, it is necessary to solve the problems of estimation, identification, forecasting and control [30–33], using the methods of research and investigations of adaptive [1, 3, 5, 34] and stochastic [30, 35–38] control systems, the optimization of intelligent control systems based on fuzzy logic and neural networks [39–50], as well as the methods for controlling stochastic systems in a fuzzy environment [51–53].

The forecasting problem of an object state and a control problem are mathematically dual problems [54]. Therefore, the extrapolation methods of random processes and sequences are widely used for the synthesis of stochastic control systems. One of the fundamental methods of a stochastic control theory is the theory of filtering and preemption developed by Wiener and Kolmogorov [1]. However, the use of the Wiener-Kolmogorov method [55, 56] presupposes the solution of integral equation

(Wiener-Hopf equation) that reduces the area of its application. The Wiener-Hopf equation has rarely shown analytical solutions for real tasks, and the solution by numerical methods is a cumbrous and laborious procedure. Kalman and Bucy [57, 58] made a significant contribution into the filtering theory. In their theory, the problem of preemption and filtering is solved by recurrent methods which allows one to use computing machinery. R. Kalman and R. Bucy's results also apply to non-stationary processes. Based on the Kalman-Bucy theory, the forecasting is carried out in the form of an output variable dynamic system when the control is carried out by observations. To determine the coefficients of a dynamic system it is necessary to solve the Riccati equation with set initial conditions. However, in spite of a number of obvious advantages, the Kalman-Bucy method has an essential shortcoming. The optimal solution of a forecasting problem based on the given method can only be obtained for random Markov processes. The most general extrapolation form for solving the problem of nonlinear extrapolation is found using the Kolmogorov-Gabor polynomial [59], but the determination of its parameters for a large number of known values and a used order of nonlinear relation is a very difficult and laborious procedure. Thereupon, different simplifications and restrictions on the properties of random sequences are used during the forming of the realizable in practice forecasting algorithms. For example, a number of suboptimal methods of nonlinear extrapolation [60] with a bounded order of a stochastic relation based on the approximation of a posteriori density of probabilities of an estimated vector by orthogonal Hermite polynomial expansion or in the form of Edgeworth series is offered by V. S. Pugachev. The solution of the non-stationary A. N. Kolmogorov equation (a particular case of R. L. Stratanovich differential equation for description of Markovian processes) is obtained, provided that a drift coefficient is a linear function of the state, and a diffusion coefficient is equal to a constant.

In most cases during the forming of stochastic control systems, a stationary or the Markov model is used. Also, limitations could be imposed on the nonlinear properties of a model that essentially restricts control accuracy. Thereupon, the aim of the work is a synthesis of control systems taking an arbitrary number of control object states into account and a random order of nonlinear relations of a random sequence of the change of the parameters of an investigated stochastic system.

A very promising approach to the study of various stochastic models in the form of random sequences is the use of the mathematical apparatus of canonical decompositions [61–66], which allows one to consider different degrees of relationships between the characteristics of random sequences.

Stochastic models of this class can successfully be used to solve the problems of predicting electricity consumption [67], estimating the pre-emergency state of computer systems of critical use [68–72], recognizing the characteristics of electrocardiograms and electroencephalograms [19, 20], etc.

The aim of this work is to improve the management methods of stochastic systems based on the predictive models of random sequences using the mathematical apparatus of canonical decomposition and taking the different degrees of nonlinear connections of random sequences into account.

2 Linear Forecasting Model for an Arbitrary Number of Stochastic System States

Let properties of a stochastic system are fully set by discretized functions $M[X(\mu)X(j)]$, $M[U(\mu)U(j)]$, $M[U(\mu)X(j)]$, $\mu, j = \overline{1, i+1}$, where $\{X\} = X(\mu)$, $\mu = \overline{1, i+1}$ is a random sequence of the values of a controlled parameter of a control object; $\{U\} = U(\mu)$, $\mu = \overline{1, i+1}$ is a random sequence of control actions; μ and j are some moments of time t_μ and t_j. Without the loss, the generality let us assume that $M[X(\mu)] = 0$, $M[U(\mu)] = 0$, $\mu = \overline{1, i+1}$. It is necessary to obtain an optimal in mean-square sense model of an investigated control system.

Taking that the system properties are fully set by correlation functions into account, the linear equation [61, 73] can be used as the model of a control system

$$X(i+1) = \sum_{\mu=1}^{i} f_\mu^{(i)}(i+1)X(\mu) + \sum_{j=1}^{i} a_\mu^{(i)}(i+1)U(\mu) + \xi_{i+1}. \quad (1)$$

where $f_\mu^{(i)}(i+1)$, $a_\mu^{(i)}(i+1)$ are non-random functions and ξ_{i+1} is a stochastic error of a control system.

Thus, the problem is reduced to the determination of optimal in mean-square sense parameters $f_\mu^{(i)}(i+1)$, $a_\mu^{(i)}(i+1)$.

The process of the system functioning can be presented by a random sequence $\{X'\} = \{X(1), U(1), X(2), U(2), \ldots, X(i), U(i), X(i+1)\}$ (a control action which influences on the following object state is formed by a known value of the parameter of an investigated object). For such a random sequence, a canonical expansion [73, 74] can be obtained in a standard way:

$$X'(j) = \sum_{\nu=1}^{j-1} V_\nu \varphi_\nu(j), \; j = \overline{1, 2i+1}, \quad (2)$$

Considering the properties of a random sequence $\{X'\}$ ($X'(j) = X(\mu)$ for $j = 2\mu - 1$, $\mu = \overline{1, i+1}$ and $X'(j) = U_\mu$ for $j = 2\mu$, $\mu = \overline{1, i}$):

$$V_\nu = X(\nu) - \sum_{\mu=1}^{\nu-1} V_\mu \varphi_\mu(\nu), \; \nu = 2l - 1, l = \overline{1, i+1}; \quad (3)$$

$$V_\nu = U(\nu) - \sum_{\mu=1}^{\nu-1} V_\mu \varphi_\mu(\nu), \; \nu = 2l, l = \overline{1, i+1}. \quad (4)$$

Canonical expansion (2) accurately describes the values of a random sequence $\{X'\}$ in each section and provides the minimum of a mean-square error of approximation in the intervals between them. The ratios used to determine the dispersions of uncorrelated random coefficients V_ν, $\nu = \overline{1, 2i+1}$ are of the form:

$$D_\nu = M\left[V_\nu^2\right] = D_x(\nu) - \sum_{\mu=1}^{\nu-1} D_\mu \varphi_\mu^2(\nu),\ \nu = 2l-1,\ l = \overline{1, i+1}; \tag{5}$$

$$D_\nu = D_U(\nu) - \sum_{\mu=1}^{\nu-1} D_\mu \varphi_\mu^2(\nu),\ \nu = 2l,\ l = \overline{1, i}. \tag{6}$$

Non-random coordinate functions are calculated with the help of the following expressions

$$\varphi_\mu(\nu) = \frac{1}{D_\mu}\left\{M[X(\mu)X(\nu)] - \sum_{j=1}^{\mu-1} D_j \varphi_j(\mu)\varphi_j(\nu)\right\},\ \mu = 2l-1,$$
$$l = \overline{1, i+1},\ \nu = 2p-1,\ p = \overline{1, i+1},\ \mu \le \nu; \tag{7}$$

$$\varphi_\mu(\nu) = \frac{1}{D_\mu}\left\{M[U(\mu)X(\nu)] - \sum_{j=1}^{\mu-1} D_j \varphi_j(\mu)\varphi_j(\nu)\right\},\ \mu = 2l,$$
$$l = \overline{1, i},\ \nu = 2p-1,\ p = \overline{1, i+1},\ \mu \le \nu; \tag{8}$$

$$\varphi_\mu(\nu) = \frac{1}{D_\mu}\left\{M[X(\mu)U(\nu)] - \sum_{j=1}^{\mu-1} D_j \varphi_j(\mu)\varphi_j(\nu)\right\},\ \mu = 2l-1,$$
$$l = \overline{1, i},\ \nu = 2p,\ p = \overline{1, i},\ \mu \le \nu; \tag{9}$$

$$\varphi_\mu(\nu) = \frac{1}{D_\mu}\left\{M[U(\mu)U(\nu)] - \sum_{j=1}^{\mu-1} D_j \varphi_j(\mu)\varphi_j(\nu)\right\},\ \mu = 2l,$$
$$l = \overline{1, i},\ \nu = 2p,\ p = \overline{1, i},\ \mu \le \nu. \tag{10}$$

Coordinate functions have the properties

$$\varphi_\mu(\nu) = \begin{cases} 1,\ \mu = \nu; \\ 0,\ \mu > \nu. \end{cases} \tag{11}$$

Let us assume that the system is in an original state $X(1) = x(1)$. Expansion (2) is true for this value, and substitution $x(1)$ into (2) concretizes ($v_1 = x(1)$) the value of a random coefficient V_1:

$$X'(j = 2i+1/X(1) = x(1)) = X(i+1/X(1) = x(1)) =$$
$$= x(1)\varphi_1(2i+1) + \sum_{\nu=2}^{2i+1} V_\nu \varphi_\nu(2i+1). \tag{12}$$

Expression (12) accurately describes the conditional random sequence $\{X'\}$ provided that $X(1) = x(1)$ and that it is a model of the system on the assumption that only one original state is known:

$$f_1^{(1)}(i+1) = \varphi_1(2i+1) = F_1^{(1)}(2i+1),$$

$$a_1^{(1)}(i+1) = 0, \quad \xi_{i+1} = \sum_{\nu=2}^{2i+1} V_\nu \varphi_\nu(2i+1).$$

Optimal in mean-square sense estimation of the parameter of a control object at the point of time t_{i+1} is determined as

$$\hat{X}(i+1/X(1) = x(1)) = \hat{X}'^{(1)}(2i+1) = x(1)\varphi_1(2i+1). \tag{13}$$

The value of control action $u(1)$ is concretized by the second coefficient $V_2 = v_2 = u(1) - x(1)\varphi_1(2)$ and expansion (2) takes form:

$$\begin{aligned} &X'(j = 2i+1/X(1) = x(1), U(1) = u(1)) = \\ &= X(i+1/X(1) = x(1), U(1) = u(1)) = \\ &= x(1)\varphi_1(2i+1) + [u(1) - x(1)\varphi_1(2)]\varphi_2(2i+1) + \sum_{\nu=3}^{2i+1} V_\nu \varphi_\nu(2i+1). \end{aligned} \tag{14}$$

or

$$\begin{aligned} X(i+1/X(1) = x(1), U(1) = u(1)) = x(1)[\varphi_1(2i+1) - \varphi_1(2)\varphi_2(2i+1)] + \\ + u(1)\varphi_2(2i+1) + \sum_{\nu=3}^{2i+1} V_\nu \varphi_\nu(2i+1). \end{aligned} \tag{15}$$

Thus, provided that only original state $x(1)$ of an object and control action $u(1)$ are used, the parameters of model (1) are determined by the expressions

$$\begin{aligned} &f_1^{(1)}(i+1) = \varphi_1(2i+1) - \varphi_1(2)\varphi_2(2i+1) = F_1^{(2)}(2i+1), \\ &a_1^{(1)}(i+1) = \varphi_2(2i+1) = F_2^{(2)}(2i+1), \\ &\xi_{i+1} = \sum_{\nu=3}^{2i+1} V_\nu \varphi_\nu(2i+1). \end{aligned}$$

At that, optimal in mean-square sense estimation of the state of a control object at the point of time t_{i+1} by known $x(1)$ and $u(1)$ takes form

$$\begin{aligned} &\hat{X}(i+1/X(1) = x(1), U(1) = u(1)) = \hat{X}'^{(2)}(2i+1) = \\ &= x(1)[\varphi_1(2i+1) - \varphi_1(2)\varphi_2(2i+1)] + u(1)\varphi_2(2i+1). \end{aligned} \tag{16}$$

The continuation of similar calculations for the consecutively increasing number of known values $x(\mu), u(\mu), \mu = \overline{1, i}$ allows one to obtain general expressions to determine the optimal parameters of an investigated stochastic control system [74]:

$$f_{\mu}^{(i)}(i+1) = F_{2\mu-1}^{(2i)}(2i+1), \mu = \overline{1, i}; \tag{17}$$

$$a_{\mu}^{(i)}(i+1) = F_{2\mu}^{(2i)}(2i+1), \mu = \overline{1, i}; \tag{18}$$

$$\xi_{i+1} = V_{2i+1}; \tag{19}$$

$$F_{\nu}^{(k)}(j) = \begin{cases} F_{\nu}^{(k-1)}(j) - F_{\nu}^{(k-1)}(k)\varphi_k(j), \nu \le k-1; \\ \varphi_k(j), \nu = k. \end{cases} \tag{20}$$

The optimal estimation of the state of a control object throughout the whole backstory is determined as well by recurrent relations

$$\hat{X}'^{(k)}(j) = \hat{X}'^{(k-1)}(j) + \left[x(k) - \hat{X}'^{(k-1)}(k)\right]\varphi_k(j), \ k = 2p-1, p = \overline{1, i}; \tag{21}$$

$$\hat{X}'^{(k)}(j) = \hat{X}'^{(k-1)}(j) + \left[u(k) - \hat{X}'^{(k-1)}(k)\right]\varphi_k(j), \ k = 2p, p = \overline{1, i}. \tag{22}$$

where

$$\hat{X}'^{(k)}(j) = x(\mu/x(1), ..., x(l); u(1), ...u(n)), \ j = 2\mu - 1, \mu = \overline{1, i+1};$$
$$l = n = k/2 \ \text{if}\{k/2\} = 0;$$

$$\hat{X}'^{(k)}(j) = u(\mu/x(1), ..., x(l); u(1), ...u(n)), \ j = 2\mu, \mu = \overline{1, i};$$
$$l = [k/2] + 1, n = [k/2] \ \text{if}\{k/2\} \neq 0.$$

The error of an obtained model of a control system is only determined by the stochastic nature of the process of the change of values of an investigated object parameter and is equal to the dispersion of a posteriori sequence $\{X'\}$, provided that the values $x(\mu), u(\mu), \mu = \overline{1, i}$ are known.

$$\varepsilon(i+1) = D_{2i+1} = D_x(2i+1) - \sum_{\mu=1}^{2i} D_{\mu}\varphi_{\mu}^2(2i+1). \tag{23}$$

Let us assume that the values of the parameter of an investigated object are measured with an error. Thus, the model of a stochastic control system takes place:

$$X(i+1) = \sum_{\mu=1}^{i} f_{\mu}^{(i)}(i+1)X(\mu) + \sum_{j=1}^{i} a_{\mu}^{(i)}(i+1)U(\mu) + \xi_{i+1}. \tag{24}$$

$$Z(i) = X(i) + Y(i), \tag{25}$$

where $Y(i)$ and $Z(i)$ are respectively an error and the result of measurement.

Taking the errors of measurement into account, the equation of the object control can be written as

$$\hat{X}_z^{(i)}(i+1) = \sum_{\mu=1}^{i} f_\mu^{(i)}(i+1)Z(\mu) + \sum_{j=1}^{i} a_\mu^{(i)}(i+1)U(\mu) \tag{26}$$

In a natural way, the inaccuracy of values $x(\mu)$, $\mu = \overline{1,i}$ determines an increased stochastic uncertainty of the system (mean-square error of control has increased by the value $\sum_{\mu=1}^{i}\sum_{j=1}^{i} f_\mu^{(i)}(i+1)f_j^{(i)}(i+1)M[Y(\mu)Y(j)]$). The decrease of influence of the given factor on the accuracy of control is possible due to the application of a linear filtering operation to the results of measurements

$$\hat{X}_f(\mu/x(1),\ldots,x(\mu-1);u(1),\ldots u(\mu-1)) = \left(1 - B_{2\mu-1}\right)\hat{X}_f^{\prime(2\mu-2)}(2\mu-1)+$$
$$+ B_{2\mu-1}z(\mu), \mu = \overline{1,i}. \tag{27}$$

The estimation (27) of the unknown value $x(\mu)$ is determined as a weighted average of forecasting result $\hat{X}_\varphi^{(2\mu-2)}(\mu)$ and measurement result $z(\mu)$ [54]. It should be noted that control actions $u(\mu)$, $\mu = \overline{1,i}$ are known precisely, which is why $B_{2\mu} = 1$, $\mu = \overline{1,i}$. The substitution of the estimation (27) into (26) gives the control equation taking into account preliminary filtering of measurements errors

$$\hat{X}_f^{(i)}(i+1) = \sum_{\mu=1}^{i} g_\mu^{(i)}(i+1)Z(\mu) + \sum_{j=1}^{i} c_\mu^{(i)}(i+1)U(\mu). \tag{28}$$

where

$$g_\mu^{(i)}(i+1) = S_{2\mu-1}^{(2i)}(2i+1), \mu = \overline{1,i}; \tag{29}$$

$$c_\mu^{(i)}(i+1) = S_{2\mu}^{(2i)}(2i+1), \mu = \overline{1,i}; \tag{30}$$

$$S_\nu^{(k)}(j) = \begin{cases} S_\nu^{(k-1)}(j) - S_\nu^{(k-1)}(k)B_k\varphi_k(j), \nu \le k-1; \\ B_k\varphi_k(j), \nu = k. \end{cases} \tag{31}$$

The values of weight coefficients are determined from the condition of the minimum of a mean-square error of filtering

$$\varepsilon_f(\mu) = M\left[\left|\hat{X}_f(\mu) - X(\mu)\right|^2\right] =$$

$$M\left[\left|\left(1 - B_{2\mu-1}\right)\left\{\sum_{\nu=1}^{\mu-1} g_\nu^{(\mu-1)}(\mu)Z(\nu) + \sum_{\nu=1}^{\mu-1} c_\nu^{(\mu-1)}(\mu)U(\nu)\right\} + B_{2\mu-1}z(\mu) - X(\mu)\right|^2\right].$$

After the differentiation of this expression by $B_{2\mu-1}$ and the solving of a corresponding equation we obtain the expression for the calculation of an optimal value of filtering coefficient $B_{2\mu-1}$

$$B_{2\mu-1} = \frac{H_{2\mu-1} + G_{2\mu-1} - L_{2\mu-1}}{H_{2\mu-1} + G_{2\mu-1} - 2L_{2\mu-1} + D_y(\mu)}, \tag{32}$$

where

$$H_{2\mu-1} = \sum_{\nu=1}^{\mu-1}\sum_{j=1}^{\mu-1} g_\nu^{(\mu-1)}(\mu) g_j^{(\mu-1)}(\mu) R_x(\nu, j) +$$
$$+ \sum_{\nu=1}^{\mu-1}\sum_{j=1}^{\mu-1} g_\nu^{(\mu-1)}(\mu) c_j^{(\mu-1)}(\mu) R_{xu}(\nu, j) +$$
$$+ \sum_{\nu=1}^{\mu-1}\sum_{j=1}^{\mu-1} c_\nu^{(\mu-1)}(\mu) c_j^{(\mu-1)}(\mu) R_u(\nu, j) - 2\sum_{\nu=1}^{\mu-1} g_\nu^{(\mu-1)}(\mu) R_x(\nu, \mu) -$$
$$- 2\sum_{\nu=1}^{\mu-1} c_\nu^{(\mu-1)}(\mu) R_{xu}(\nu, \mu) + D_x(\mu),$$

$$G_{2\mu-1} = \sum_{\nu=1}^{\mu-1}\sum_{j=1}^{\mu-1} g_\nu^{(\mu-1)}(\mu) g_j^{(\mu-1)}(\mu) R_y(\nu, j),$$

$$L_{2\mu-1} = \sum_{\nu=1}^{\mu-1} g_\nu^{(\mu-1)}(\mu) R_y(\nu, \mu).$$

Each element of the formula (31) has obvious physical meaning: summand $H_{2\mu-1}$ determines the contribution into a resulting error made by a stochastic nature of the control system, summands $G_{2\mu-1}$ and $L_{2\mu-1}$ are connected to the errors of prior measurements, and summand $D_y(\mu)$ is a dispersion of the error of the last measurement.

In case the errors are uncorrelated $M[Y(\nu)Y(\mu)] = 0$ at $\nu \neq \mu$ expression (31) is simplified to the form

$$B_{2\mu-1} = \frac{H_{2\mu-1} + G'_{2\mu-1}}{H_{2\mu-1} + G'_{2\mu-1} + D_y(\mu)}, \tag{33}$$

where

$$G'_{2\mu-1} = \sum_{\nu=1}^{\mu-1} \left[g_{\nu}^{(\mu-1)}(\mu) \right]^2 D_y(\nu).$$

The error of control, with the help of Eq. (28), can be written as

$$\Delta^{(k)}(i+1) = \hat{X}_f^{(i)}(i+1) - X^{(i)}(i+1). \tag{34}$$

Substitution of expressions (28) and (1) instead of $\hat{X}_f^{(i)}(i+1)$ and $X^{(i)}(i+1)$ respectively into (34) gives

$$\Delta^{(k)}(i+1) = \sum_{\mu=1}^{i} \left[g_{\mu}^{(i)}(i+1) - f_{\mu}^{(i)}(i+1) \right] X(\mu) + \sum_{\mu=1}^{i} g_{\mu}^{(i)}(i+1) Y(\mu) -$$
$$- \sum_{\mu=1}^{i} \left[c_{\mu}^{(i)}(i+1) - a_{\mu}^{(i)}(i+1) \right] U(\mu) - \sum_{\mu=1}^{i} f_{\mu}^{(i)}(i+1) X(\mu) - W_{i+1}. \tag{35}$$

Taking (35) into account a mean-square error of a stochastic control system is determined from expression

$$\varepsilon_f(i+1) = \sum_{\mu=1}^{i} \sum_{j=1}^{i} \left[g_{\mu}^{(i)}(i+1) - f_{\mu}^{(i)}(i+1) \right] \left[g_{j}^{(i)}(i+1) - f_{j}^{(i)}(i+1) \right] R_x(\mu, j) +$$
$$+ \sum_{\mu=1}^{i} \sum_{j=1}^{i} g_{\mu}^{(i)}(i+1) g_{j}^{(i)}(i+1) R_y(\mu, j) + \sum_{\mu=1}^{i} \sum_{j=1}^{i} f_{\mu}^{(i)}(i+1) f_{j}^{(i)}(i+1) R_x(\mu, j) +$$
$$+ \sum_{\mu=1}^{i} \sum_{j=1}^{i} \left[c_{\mu}^{(i)}(i+1) - a_{\mu}^{(i)}(i+1) \right] \left[c_{j}^{(i)}(i+1) - a_{j}^{(i)}(i+1) \right] R_u(\mu, j) +$$
$$+ \sum_{\mu=1}^{i} \sum_{j=1}^{i} \left[g_{\mu}^{(i)}(i+1) - f_{\mu}^{(i)}(i+1) \right] \left[c_{j}^{(i)}(i+1) - a_{j}^{(i)}(i+1) \right] R_{xu}(\mu, j) +$$
$$+ \sum_{\mu=1}^{i} \sum_{j=1}^{i} \left[g_{\mu}^{(i)}(i+1) - f_{\mu}^{(i)}(i+1) \right] f_{j}^{(i)}(i+1) R_x(\mu, j) +$$
$$+ \sum_{\mu=1}^{i} \sum_{j=1}^{i} \left[c_{\mu}^{(i)}(i+1) - a_{\mu}^{(i)}(i+1) \right] f_{j}^{(i)}(i+1) R_{xu}(\mu, j) + D_{2i+1}. \tag{36}$$

3 Generalized Nonlinear Model of a Stochastic Control System

Let us assume that a stochastic system has nonlinear relations and its properties are fully set at a discrete number of points t_i, $i = \overline{1, I}$ by moment functions $M\left[X^{\lambda}(\nu)X^{s}(i)\right]$, $M\left[U^{\lambda}(\nu)X^{s}(i)\right]$, $M\left[U^{\lambda}(\nu)U^{s}(i)\right]$, $\lambda, s=\overline{1, N}$; $\nu, j = \overline{1, I}$.

To obtain a canonical model of a vector random sequence $\{X(i), U(i)\}$, $i = \overline{1, I}$ taking nonlinear relations into account, we consider an array of random values

$$\begin{pmatrix} X(1) & X(2) & \ldots & X(I-1) & X(I) \\ X^2(1) & X^2(2) & \ldots & X^2(I-1) & X^2(I) \\ \ldots & \ldots & \ldots & \ldots & \ldots \\ X^N(1) & X^N(2) & \ldots & X^N(I-1) & X^N(I) \\ U(1) & U(2) & \ldots & U(I-1) & U(I) \\ U^2(1) & U^2(2) & \ldots & U^2(I-1) & U^2(I) \\ \ldots & \ldots & \ldots & \ldots & \ldots \\ U^N(1) & U^N(2) & \ldots & U^N(I-1) & U^N(I) \end{pmatrix} \tag{37}$$

The correlation moments of array elements (37) fully describe the probabilistic relations of sequence $\{X(i), U(i)\}, i = \overline{1, I}$ at an investigated number of points $t_i, i = \overline{1, I}$. Therefore, the application of vector linear canonical expansion [73] to the first line $X(i)$, $i = \overline{1, I}$ allows one to obtain a canonical expansion with a full taking into account of a priori information for each component:

$$X(i) = M[X(i)] + \sum_{\nu=1}^{i-1}\sum_{l=1}^{2}\sum_{\lambda=1}^{N} W_{\nu l}^{(\lambda)} \beta_{l\lambda}^{(1,1)}(\nu, i) + W_{i1}^{(1)},\ i = \overline{1, I}; \tag{38}$$

$$W_{\nu 1}^{(\lambda)} = X^{\lambda}(\nu) - M\left[X^{\lambda}(\nu)\right] - \sum_{\mu=1}^{\nu-1}\sum_{m=1}^{2}\sum_{j=1}^{N} W_{\mu m}^{(j)} \beta_{mj}^{(1,\lambda)}(\mu, \nu) -$$

$$- \sum_{j=1}^{\lambda-1} W_{\nu 1}^{(j)} \beta_{1j}^{(1,\lambda)}(\nu, \nu),\ \lambda = \overline{1, N}, \nu = \overline{1, I}; \tag{39}$$

$$W_{\nu 2}^{(\lambda)} = U^{\lambda}(\nu) - M\left[U^{\lambda}(\nu)\right] - \sum_{\mu=1}^{\nu-1}\sum_{m=1}^{2}\sum_{j=1}^{N} W_{\mu m}^{(j)} \beta_{mj}^{(2,\lambda)}(\mu, \nu) -$$

$$- \sum_{j=1}^{N} W_{\nu 1}^{(j)} \beta_{1j}^{(2,\lambda)}(\nu, \nu) - \sum_{j=1}^{\lambda-1} W_{\nu 2}^{(j)} \beta_{2j}^{(2,\lambda)}(\nu, \nu),\ \lambda = \overline{1, N}, \nu = \overline{1, I}; \tag{40}$$

$$D_{1,\lambda}(\nu) = M\left[\left\{W_{\nu 1}^{(\lambda)}\right\}^2\right] = M\left[X^{2\lambda}(\nu)\right] - M^2\left[X^{\lambda}(\nu)\right] -$$

$$-\sum_{\mu=1}^{\nu-1}\sum_{m=1}^{2}\sum_{j=1}^{N} D_{mj}(\mu)\Big\{\beta_{mj}^{(1,\lambda)}(\mu,\nu)\Big\}^{2}-$$

$$-\sum_{j=1}^{\lambda-1} D_{1j}(\nu)\Big\{\beta_{1j}^{(1,\lambda)}(\nu,\nu)\Big\}^{2},\ \lambda=\overline{1,N},\ \nu=\overline{1,I}; \tag{41}$$

$$D_{2,\lambda}(\nu)=M\left[\Big\{W_{\nu 2}^{(\lambda)}\Big\}^{2}\right]=M\big[U^{2\lambda}(\nu)\big]-M^{2}\big[U^{\lambda}(\nu)\big]-$$

$$-\sum_{\mu=1}^{\nu-1}\sum_{m=1}^{H}\sum_{j=1}^{N} D_{mj}(\mu)\Big\{\beta_{mj}^{(l,\lambda)}(\mu,\nu)\Big\}^{2}-\sum_{j=1}^{N} D_{1j}(\nu)\Big\{\beta_{1j}^{(2,\lambda)}(\nu,\nu)\Big\}^{2}-$$

$$-\sum_{j=1}^{\lambda-1} D_{2j}(\nu)\Big\{\beta_{2j}^{(2,\lambda)}(\nu,\nu)\Big\}^{2},\ \lambda=\overline{1,N},\ \nu=\overline{1,I}; \tag{42}$$

$$\beta_{1\lambda}^{(1,s)}(\nu,i)=\frac{M\Big[W_{\nu 1}^{(\lambda)}\big(X^{s}(i)-M[X^{s}(i)]\big)\Big]}{M\left[\Big\{W_{\nu 1}^{(\lambda)}\Big\}^{2}\right]}=$$

$$=\frac{1}{D_{1\lambda}(\nu)}\Big(M\big[X^{\lambda}(\nu)X^{s}(i)\big]-M\big[X^{\lambda}(\nu)\big]M\big[X^{s}(i)\big]-$$

$$-\sum_{\mu=1}^{\nu-1}\sum_{m=1}^{2}\sum_{j=1}^{N} D_{mj}(\mu)\beta_{mj}^{(l,\lambda)}(\mu,\nu)\beta_{mj}^{(h,s)}(\mu,i)-$$

$$-\sum_{j=1}^{\lambda-1} D_{1j}(\nu)\beta_{1j}^{(1,\lambda)}(\nu,\nu)\beta_{1j}^{(1,s)}(\nu,i),\ \lambda,s=\overline{1,N},\ \nu=\overline{1,i}. \tag{43}$$

$$\beta_{2\lambda}^{(1,s)}(\nu,i)=\frac{M\Big[W_{\nu 2}^{(\lambda)}\big(X^{s}(i)-M[X^{s}(i)]\big)\Big]}{M\left[\Big\{W_{\nu 2}^{(\lambda)}\Big\}^{2}\right]}=$$

$$=\frac{1}{D_{2\lambda}(\nu)}\big(M\big[U^{\lambda}(\nu)X^{s}(i)\big]-M\big[U^{\lambda}(\nu)\big]M\big[X^{s}(i)\big]-$$

$$-\sum_{\mu=1}^{\nu-1}\sum_{m=1}^{2}\sum_{j=1}^{N} D_{mj}(\mu)\beta_{mj}^{(2,\lambda)}(\mu,\nu)\beta_{mj}^{(1,s)}(\mu,i)-$$

$$-\sum_{j=1}^{N} D_{1j}(\nu)\beta_{1j}^{(2,\lambda)}(\nu,\nu)\beta_{1j}^{(1,s)}(\nu,i)-$$

$$-\sum_{j=1}^{\lambda-1} D_{2j}(\nu)\beta_{2j}^{(2,\lambda)}(\nu,\nu)\beta_{2j}^{(1,s)}(\nu,i),\ \lambda,s=\overline{1,N},\ \nu=\overline{1,i}. \tag{44}$$

Random sequence $\{X(i), U(i)\}, i = \overline{1, I}$ is represented with the help of $2 \times N$ arrays $\{W_{\nu l}^{(\lambda)}\}$, $\lambda = \overline{1, N}$; $l = \overline{1, 2}$ of uncorrelated centered random coefficients $W_{\nu l}^{(\lambda)}$, $\nu = \overline{1, I}$. Each of these coefficients contains information about the corresponding values $X^{(\lambda)}(\nu)$, $U^{(\lambda)}(\nu)$ and coordinate functions $\beta_{l\lambda}^{(h,s)}(\nu, i)$ describe probabilistic relations of $\lambda + s$ order between components $X(t)$ and $U(t)$ in the sections t_ν and t_i.

The block diagram of the algorithm to calculate parameters $D_{l,\lambda}(\nu)$, $l = \overline{1, 2}$, $\lambda = \overline{1, N}$, $\nu = \overline{1, I}$ and $\beta_{l\lambda}^{(h,s)}(\nu, i)$, $l, h = \overline{1, 2}$, $\lambda, s = \overline{1, N}$, $\nu, i = \overline{1, I}$ of canonical expansion (38) is presented in Fig. 1.

Let us assume that as a result of a measurement the first value $x(1)$ of the sequence at point t_1 is known. Consequently, the values of coefficients $W_{11}^{(\lambda)}$, $\lambda = \overline{1, N}$ are known:

$$w_{11}^{(\lambda)} = x^\lambda(1) - M\big[X^\lambda(1)\big] - \sum_{j=1}^{\lambda-1} w_{11}^{(j)} \beta_{1j}^{(1,\lambda)}(1, 1),\ \lambda = \overline{1, N} \tag{45}$$

Substituting $w_{11}^{(1)}$ into (38) allows one to obtain a posteriori canonical expansion of the first component $\big\{X^{(1,1)}\big\} = X(i/x_1(1))$ of a random sequence $\{X(i), U(i)\}, i = \overline{1, I}$:

$$\begin{aligned} X^{(1,1)}(i) = X(i/x(1)) = M[X(i)] + (x(1) - M[X(1)])\beta_{11}^{(1,1)}(1, i) + \\ + \sum_{\lambda=2}^{N} W_{11}^{(\lambda)} \beta_{1\lambda}^{(1,1)}(1, i) + \sum_{\lambda=1}^{N} W_{12}^{(\lambda)} \beta_{1\lambda}^{(1,1)}(1, i) + \\ + \sum_{\nu=2}^{i-1} \sum_{l=1}^{2} \sum_{\lambda=1}^{N} W_{\nu l}^{(\lambda)} \beta_{l\lambda}^{(1,1)}(\nu, i) + W_{i1}^{(1)},\ i = \overline{1, I}. \end{aligned} \tag{46}$$

The application of the operation of mathematical expectation to (46) gives an optimal (by the criterion of minimum of mean-square error of extrapolation) estimation of future values of sequence $\{X\}$ provided that one value $x(1)$ is used to determine the given estimation:

$$m_{1,1}^{(1,1)}(1, i) = M[X(i/x(1))] = M[X(i)] + (x(1) - M[X(1)])\beta_{11}^{(1,1)}(1, i),\ i = \overline{1, I}. \tag{47}$$

Taking the fact that coordinate functions $\beta_{l\lambda}^{(h,s)}(\nu, i)$, $l, h = \overline{1, 2}$, $\lambda, s = \overline{1, N}$, $\nu, i = \overline{1, I}$ are determined from the condition of minimum of a mean-square error of approximation in the intervals between arbitrary values $X^\lambda(\nu)$ and $U^h(i)$ into account, expression (47) can be generalized in case of the forecasting $x^s(i)$, $s = \overline{1, N}$, $i = \overline{1, I}$:

$$m_{1,1}^{(1,1)}(s, i) = M\big[X^s(i/x(1))\big] = M\big[X^s(i)\big] + (x(1) - M[X(1)])\beta_{11}^{(1,s)}(1, i). \tag{48}$$

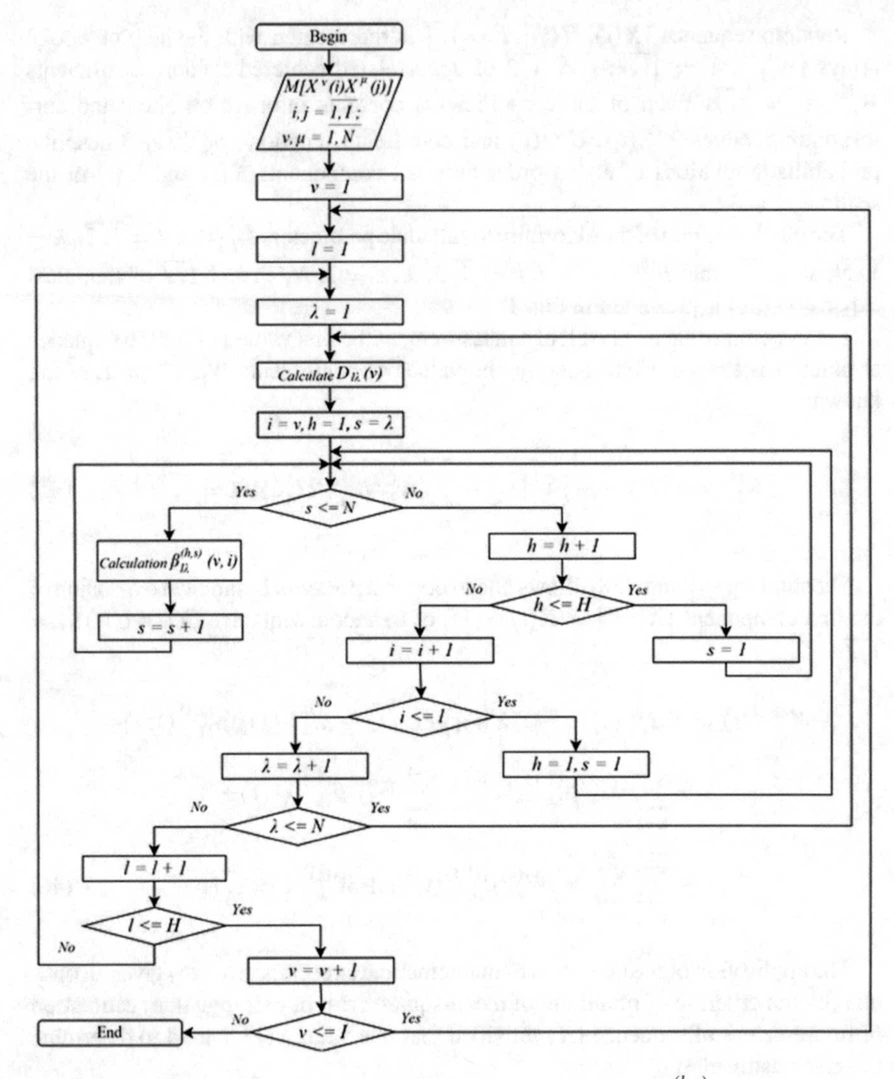

Fig. 1 Block diagram of the algorithm to calculate parameters $D_{l,\lambda}(\nu)$, $\beta_{l\lambda}^{(h,s)}(\nu, i)$ of a canonical expansion (38) ($H = 2$)

where $m_{1,1}^{(1,1)}(s, i)$ is optimal estimation of a future value $x^s(i)$ provided that value $x(1)$ is used for forecasting.

Specification of the second value $w_{11}^{(2)}$ in (46) gives canonical expansion of a posteriori sequence $\left\{X^{(1,2)}\right\} = X\left(i/x_1(1), x_1(1)^2\right)$:

$$X^{(1,2)}(i) = X\left(i/x(1), x(1)^2\right) = M[X(i)] + (x(1) - M[X(1)])\beta_{11}^{(1,1)}(1, i) +$$
$$+ \left[x^2(1) - (x(1) - M[X(1)])\beta_{11}^{(1,2)}(1, 1)\right]\beta_{12}^{(1,1)}(i) +$$

$$+\sum_{\lambda=3}^{N} W_{11}^{(\lambda)}\beta_{1\lambda}^{(1,1)}(1,i)+\sum_{\lambda=1}^{N} W_{12}^{(\lambda)}\beta_{1\lambda}^{(1,1)}(1,i)+$$
$$+\sum_{\nu=2}^{i-1}\sum_{l=1}^{2}\sum_{\lambda=1}^{N} W_{\nu l}^{(\lambda)}\beta_{l\lambda}^{(1,1)}(\nu,i)+W_{i1}^{(1)},\ i=\overline{1,I}. \tag{49}$$

Application of an operation of mathematical expectation to (49) allows to obtain the algorithm of extrapolation by two values $x(1),\ x(1)^2$ using expression (48):

$$m_{1,1}^{(1,2)}(s,i)=M\left[X^s\left(i/x(1),x(1)^2\right)\right]=$$
$$=m_{1,1}^{(1,1)}(s,i)+\left[x^2(1)-m_{1,1}^{(1,1)}(2,1)\right]\beta_{12}^{(1s)}(1,i),\ i=\overline{1,I}. \tag{50}$$

After N iterations, the value of the random coefficient $W_{12}^{(1)}=w_{12}^{(1)}$ can be calculated based on the information about control action $U(1)=u(1)$:

$$w_{12}^{(1)}=u(1)-\sum_{\lambda=1}^{N} w_{11}^{(\lambda)}\beta_{1\lambda}^{(2,1)}(1,1) \tag{51}$$

and forecasting algorithm with the use of values $x(1),\ x(1)^2,\ldots,x(1)^N,u(1)$ takes form

$$m_{2,1}^{(1,1)}(s,i)=m_{1,1}^{(1,N)}(s,i)+\left(u(1)-m_{1,2}^{(1,N)}(1,1)\right)\beta_{2,1}^{(1,s)}(1,i). \tag{52}$$

A consequent fixation of known values and their consecutive substitution into a canonical expansion (38) allows one to obtain the model of a stochastic control system for an arbitrary number of known values $x(\mu),u(\mu)$:

$$m_{j,h}^{(\mu,p)}(s,i)=\begin{cases} M[X(i)],\ \text{if } \mu=0,\\ m_{j,h}^{(\mu,p-1)}(s,i)+\left(u^p(\mu)-m_{j,j}^{(\mu,p-1)}(p,\mu)\right)\beta_{j,p}^{(h,s)}(\mu,i),\\ \text{if } p>1,\ j=2\\ m_{j-1,h}^{(\mu,N)}(s,i)+\left(u(\mu)-m_{j-1,j}^{(\mu,N)}(1,\mu)\right)\beta_{j,p}^{(h,s)}(\mu,i),\\ \text{if } p=1,\ j=2\\ m_{j,h}^{(\mu,p-1)}(s,i)+\left(x^p(\mu)-m_{j,j}^{(\mu,p-1)}(p,\mu)\right)\beta_{j,p}^{(h,s)}(\mu,i),\\ \text{if } p>1,\ j=1\\ m_{2,h}^{(\mu-1,N)}(s,i)+\left(x(\mu)-m_{2,1}^{(\mu-1,N)}(1,\mu)\right)\beta_{j,p}^{(h,s)}(\mu,i),\ \text{for}\\ p=1,j=1, \end{cases} \tag{53}$$

where $m_{2,1}^{(\mu,N)}(1,i)=M\left[X(i)/x^n(\nu),u^n(\nu),\ n=\overline{1,N},\ \nu=\overline{1,\mu}\right]$ is an optimal in a mean-square sense estimation of future values of an investigated random sequence provided that posteriori information $x^n(\nu), u^n(\nu), n = \overline{1, N}, \nu = \overline{1, \mu}$ is applied for the forecasting.

The diagram on Fig. 2 reflects the peculiarities of a calculation process when using extrapolator (53).

The expression for the mean-square error of the extrapolation using algorithm (53) by known values $x^n(\nu), u^n(\nu), n = \overline{1, N}, \nu = \overline{1, \mu}$ is of the form

$$E^{(\mu,N)}(i) = M\left[X^2(i)\right] - M^2\left[X(i)\right] - \\ -\sum_{k=1}^{\mu}\sum_{m=1}^{2}\sum_{j=1}^{N} D_{mj}(\mu)\left\{\beta_{mj}^{(1,\lambda)}(k,\nu)\right\}^2, i = \overline{\mu + 1, I}. \quad (54)$$

The mean-square error of extrapolation $E^{(\mu,N)}(i)$ is equal to the dispersion of the posteriori random sequence

$$X^{(\mu,N)}(i) = X\left(i/x^n(\nu), u^n(\nu), n = \overline{1, N}, \nu = \overline{1, \mu}\right) = m_{2,1}^{(\mu,N)}(1, i) + \\ + \sum_{\nu=\mu+1}^{i-1}\sum_{l=1}^{2}\sum_{\lambda=1}^{N} W_{\nu l}^{(\lambda)}\beta_{l\lambda}^{(1,1)}(\nu, i) + W_{i1}^{(1)}, i = \overline{\mu + 1, I}.$$

The synthesis and application of mathematical model (53) of a stochastic control system presuppose the realization of the following stages:

Stage 1. The gathering of data on an investigated random sequence $\{X(i), U(i)\}, i = \overline{1, I}$;

Stage 2. The estimation of moment functions $M\left[X^\lambda(\nu)X^s(i)\right]$, $M\left[U^\lambda(\nu)X^s(i)\right]$, $M\left[U^\lambda(\nu)U^s(i)\right]$, $\lambda, s = \overline{1, N}$; $\nu, j = \overline{1, I}$ based on the accumulated realizations of a random sequence $\{X(i), U(i)\}, i = \overline{1, I}$;

Stage 3. The forming of canonical expansion (38);

Stage 4. The calculation of the parameters of extrapolation algorithm (53);

Stage 5. The estimation of the future values of extrapolated realization based on expression (53);

Stage 6. The estimation of the quality of solving the forecasting problem for an investigated sequence using expression (54).

By analogy with (26) an optimal nonlinear model for the case of the measurement with errors of system parameters can be obtained.

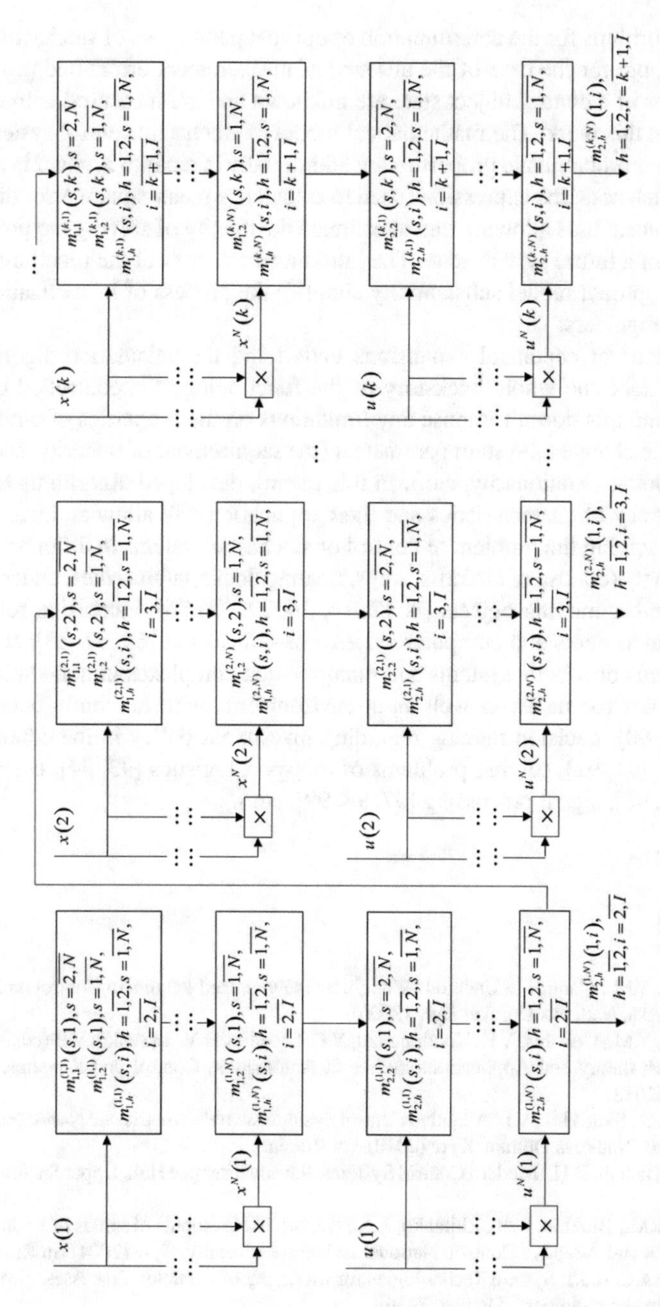

Fig. 2 The diagram of the procedure of forming the future values of a random sequence using algorithm (53)

4 Conclusion

The linear algorithms for the determination of optimal parameters of stochastic control systems, both for the case of the absence of measurement errors and provided that true values of a control object state are unknown and are measured with errors are obtained in this work. The mathematical model of stochastic control systems in which nonlinear probabilistic properties are additionally taken into account is added as well. For each case, the expressions used to calculate a mean-square error of control are introduced, thus allowing one to estimate the quality of solving the problem of estimation of a future system state. The introduced schemes of the functioning of the nonlinear optimal model substantially simplify the process of its realization by development engineers.

The apparatus of canonical expansions underlying the calculation algorithms allows one to take the whole backstory of the functioning of a controlled object into account and this doesn't impose any limitations on the properties of a random sequence of the change of system parameters (the requirement of linearity, Markov property, monotony, stationarity, etc.). In this regard, developed algorithms have a maximal accuracy of characteristics and their application will allow an increase of the quality of solving the problem of control of stochastic systems in different areas of human activity (control of aircrafts, ships, floating docks, submarines, underwater vehicles and other moving objects [8, 12, 14, 26, 27, 75, 76]; increasing reliability of technical systems and computer systems of critical use [68, 77–83]; solving control problems of robotic systems and manipulation complexes in non-stationary and dynamic environments, as well as in environments with randomly occurring obstacles [84–90]; decision making regarding investment policy in the conditions of uncertainty [91, 92]; solving problems of transport logistics [93, 94]; education processes [95, 96]; signal processing [97, 98, 99], etc.).

References

1. Kuntsevich, V.M.: Control in Uncertain Conditions: Guaranteed Results in Control and Identification Tasks. Naukova Dumka, Kyiv (2006)
2. Kuntsevich, V.M., Gubarev, V.F., Kondratenko, Y.P., Lebedev, D.V., Lysenko, V.P. (eds.): Control Systems: theory and Applications. Series in Automation, Control and Robotics. River Publishers (2018)
3. Zhiteckij, L.S., Skurikhin, V.I.: Adaptive Control Systems with Parametric and Non-Parametric Uncertainties. Naukova Dumka, Kyiv (2010). (in Russian)
4. Dorf, R.C., Bishop, R.H.: Modern Control Systems, 9th edn. Prentice Hall, Upper Saddle River, NJ (2001)
5. Azarskov, V.M., Blokhin, L.N., Zhiteckij, L.S., Kussul, N.N.: Robust Methods of Estimation, Identification and Adaptive Control. National Aviation University, Kyiv (2004) (in Russian)
6. Mikhalyov, A.I., (ed.): System Technologies for Modeling of Complex Processes. NMetAU-CPI "System Technologies", Dnipro (2016)

7. Kuntsevich, V.M., Kuntsevich, A.V.: Analysis of the pursuit-evasion process for moving plants under uncertain observation errors dependent on states. In: 15th IFAC World Congress, Preprints. Barcelona, Spain (2002)
8. Dorf, R.C.: Modern Control Systems, 5th edn. Addison-Wesley Publishing Company, Reading, Massachusetts (1990)
9. Goodwin, G.C., Graebe, S.F., Salgado, M.E.: Control Systems Design. Prentice Hall, Upper Saddle River, NJ (2001)
10. Ladanyuk, A.P., Kyshen'ko, V.D., Luts'ka, N.M., Ivashchuk, V.V.: Methods of Modern Control Theory. NUKhT, Kyiv (2010) (in Ukrainian)
11. Krak, Y.V., Levoshych, O.L.: Control Theory. Taras Shevchenko KNU, Kyiv (2001)
12. Osadchiy, S.I., Kalich, V.M., Didyk, O.K.: Structural identification of unmanned supercavitation vehicle based on incomplete experimental data. In: IEEE 2nd International Conference on Actual Problems of Unmanned Air Vehicles Developments, Kiev, Ukraine, pp. 93–95, 15–17 Oct 2013. https://doi.org/10.1109/APUAVD.2013.6705294
13. Burl, J.B.: Linear Optimal Control. Addison-Wesley, Menlo Park, California (1999)
14. Timchenko, V.L., Kondratenko, Y.P.: Robust stabilization of marine mobile objects on the Basis of systems with variable structure of feedbacks. J. Autom. Inf. Sci. **43**(6), 16–29 (2011). https://doi.org/10.1615/jautomatinfscien.v43.i6.20
15. Kondratenko, Y.P., Kozlov, O.V., Topalov, A.M.: Fuzzy controllers for increasing efficiency of the floating dock's operations: design and optimization. In: Kuntsevich, V.M., Gubarev, V.F., Kondratenko, Y.P., Lebedev, D.V., Lysenko, V.P. (eds.) Control Systems: Theory and Applications. Series in Automation, Control and Robotics, , pp. 197–232. River Publisher (2018)
16. Prokopenya, A.N.: Symbolic-numerical analysis of the relative equilibria stability in the planar circular restricted four-body problem. In: Gerdt, V., Koepf, W., Seiler, W., Vorozhtsov, E. (eds.) Computer Algebra in Scientific Computing. CASC 2017. Lecture Notes in Computer Science, vol. 10490. Springer, Cham, pp. 329–345 (2017). https://doi.org/10.1007/978-3-319-66320-3_24
17. Atamanyuk, I.P., Kondratenko, Y.P., Sirenko, N.N.: Forecasting economic indices of agricultural enterprises based on vector polynomial canonical expansion of random sequences. In: Ermolayev, V. et al. (eds.) Proceedings of the 12th International Conference on Information and Communication Technologies in Education, Research, and Industrial Application. Integration, Harmonization and Knowledge Transfer, ICTERI'2016, CEUR-WS, 21–24 June 2016, vol-1614, pp. 458–468, Kyiv, Ukraine (2016)
18. Atamanyuk, I.P., Kondratenko, Y.P., Sirenko, N.N.: Management system for agricultural enterprise on the basis of its economic state forecasting, complex systems: solutions and challenges in economics, management and engineering, In: Berger-Vachon, C., Lafuente, A.M.G., Kacprzyk, J., Kondratenko, Y., Merigó, J.M., Morabito, F.C. (eds.) Book Series: Studies in Systems, Decision and Control, vol. 125, pp. 453–470. Springer International Publishing, Berlin, Heidelberg (2018). https://doi.org/10.1007/978-3-319-69989-9_27
19. Atamanyuk, I., Kondratenko, Y.: Calculation method for a computer's diagnostics of cardiovascular diseases based on canonical decompositions of random sequences. In: Batsakis, S. et al. (eds.) ICT in Education, Research and Industrial Applications: Integration, Harmonization and Knowledge Transfer. Proceedings of the 11th International Conference ICTERI-2015, CEUR-WS, vol. 1356, pp. 108–120, Lviv, Ukraine, May 14–16 2015
20. Shebanin, V., Atamanyuk, I., Kondratenko, Y., Volosyuk, Y.: Canonical mathematical model and information technology for cardio-vascular diseases diagnostics. In: 14th International Conference The Experience of Designing and Application of CAD Systems in Microelectronics, CADSM 2017—Proceedings Open Access, pp. 438–440 (2017). https://doi.org/10.1109/cadsm.2017.7916170
21. Petunin, V.I.: Application of the Theory of Random Processes in Biology and Medicine. Naukova dumka, Kyiv (1981). (in Russian)
22. Chikrii, A.A.: Conflict-Controlled Processes. Kluwer Academic Publishers, Boston-London-Dordrecht (1997)

23. Chikrii, A.A.: Control of moving objects in condition of conflict. In: Kuntsevich, V.M., Gubarev, V.F., Kondratenko, Y.P., Lebedev, D.V., Lysenko, V.P. (eds.) Control Systems: Theory and Applications. Series in Automation, Control and Robotics, pp. 17–22. River Publishers (2018)
24. Chen, G., Chen, G., Hsu, S.-H.: Linear Stochastic Control Systems. CRC Press, New York, 464 p (2000)
25. Solesvik, M., Kondratenko, Y.: Architecture for collaborative digital simulation for the polar regions. In: Kharchenko, V., Kondratenko, Y., Kacprzyk, J. (eds.) Green IT Engineering: Social, Business and Industrial Applications. Studies in Systems, Decision and Control, vol 171. Springer, Cham, pp. 517–531 (2019). https://doi.org/10.1007/978-3-030-00253-4_22
26. Sharman, R., Lane, T. (eds.): Aviation Turbulence: Processes, Detection, Prediction. Edition, Kindle Edition, 523 p (2016)
27. Kazak, V.M.: Systemic Methods of Restoration of Survivability of Aircraft in Special Situations in Flight. Publications National Aviation University "NAU-druk" (2010) (in Ukrainian)
28. Lebedev, D.V., Tkachenko, A.I.: Navigation and Orientation Control of Small Spacecraft. Naukova Dumka, Kyiv (2006). (in Russian)
29. Fillipova, T.A., Sidorkin, YuM, Rusina, A.G.: Optimization of Regimes of Electric Power Stations and Power Systems. Publ, NSTU, Novosibirsk (2007). (in Russian)
30. Pupkov, K.A., Yegupov, N.D. (eds.): Methods of classical and modern theory of automatic control. In: Bauman, N.E. (ed.) Statistical Dynamics and Identification of Automatic Control Systems, vol. 2. Publications MSTU n.a., Moscow (2004)
31. Kondratenko, Y.P., Kozlov, O.V., Kondratenko, G.V., Atamanyuk, I.P.: Mathematical model and parametrical identification of ecopyrogenesis plant based on soft computing techniques, complex systems: solutions and challenges in economics, management and engineering, In: Berger-Vachon, C., Lafuente, A.M.G., Kacprzyk, J., Kondratenko, Y., Merigó, J.M., Morabito, F.C. (eds.) Book Series: Studies in Systems, Decision and Control, vol. 125, pp. 201–233. Springer International Publishing, Berlin, Heidelberg (2018). https://doi.org/10.1007/978-3-319-69989-9_13
32. Gubarev, V.F., Gummel, A.V., Kryshtal, A.A., Oles V.Y.: On the structural and parametric identification under the limited uncertainty and approximating models of multidimensional systems. J. Autom. Inf. Sci. **43**(6), 1–15 (2011)
33. Kuntsevich, V.M.: Estimation of impact of bounded perturbations on nonlinear discrete systems. In: Kuntsevich, V.M., Gubarev, V.F., Kondratenko, Y.P., Lebedev, D.V., Lysenko, V.P. (eds.) Control Systems: Theory and Applications. Series in Automation, Control and Robotics, pp. 3–16. River Publishers (2018)
34. Kuntsevich, V.M., Luchak, M.M.: Guaranteed Estimates, Adaptation and Robustness in Control Systems. Lecture Notes in Control and Information Sciences. Springer-Verlag, Heidelberg, vol. 169 (1992)
35. Simon, D.: Optimal State Estimation: Kalman, H-infinity, and Nonlinear Approaches. Wiley (2006)
36. Åström, K.J.: Introduction to Stochastic Control Theory. Academic Press, NewYork, London (1970)
37. Aoki, M.: Optimisation of Stochastic Systems. 2nd ed. Academic Press (1989)
38. Rosin, M.F., Bulygin, V.S.: Statistic Dynamics and Effectiveness Theory of the Control Systems. Mashinostroyeniye, Moscow (1981). (in Russian)
39. Lewis, F.L., Campos, J., Selmic, R.: Neuro-Fuzzy Control of Industrial Systems with Actuator Nonlinearities. SIAM, Philadelphia (2002)
40. Kondratenko, Y., Gordienko E.: Implementation of the neural networks for adaptive control system on FPGA. In: Katalinic B. (ed.) Annals of DAAAM for 2012 & Proceeding of the 23th International DAAAM Symposium "Intelligent Manufacturing and Automation", vol. 23, no. 1, pp. 0389–0392. Published by DAAAM International, Vienna, Austria, EU (2012)
41. Kondratenko, Y.P., Kozlov, O.V., Gerasin, O.S., Zaporozhets, Y.M.: Synthesis and research of neuro-fuzzy observer of clamping force for mobile robot automatic control system. In: 2016 IEEE First International Conference on Data Stream Mining & Processing (DSMP), pp. 90–95 (2016). https://doi.org/10.1109/dsmp.2016.7583514

42. Kondratenko, Y.P., Simon, D.: Structural and Parametric optimization of fuzzy control and decision making systems. In: Zadeh, L. et al. (eds.) Recent Developments and the New Direction in Soft-Computing Foundations and Applications. Studies in Fuzziness and Soft Computing, vol. 361. Springer International Publishing AG. Part of Springer Nature (2018). https://doi.org/10.1007/978-3-319-75408-6_22
43. Gomolka, Z., Dudek-Dyduch, E., Kondratenko, Y.P.: From homogeneous network to neural nets with fractional derivative mechanism. In: Rutkowski, L. et al. (eds.) International Conference on Artificial Intelligence and Soft Computing, ICAISC-2017, Part I, Zakopane, Poland, 11–15 June 2017, LNAI 10245, Springer, Cham, pp. 52–63 (2017). https://doi.org/10.1007/978-3-319-59063-9_5
44. Solesvik, M., Kondratenko, Y., Kondratenko, G., Sidenko, I., Kharchenko, V., Boyarchuk, A.: Fuzzy decision support systems in marine practice. In: IEEE International Conference on Fuzzy Systems (FUZZ-IEEE), pp. 1–6, Naples, Italy (2017). https://doi.org/10.1109/fuzzieee.2017.8015471
45. Kondratenko, Y., Kondratenko, V.: Soft computing algorithm for arithmetic multiplication of fuzzy sets based on universal analytic models. In: Ermolayev, V. et al. (eds.) Information and Communication Technologies in Education, Research, and Industrial Application. Communications in Computer and Information Science, ICTERI'2014, vol. 469, pp. 49–77. Springer International Publishing Switzerland (2014) https://doi.org/10.1007/978-3-319-13206-8_3
46. Kondratenko Y.P., Kondratenko N.Y.: Synthesis of analytic models for subtraction of fuzzy numbers with various membership function's shapes. In: Gil-Lafuente, A., Merigó, J., Dass, B., Verma, R. (eds.) Applied Mathematics and Computational Intelligence. FIM 2015. Advances in Intelligent Systems and Computing, vol 730. Springer, Cham, pp. 87–100 (2018). https://doi.org/10.1007/978-3-319-75792-6_8
47. Shebanin, V., Atamanyuk, I., Kondratenko, Y., Volosyuk Y.: Application of fuzzy predicates and quantifiers by matrix presentation in informational resources modeling. In: Proceedings of XII International Conference "MEMSTECH 2016", 22–24 Apr 2016, pp. 146–149, Lviv-Poljana (2016)
48. Kondratenko, Y.P., Kondratenko, N.Y.: Soft computing analytic models for increasing efficiency of fuzzy information processing in decision support systems. In: Hudson R. (ed.) Decision Making: Processes, Behavioral Influences and Role in Business Management. Nova Science Publishers, New York, pp. 41–78 (2015)
49. Kondratenko, Y.P., Al Zubi, E.Y.M.: The optimisation approach for increasing efficiency of digital fuzzy controllers. In: Annals of DAAAM for 2009 & Proceeding of the 20th Int. DAAAM Symp. "Intelligent Manufacturing and Automation", pp. 1589–1591. Published by DAAAM International, Vienna, Austria (2009)
50. Kacprzyk, J.: Multistage Fuzzy Control: A Prescriptive Approach. Wiley, New York, NY, USA (1997)
51. Kacprzyk, J.: A new approach to the control of stochastic systems in a fuzzy environment. Archiwum Automatyki i Telemechaniki **25**, 443–444 (1980)
52. Kacprzyk, J., Safteruk, K., Staniewski, P.: On The Control of Stochastic Systems in a Fuzzy Environment over Infinite Horizon. Systems Science **7**, 121–131 (1981)
53. Esogbue, A.O., Fedrizzi, M., Kacprzyk, J.: Fuzzy dynamic programming with stochastic systems. In: Kacprzyk, J., Fedrizzi, M. (eds.) Combining Fuzzy Imprecision with Probabilistic Uncertainty in Decision Making. Lecture Notes in Economics and Mathematical Systems, vol 310. Springer, Berlin, Heidelberg, pp. 266–285 (1988). https://doi.org/10.1007/978-3-642-46644-1_19
54. Kvakernak, H.: Sivan River: Linear optimum control systems, 650 p. World, Moscow, (1977)
55. Wiener, N.: Extrapolation, Interpolation, and Smoothing of Stationary Time Series: With Engineering Applications, 250 p. MIT Press, New-York (1949)
56. Kolmogorov, A.N.: Interpolation and extrapolation of stationary random sequences. J. Proc. Acad. Sci. USSR. Math. Ser. **5**, 3–14 (1941)
57. Kalman, R.E.: A new approach to linear filtering and prediction problems. Trans. ASME Ser. D, J. Basic Eng **82** (Series D), 35–45 (1960)

58. Simon, D.: Training fuzzy systems with the extended Kalman filter. Fuzzy Sets Syst. **132**, 189–199 (2002)
59. Box, G.E.P., Jenkins G.M.: Time—Series Analysis, Forecasting and Control. Holden–Day, San Francisco, 575 p. (1970)
60. Pugachev, V.S.: The Theory of Random Functions and its Application, 720 p. Fitmatgiz, Moscow (1962)
61. Kudritsky, V.D.: Filtering, Extrapolation and Recognition Realizations of Random Functions, 176 p. FADA Ltd., Kyiv (2001)
62. Atamanyuk, I.P., Kondratenko, Y.P.: Method of generating realizations of random sequence with the specified characteristics based on nonlinear canonical decomposition. J. Autom. Inf. Sci. **48**(10), 31–48 (2016). https://doi.org/10.1615/JAutomatInfScien.v48.i10.40
63. Shebanin, V.S., Atamanyuk, I.P., Kondratenko, Y.P.: Simulation of vector random sequences based on polynomial degree canonical decomposition. East.-Eur. J. Enterp. Technol. **5**(4(83)), 4–12 (2016). http://dx.doi.org/10.15587/1729-4061.2016.80786
64. Belan, V.V., Osadchiy, S.I.: Using canonical decomposition of spectral matrices to factor them. J. Autom. Inf. Sci. **27**(2), 57–62 (1995)
65. Shebanin, V.S., Kondratenko, Y.P., Atamanyuk, I.P.: The method of optimal nonlinear extrapolation of vector random sequences on the basis of polynomial degree canonical expansion. In: Gil-Lafuente, A., Merigó, J., Dass, B., Verma, R. (eds.) Applied Mathematics and Computational Intelligence. FIM 2015. Advances in Intelligent Systems and Computing, vol 730, pp. 14–25. Springer, Cham (2018). https://doi.org/10.1007/978-3-319-75792-6_2
66. Atamanyuk, I., Shebanin, V., Kondratenko, Y., Havrysh, V., Volosyuk, Y.: Method of an optimal nonlinear extrapolation of a noisy random sequence on the basis of the apparatus of canonical expansions. In: Chertov, O., Mylovanov, T., Kondratenko, Y., Kacprzyk, J., Kreinovich, V., Stefanuk, V. (eds.) Recent Developments in Data Science and Intelligent Analysis of Information. Proceedings of the XVIII International Conference on Data Science and Intelligent Analysis of Information, ICDSIAI 2018, Advances in Intelligent Systems and Computing, 4–7 June 2018, Kyiv, Ukraine. vol. 836, pp. 329–337. Springer International Publishing (2019). https://doi.org/10.1007/978-3-319-97885-7_32
67. Atamanyuk, I., Kondratenko, V., Kondratenko Y., Shebanin, V., Solesvik, M.: Models and algorithms for prediction of electrical energy consumption based on canonical expansions of random sequences. In: Kharchenko, V., Kondratenko, Y., Kacprzyk, J. (eds.) Green IT Engineering: Social, Business and Industrial Applications. Studies in Systems, Decision and Control, vol 171, pp. 397–421. Springer, Cham (2019). https://doi.org/10.1007/978-3-030-00253-4_17
68. Drozd, J., Drozd, A.: Models, methods and means as resources for solving challenges in co-design and testing of computer systems and their components. In: Proceedings of the 9th International Conference on Digital Technologies, pp. 225–230 (2013)
69. Trunov, A.N.: An adequacy criterion in evaluating the effectiveness of a model design process. East.-Eur. J. Enterp. Technol. **1**(4(73)), 36–41 (2015)
70. Atamanyuk, I., Kondratenko, Y.: Computer's Analysis Method and Reliability Assessment of Fault-Tolerance Operation of Information Systems. In: Batsakis, S. et al. (eds.) ICT in Education, Research and Industrial Applications: Integration, Harmonization and Knowledge Transfer. Proceedings of the 11th International Conference ICTERI-2015, CEUR-WS, vol. 1356, pp. 507–522, Lviv, Ukraine, 14–16 May 2015
71. Atamanyuk, I., Shebanin, V., Volosyuk, Y., Kondratenko, Y.: Generalized method for prediction of the electronic devices and information systems' state. In: 2018 14th International Conference on Perspective Technologies and Methods in MEMS Design, MEMSTECH 2018—Proceedings, 18–22 Apr 2018, Lviv, Ukraine (2018). https://doi.org/10.1109/MEMSTECH.2018.8365709
72. Atamanyuk, I.P., Kondratenko, Y.P., Shebanin, V.S.: Calculation methods of the prognostication of the computer systems state under different level of information uncertainty. In: Ermolayev, V. et al. (eds.) Proceedings of the 12th International Conference on Information and Communication Technologies in Education, Research, and Industrial Application. Integration,

Harmonization and Knowledge Transfer, ICTERI'2016, CEUR-WS, 21–24 June 2016, vol. 1614, pp. 292–307 Kyiv, Ukraine (2016)
73. Atamanyuk, I.P.: Algorithm of extrapolation of a nonlinear random process on the basis of its canonical decomposition. J. Kibernetika i Sistemnyj Analiz, no. 2, pp. 131–138 (2005)
74. Atamanyuk, I.P.: Polynomial algorithm of optimal extrapolation of stochastic system parameters. J. Upravlyayushchie Sistemy i Mashiny, (1), 16–19 (2002)
75. Oskin, A.A., Dyda, A.A., Longhi, S., Monteriu, A.: Underwater robot intelligent control based on multilayer neural network. In: Duro, R., Kondratenko, Y. (eds.) Advances in Intelligent Robotics and Collaborative Automation. Series in Automation, Control and Robotics, pp. 147–166. River Publishers (2015)
76. Kondratenko, Y.P., Kozlov, O.V., Korobko, O.V., Topalov, A.M.: Synthesis and optimization of fuzzy control system for floating Dock's docking operations. In: Santos, W. (eds.) Fuzzy Control Systems: Design, Analysis and Performance Evaluation, pp. 141–215. Nova Science Publishers, New York City, USA (2017)
77. Nesterov, M., Skarga-Bandurova, I.: Troubleshooting and performance methodology for business critical systems. In: Proceedings of 2018 IEEE 9th International Conference on Dependable Systems, Services and Technologies, DESSERT 2018, IEEE, Kiev, Ukraine (2018). https://doi.org/10.1109/DESSERT.2018.8409188
78. Atamanyuk, I., Kondratenko, Y., Shebanin, V., Mirgorod, V.: Method of polynomial predictive control of fail-safe operation of technical systems. In: Proceedings of the XIIIth International Conference "The Experience of Designing and Application of CAD Systems in Microelectronics", CADSM 2015, 19–23 Feb 2015, pp. 248–251, Polyana, Svalyava, Ukraine (2015). https://doi.org/10.1109/cadsm.2015.7230848
79. Shcherbovskykh, S., Spodyniuk, N., Stefanovych, T., Zhelykh, V., Shepitchak, V.: Development of a reliability model to analyze the causes of a poultry module failure. East. Eur. J. Enterp. Technol. **4**(3(82)), 4–9 (2016)
80. Shcherbovskykh, S., Stefanovych, T.: Reliability model developing for protective fittings taking into account load-sharing effect. East. Eur. J. Enterp. Technol. **1**(3(73)), pp. 37–44 (2015)
81. Lobur, M., Shcherbovskykh, S., Stefanovych, T.: Modelling of type I and II errors of switching device for systems with hot and cold redundancy based on two-terminal dynamic fault tree. In: Proceedings of the 14th International Conference on the Experience of Designing and Application of CAD Systems in Microelectronics (CADSM'2017), pp. 19–21. IEEE, Lviv, Ukraine (2017). https://doi.org/10.1109/CADSM.2017.7916075
82. Palagin, A.V., Opanasenko, V.N.: Design and application of the PLD-based reconfigurable devices. In: Adamski, M., Barkalov, A., Wegrzyn, M. (eds.) Design of Digital Systems and Devices. Lecture Notes in Electrical Engineering, vol. 79, pp. 59–91. Springer, Verlag, Berlin, Heidelberg (2011)
83. Maevsky, D., Maevskaya, E., Shapa, L.: Software reliability growth model‘s assumptions in context of the secondary faults. In: ICTERI 2017. ICT in Education, Research and Industrial Applications. Integration, Harmonization and Knowledge Transfer. CEUR Workshop Proceedings, vol. 1844, pp. 645–653 (2017). http://ceur-ws.org/Vol-1844/10000645.pdf
84. Kondratenko, Y.P.: Robotics, automation and information systems: future perspectives and correlation with culture, sport and life science. In: Gil-Lafuente, A.M., Zopounidis, C. (eds.) Decision Making and Knowledge Decision Support Systems. Lecture Notes in Economics and Mathematical Systems, vol. 675, pp. 43–56. Springer International Publishing Switzerland (2015). https://doi.org/10.1007/978-3-319-03907-7_6
85. Kondratenko, Y., Gerasin, O., Topalov, A.: A simulation model for robot's slip displacement sensors. Int. J. Comput. **15**(4) 224–236 (2016)
86. Tkachenko, A.N., Brovinskaya, N.M., Kondratenko, Y.P.: Evolutionary adaptation of control processes in robots operating in non-stationary environments. Mechan. Mach. Theory. **18**(4), 275–278 (1983). https://doi.org/10.1016/0094-114x(83)90118-0
87. Atamanyuk, I.P., Kondratenko, V.Y., Kozlov O.V., Kondratenko, Y.P.: The algorithm of optimal polynomial extrapolation of random processes. In: Engemann, K.J., Gil-Lafuente, A.M., Merigo, J.L. (eds.) Modeling and Simulation in Engineering, Economics and Management,

International Conference MS 2012, Proceedings. Lecture Notes in Business Information Processing, (May 30–June 1, 2012), vol. 115, pp. 78–87. Springer, New Rochelle, NY, USA (2012). https://doi.org/10.1007/978-3-642-30433-0_9
88. Kondratenko, Y., Khademi, G., Azimi, V., Ebeigbe, D., Abdelhady, M., Fakoorian, S.A., Barto, T., Roshanineshat, A.Y., Atamanyuk, I., Simon, D.: Robotics and prosthetics at Cleveland State University: modern information, communication, and modeling technologies. In: Ginige, A. et al. (eds.), Information and Communication Technologies in Education, Research, and Industrial Applications: ICTERI'2016, CCIS 783, pp. 133–155. Springer, Cham (2017). https://doi.org/10.1007/978-3-319-69965-3_8
89. Kryvonos, Y.G., Krak, Y.V., Kyrychenko, M.F.: Modeling, Analysis and Synthesis of Manipulation Systems. Naukova Dumka, Kyiv (2006) (in Ukrainian)
90. Duro, R., Kondratenko, Y. (eds.): Advances in Intelligent Robotics and Collaborative Automation. Series in Automation, Control and Robotics. River Publishers (2015)
91. Gil-Aluja, J.: Investment in Uncertainty. Kluwer Academic Publishers, Dordrecht, Boston, London (1999)
92. Kalinichenko, A., Havrysh, V., Perebyynis, V.: Sensitivity analysis in investment project of biogas plant. J. Appl. Ecol. Environ. Res. **15**(4), 969–985 (2017)
93. Kondratenko, Y.P., Encheva, S.B., Sidenko, E.V.: Synthesis of intelligent decision support systems for transport logistic. In: Proceeding of the 6th IEEE International Conference on Intelligent Data Acquisition and Advanced Computing Systems: Technology and Applications, IDAACS'2011. vol. 2 (Sept 15–17 2011, Prague, Czech Republic), pp. 642–646. https://doi.org/10.1109/idaacs.2011.6072847
94. Encheva, S., Kondratenko, Y., Solesvik, M.Z., Tumin, S.: Decision support systems in logistics. In: AIP Conference Proceedings, vol. 1060, pp. 254–256 (2008). https://doi.org/10.1063/1.3037065
95. Kondratenko, Y., Simon, D., Atamanyuk I.: University curricula modification based on advancements in information and communication technologies. In: Ermolayev, V. et al. (eds.) Proceedings of the 12th International Conference on Information and Communication Technologies in Education, Research, and Industrial Application. Integration, Harmonization and Knowledge Transfer, ICTERI'2016, CEUR-WS, 21–24 June 2016, vol. 1614, pp. 184–199, Kyiv, Ukraine (2016)
96. Kondratenko, G., Kondratenko, Y., Sidenko, I.: Fuzzy decision making system for model-oriented academia/industry cooperation: university preferences, complex systems: solutions and challenges in economics, management and engineering. In: Berger-Vachon, C. et al. (eds.) Studies in Systems, Decision and Control, vol. 125, pp. 109–124. Springer International Publishing, Berlin, Heidelberg (2018). https://doi.org/10.1007/978-3-319-69989-9_7
97. Sidenko, I., Filina, K., Kondratenko, G., Chabanovskyi, D., Kondratenko, Y.: Eye-tracking technology for the analysis of dynamic data. In: Proceedings of 2018 IEEE 9th International Conference on Dependable Systems, Services and Technologies, DESSERT 2018, 24–27 May 2018, Kiev, Ukraine, pp. 509–514. https://doi.org/10.1109/dessert.2018.8409181
98. Kondratenko, Y., Kondratenko, N.: Computational library of the direct analytic models for real-time fuzzy information processing. In: Proceedings of the 2018 IEEE Second International Conference on Data Stream Mining & Processing (DSMP), pp. 38–43, Lviv, Ukraine, 21–25 Aug 2018. https://doi.org/10.1109/DSMP.2018.8478518, ISBN 978-1-5386-2875-1
99. Kozina, G., Kudermetov, R.: Computer network design under uncertain input parameters. In: The Experience of Designing and Application of CAD Systems in Microelectronics, 2003. CADSM-2003. Proceedings of the 7th International Conference. IEEE, Slavske, Ukraine (2003). https://doi.org/10.1109/CADSM.2003.1255030

Program Iterations Method and Relaxation of a Pursuit-Evasion Differential Game

Alexander Chentsov and Daniel Khachay

Abstract We consider special case of non-linear zero-sum pursuit-evasion differential game. This game is defined by two closed sets - target set and one defining state constraints. We find an optimal non-anticipating strategy for player I (the pursuer). Namely, we construct his successful solvability set specified by limit function of the iterative procedure in space of positions. For positions outside of the successful solvability set, we consider relaxation of our game by determining the smallest size of a neighborhoods of two mentioned sets, for which the pursuer can solve his problem. Then, we construct his successful solvability set in terms of those neighborhoods.

1 Introduction

In the past few decades, differential game theory has been an actively developing field in operations research and control theory. The first mention of the differential game goes back to R. Isaacs. In well-known study [1], Isaacs overviewed a number of applications that can be reduced to a model of differential game. These applications are of great practical importance. Later, this theory was significantly developed by soviet mathematicians Pontryagin [2], Krasovskii [3], Pshenichny [4] and improved by A. B. Kurzhanski, Yu. S. Osipov and A. N. Subbotin. An important part of control theory with indeterminacy, namely identification, control, and observation under uncertainty is studied by Kuntsevich [5–7].

In this paper, we consider a non-linear zero-sum pursuit-evasion differential game, defined by two closed sets in space of positions. This setting corresponds to [8, 9]. For considered differential game in [8], the fundamental theorem of alternative was established by N. N. Krasovskii and A. I. Subbotin. Important generalization of this

A. Chentsov (✉) · D. Khachay
Krasovskii Institute of Mathematics and Mechanics, Ekaterinburg, Russia
e-mail: chentsov@imm.uran.ru

A. Chentsov · D. Khachay
Ural Federal University, Ekaterinburg, Russia
e-mail: daniil.khachay@gmail.com

Y. P. Kondratenko et al. (eds.), *Advanced Control Techniques in Complex Engineering Systems: Theory and Applications*, Studies in Systems, Decision and Control 203, https://doi.org/10.1007/978-3-030-21927-7_7

result was obtained by Kryazhimskiy [10]. According to theorem of alternative, the set defining state constraints, can be split into two disjoint subsets, specifying successful solvability sets for each player. If differential game satisfies Isaacs condition, then the alternative can be implemented in terms of pure positional strategies and, of course, in terms of non-anticipating strategies (see [11, 12]). Those strategies are also known in literature as Elliott-Kalton strategies [12] and quasi-strategies [13–15]. Multivalued cases of non-anticipating strategies were considered in [13–15]. The problem of construction of an alternative partition can be reduced to problem of finding the set of successful solvability of a pursuer (player I) (that is, the maximum stable bridge by N. N. Krasovskii), who is interested in the guaranteed feasibility of approach. Various methods were used to build this set. In [16–18], V. N. Ushakov and his students considered special kind of procedures constructed by dynamic programming. The construction of guaranteed solutions to problems in the theory of differential games was studied by Chikrii [19].

The program constructions were used to find solutions for non-linear differential game (see [9, 20, 21] and others). For general case of differential game, the program iterations method was proposed (see [13, 14, 22–24]). In our case, we use modification of program iterations method, which called stability iterations [25, 26]. With this modification, it is possible to solve pursuit-evasion games with a restriction on the number of switchings (see [26–28]). More precisely, at each stage of the iterative procedure, we construct the successful solvability set.

In this study, we relax the initial setting of considered pursuit-evasion differential game by analyzing possibilities of player I in terms of reachability of closed neighborhoods of target set within corresponding neighborhoods of state constraints set. Moreover, for each fixed position, we aim to find the smallest "size" of mentioned neighborhood, for which pursuer can successfully solve the problem. This "size" also estimates possibilities of player II (evader) in following way: for smaller neighborhoods, he can perform evasion procedure with finite number of switchings. Based on this approach, we can define minimax function, which values can be represented as guaranteed result for special payoff. To define this function, we construct special iterative procedure in space of positions; the function itself is a fixed point of special conversion operator.

The structure of the paper is as follows. In Sect. 1, we review some important notation from set theory and general topology. In Sect. 2, we summarize generalized controls for our setting. In Sect. 3, we consider the statement of our problem, local properties of pursuit-evasion game and find minimax in terms of non-anticipating strategies. In Sect. 4, we construct iterative procedure $\varepsilon_0^{(0)}, \varepsilon_0^{(1)}, \varepsilon_0^{(2)}, \ldots$, which implements limit function ε_0. In Sect. 5, we find the conversion operator Γ and show that $\varepsilon_0^{(k+1)} = \Gamma(\varepsilon_0^{(k)})$, $k \in \{0; 1; 2; 3; \ldots\}$. Finally, in Sect. 6, we prove that ε_0 is the fixed point of operator Γ. This paper is continuation of [29].

2 Preliminaries

In this paper we use standard notation from set theory. Set is called family if all its elements are sets. To each object z we assign singleton $\{z\}$ which contains z: $z \in \{z\}$. We also assign to the set X a family $\mathcal{P}(X)$ of all its subsets and assume that $\mathcal{P}'(X) \triangleq \mathcal{P}(X) \setminus \{\emptyset\}$.

Let us define the trace of a family. Let $\mathcal{A}$ - non-empty family and B - set, then trace of family $\mathcal{A}$ on set B is defined as follows:

$$\mathcal{A}|_B \triangleq \{A \cap B : A \in \mathcal{A}\} \in \mathcal{P}'(\mathcal{P}(B)).$$

If A and B — nonempty sets then B^A is the set of all mappings from A to B.

Assume (X, τ) is topological space, then $\mathbb{C}_X[\tau]$ — is the family of all closed subsets $X : \mathbb{C}_X[\tau] \triangleq \{X \setminus G : G \in \tau\}$. Hereinafter, we will use only metrizable spaces, that is, topological spaces generated by metrics. Thus, we will use ε-neighborhoods of non-empty sets, according to proper metric.

If E is a set and $\mathcal{E} \in \mathcal{P}'(\mathcal{P}(E))$, then as $\sigma_E^0(\mathcal{E})$ we define the σ-algebra of subsets of E, also generated by family [30] $\mathcal{E}$. Moreover, if $\mathcal{E}$ is a topology on E, then $\sigma_E^0(\mathcal{E})$ is the σ-algebra of Borel subsets of E. For any σ-algebra ξ, we have standard measurable space (E, ξ). Let us define the set of all non-negative countably additive measures on σ-algebra ξ as $(\sigma - \text{add})_+[\xi]$. In case when ξ is by definition the σ-algebra of Borel sets in some topological space, then measures from $(\sigma - \text{add})_+[\xi]$ are called Borel measures. Finally, if topology is generated by metric, then [31] considered measures are regular.

Later on, we fix number $n \in \mathbb{N} \triangleq \{1; 2; \ldots\}$ and time interval $T \triangleq [t_0, \theta_0]$, where $t_0 \in \mathbb{R}$, $\theta_0 \in \mathbb{R}$, $t_0 < \theta_0$. We consider $T \times \mathbb{R}^n$ as space of positions with metric ρ defined as follows:

$$((t_1, x_1), (t_2, x_2)) \longmapsto \sup(\{|t_1 - t_2|; \|x_1 - x_2\|\}) : (T \times \mathbb{R}^n) \times (T \times \mathbb{R}^n) \to [0, \infty[,$$

where (here and later) $\|\cdot\|$ is an Euclidean norm in $\mathbb{R}^n$. Note that metric ρ generates the topology of coordinate-wise convergence, designated as $\mathbf{t}$; thus, we have metrizable topological space

$$(T \times \mathbb{R}^n, \mathbf{t}). \tag{2.1}$$

If $H \in \mathcal{P}'(T \times \mathbb{R}^n)$, then as $\rho(\cdot, H) \triangleq (\rho(z, H))_{z \in T \times \mathbb{R}^n}$ we will designate a function of distance in $(T \times \mathbb{R}^n, \rho)$ to set H; see [26, p. 289]. Then for $H \in \mathcal{P}'(T \times \mathbb{R}^n)$ and $\varepsilon \in]0, \infty[$ we obtain

$$S_0(H, \varepsilon) \triangleq \{z \in T \times \mathbb{R}^n | \rho(z, H) \leqslant \varepsilon\} \tag{2.2}$$

closed ε-neighborhood of H, i.e. $S_0(H, \varepsilon) \in \mathcal{F}$, where $\mathcal{F}$ is the family of all subsets of $T \times \mathbb{R}^n$, closed in topological space (2.1). In other words, $\mathcal{F} = \mathbb{C}_{T \times \mathbb{R}^n}(\mathbf{t})$ is by definition the family of all closed subsets of $T \times \mathbb{R}^n$, also $\mathcal{F} \setminus \{\emptyset\} \subset \mathcal{P}'(T \times \mathbb{R}^n)$.

The definition (2.2) can be extended in a following way: if $H \in \mathcal{P}'(T \times \mathbb{R}^n)$, then assume that

$$S_0(H, 0) \triangleq \{z \in T \times \mathbb{R}^n | \ \rho(z, H) \leqslant 0\} = \{z \in T \times \mathbb{R}^n | \ \rho(z, H) = 0\}, \quad (2.3)$$

thus, we have the closure of H in topological space (2.1). Now we have the following relation

$$\varepsilon \longmapsto S_0(H, \varepsilon) : [0, \infty[\to \mathcal{F}. \quad (2.4)$$

Note that (see (2.3)) if $H \in \mathcal{F} \setminus \{\emptyset\}$, then $S_0(H, 0) = H$.

3 Generalized Controls and Trajectories

This section summarizes the rather traditional construction of the extension of the control problem [15, Chap. 4], [26, Sect. 3]. For an arbitrary $t \in T$, we define finite-dimensional compacta $[t, \theta_0]$, $Y_t \triangleq [t, \vartheta_0] \times P$ and $\Omega_t \triangleq [t, \theta_0] \times P \times Q$, with σ-algebras of Borel subsets $\mathcal{T}_t$, $\mathcal{K}_t$ and $\mathcal{C}_t$ respectively; as λ_t we will designate restriction of Lebesgue measure on σ-algebra $\mathcal{T}_t$. Also [32, p. 17] $\Omega_t = Y_t \times Q$. If $K \in \mathcal{P}(Y_t)$, then $K \times Q \subset \Omega_t$; with $K \in \mathcal{K}_t$ we have cylinder $K \times Q \in \mathcal{C}_t$, and with $\Gamma \in \mathcal{T}_t$ we have $\Gamma \times P \times Q \in \mathcal{C}_t$ and $\Gamma \times P \in \mathcal{K}_t$. Then required variants of generalized controls are characterized in [26, (3.1)–(3.3)]:

$$\mathcal{H}_t \triangleq \{\eta \in (\sigma - \text{add})_+[\mathcal{C}_t] \mid \eta(\Gamma \times P \times Q) = \lambda_t(\Gamma) \ \forall \Gamma \in \mathcal{T}_t\}, \quad (3.1)$$

$$\mathcal{R}_t \triangleq \{\mu \in (\sigma - \text{add})_+[\mathcal{K}_t] \mid \mu(\Gamma \times P) = \lambda_t(\Gamma) \ \forall \Gamma \in \mathcal{T}_t\}, \quad (3.2)$$

$$\pi_t(\mu) \triangleq \{\eta \in \mathcal{H}_t \mid \eta(K \times Q) = \mu(K) \ \forall K \in \mathcal{K}_t\} \ \forall \mu \in \mathcal{R}_t. \quad (3.3)$$

In terms of subject domain, generalized controls from sets (3.1)–(3.3) corresponds to [26, p. 288]. At the moment we will note that generalized controls $\eta \in \mathcal{H}_t$ are similar to pairs $(u(\cdot), v(\cdot))$ of open-loop control system.

Each of sets (3.1)–(3.3) is non-empty metrizable compactum with weak* topology (see [26, Sect. 3]).

As $\mathcal{B}$ further we define σ-algebra of Borel subsets of Q; if $v \in Q$, then we define $\delta_v \in (\sigma - \text{add})_+[\mathcal{B}]$ as restriction of Dirac measure on $\mathcal{B}$. For more precise definition of required generalized controls, we note that with $t \in T$, semi-algebra $\mathcal{K}_t\{\times\}\mathcal{B} \triangleq \{K \times B : K \in \mathcal{K}_t, B \in \mathcal{B}\}$ of measurable rectangles generates σ-algebra $\mathcal{C}_t$ [31, p. 307–308]. Then with $\mu \in \mathcal{R}_t$ and $v \in Q$ in the form of $\mu \otimes v$ we define multiplication of measures μ and δ_v, also $\mu \otimes v \in \pi_t(\mu)$.

We will proceed to the construction of generalized trajectories, following the conditions of Kryazhimsky [10]. Let's fix a following continuous mapping (vector function) $f : T \times \mathbb{R}^n \times P \times Q \to \mathbb{R}^n$. Assuming that with $t \in T$, $C_n([t, \theta_0])$ is by definition the set of all continuous mappings from $[t, \theta_0]$ in $\mathbb{R}^n$ (called n-vector functions hereinafter), given that $x(\cdot) \triangleq (x(\xi))_{\xi\in[t,\theta_0]} \in C_n([t,\theta_0])$ in the form of $(\xi, u, v) \longmapsto f(\xi, x(\xi), u, v) : \Omega_t \to \mathbb{R}^n$ we have continuous (and even uniformly continuous) mapping on metrizable compactum. Provided with $\eta \in \mathcal{H}_t$ and $\theta \in [t, \theta_0]$ the following integral is defined component-wise: $\int_{[t,\theta[\times P\times Q} f(\xi, x(\xi), u, v)\eta(d(\xi, u, v)) \in \mathbb{R}^n$. Moreover, if $(t_*, x_*) \in T \times \mathbb{R}^n$ and $\eta \in \mathcal{H}_{t_*}$, then

$$\Phi(t_*, x_*, \eta) \triangleq \{x(\cdot) \in C_n([t_*, \theta_0]) \mid x(t) = \\ = x_* + \int_{[t_*,t[\times P\times Q} f(\xi, x(\xi), u, v)\eta(d(\xi, u, v)) \; \forall t \in [t_*, \theta_0]\} \tag{3.4}$$

is generalized integral funnel of equation

$$\dot{x} = f(t, x, u, v), \tag{3.5}$$

according to fixed position (t_*, x_*) and control η. Following [10], assume that for any selection of $(t_*, x_*) \in T \times \mathbb{R}^n$ and $\eta \in \mathcal{H}_{t_*}$ feasible solution set (3.4) is singleton. Thus, assume that

$$\varphi(\cdot, t_*, x_*, \eta) = (\varphi(t, t_*, x_*, \eta))_{t\in[t_*,\theta_0]} \in C_n([t_*, \theta_0]), \tag{3.6}$$

where $\Phi(t_*, x_*, \eta) = \{\varphi(\cdot, t_*, x_*, \eta)\}$. We also suppose, like in [10], that generalized trajectories (3.6) are uniformly bounded. To formulate this condition let us make a following definition: if $\varkappa \in [0, \infty[$, then assume that $\mathbb{B}_n(\varkappa) \triangleq \{x \in \mathbb{R}^n | \; \|x\| \leqslant \varkappa\}$. Then, considered condition can be rewritten: $\forall a \in [0, \infty[\; \exists b \in [0, \infty[: \varphi(\xi, t, x, \eta) \in \mathbb{B}_n(b) \; \forall t \in T \; \forall x \in \mathbb{B}_n(a) \; \forall \eta \in \mathcal{H}_t \; \forall \xi \in [t, \theta_0]$. According to [26, (3.7)] we have that following mapping $(x, \mu) \to \varphi(\cdot, t, x, \mu \otimes v) : \mathbb{R}^n \times \mathcal{R}_t \to C_n([t, \theta_0])$ is continuous in terms of natural topology defined by multiplication of $\mathbb{R}^n$ with coordinate-wise convergence topology and $\mathcal{R}_t$ with weak* topology. As a result, for $(t_*, x_*) \in T \times \mathbb{R}^n$ and $v \in Q$

$$\mathcal{X}_\pi(t_*, x_*, v) \triangleq \{\varphi(\cdot, t_*, x_*, \mu \otimes v) : \mu \in \mathcal{R}_{t_*}\} \tag{3.7}$$

we have a non-empty compactum in $C_n([t_*, \theta_0])$ with topology of uniform convergence. Following vector-functions $\varphi(\cdot, t_*, x_*, \mu \otimes v)$, where $t_* \in T, x_* \in \mathbb{R}^n, \mu \in \mathcal{R}_{t_*}, v \in Q$, are elements of compactum (3.7), we define them as v-trajectories, which start from position (t_*, x_*).

4 Local Properties of Pursuit-Evasion Differential Game and Minimax in Terms of Non-anticipating Strategies

In this section, we use notation from [26, Sect. 4], slightly changing it for considered special case of differential game. We fix following sets

$$(\mathbf{M} \in \mathcal{F}) \; \& \; (\mathbf{N} \in \mathcal{F}). \tag{4.1}$$

First one is a target set for first player and second one defines state constraints in terms of set cross-sections: $\mathbf{N}\langle t \rangle \triangleq \{x \in \mathbb{R}^n | \; (t, x) \in \mathbf{N}\}, \;\; \forall t \in T$. Hereinafter we assume that $\mathbf{M} \neq \emptyset$ and $\mathbf{M} \subset \mathbf{N}$. From (4.1) follows that $\mathbf{M} = \mathbf{M} \cap \mathbf{N} \in \mathcal{F} \setminus \{\emptyset\}$.

Let us consider the following problem. We need to construct approximation to $\mathbf{M}$ under state constraints $\mathbf{N}\langle t \rangle$, $t \in T$. Most of the time $(\mathbf{M}, \mathbf{N})$-approximation cannot be constructed, thus problem is intractable. However, we can relax this problem by defining some quality guarantee. Namely, let us consider $(S_0(\mathbf{M}, \varepsilon), S_0(\mathbf{N}, \varepsilon))$-approximation with guarantee $\varepsilon \in [0, \infty[$. For fixed position (t_*, x_*), we find the smallest number $\varepsilon_* \in [0, \infty[$, for which $(S_0(\mathbf{M}, \varepsilon_*), S_0(\mathbf{N}, \varepsilon_*))$-approximation can be constructed in terms of admissible control procedures. Later, we show that this number exists.

Thus, according to such admissible procedure, for every trajectory $x(\cdot) = (x(t), t_* \leqslant t \leqslant \vartheta_0)$, generated by given procedure, the following property

$$((\theta, x(\theta)) \in S_0(\mathbf{M}, \varepsilon_*)) \; \& \; ((t, x(t)) \in S_0(\mathbf{N}, \varepsilon_*) \; \forall t \in [t_*, \theta]) \tag{4.2}$$

where $\theta \in [t_*, \vartheta_0]$, holds. Moreover, for trajectory $x(\cdot)$ in (4.2), the equality $x(t_*) = x_*$ is fulfilled. Therefore, we can estimate ε_* by two following values $\rho((t_*, x_*), \mathbf{M})$ and $\rho((t_*, x_*), \mathbf{N})$. In addition, by choice of $\mathbf{M}$ and $\mathbf{N}$

$$\rho((t_*, x_*), \mathbf{N}) \leqslant \rho((t_*, x_*), \mathbf{M}). \tag{4.3}$$

We suppose that $\varepsilon_*^{(1)} \triangleq \rho((t_*, x_*), \mathbf{N})$ and $\varepsilon_*^{(2)} \triangleq \rho((t_*, x_*), \mathbf{M})$. Then, by (4.3), $\varepsilon_*^{(1)} \leqslant \varepsilon_*^{(2)}$. Therefore, it is obvious that $(S_0(\mathbf{M}, \varepsilon_*^{(2)}), S_0(\mathbf{N}, \varepsilon_*^{(2)}))$-approximation can be constructed. By the choice of ε_*, we obtain that $\varepsilon_* \leqslant \varepsilon_*^{(2)}$.

On the other hand, from (4.2), for admissible procedure, which constructs $(S_0(\mathbf{M}, \varepsilon_*), S_0(\mathbf{N}, \varepsilon_*))$-approximation, we obtain that $\varepsilon_*^{(1)} \leqslant \varepsilon_*$. This follows from (4.2) and definition of $\varepsilon_*^{(1)}$. Therefore, we obtain the chain of inequalities $\varepsilon_*^{(1)} \leqslant \varepsilon_* \leqslant \varepsilon_*^{(2)}$. Using definition of $\varepsilon_*^{(1)}$ and $\varepsilon_*^{(2)}$, we have that

$$\rho((t_*, x_*), \mathbf{N}) \leqslant \varepsilon_* \leqslant \rho((t_*, x_*), \mathbf{M}). \tag{4.4}$$

Thus, by (4.4), we defined the range for optimal value of ε_* for every position from $T \times \mathbb{R}^n$. Therefore, $\rho((\cdot), \mathbf{N})$ and $\rho((\cdot), \mathbf{M})$ - are two boundary functions which define corresponding range in functional space.

In regards to [26, (4.2),(4.3)], we define stability operators

$$\mathbb{A}[M] : \mathcal{P}(T \times \mathbb{R}^n) \to \mathcal{P}(T \times \mathbb{R}^n), \tag{4.5}$$

where $M \in \mathcal{F}$. As M we will use $\mathbf{M}$ or $S_0(\mathbf{M}, \varepsilon)$, where $\varepsilon \in]0, \infty[$. If $M \in \mathcal{F}$ and $F \in \mathcal{F}$, then in terms of [26, (4.3)] and (3.7) we obtain

$$\begin{aligned}\mathbb{A}[M](F) = \{(t, x) \in F \mid \forall v \in Q\ \exists x(\cdot) \in \mathcal{X}_\pi(t, x, v)\\ \exists \theta \in [t, \theta_0] : ((\theta, x(\theta)) \in M)\&((\xi, x(\xi)) \in F\ \forall \xi \in [t, \theta])\}.\end{aligned} \tag{4.6}$$

Moreover, if $M \in \mathcal{F},\ N \in \mathcal{F}$, and $F \in \mathcal{F}|_N$, then

$$\mathbb{A}[M](F) \in \mathcal{F}|_N, \tag{4.7}$$

where $\mathcal{F}|_N = \{\tilde{F} \in \mathcal{F} \mid \tilde{F} \subset N\} \in \mathcal{P}'(\mathcal{F})$. Regarding (4.7) we have that

$$\mathbb{A}[M](F) \in \mathcal{F}\ \forall M \in \mathcal{F}\ \forall N \in \mathcal{F}\ \forall F \in \mathcal{F}|_N. \tag{4.8}$$

From (4.6) follows

$$\mathbb{A}[M](F) = \mathbb{A}[M \cap F](F)\ \forall M \in \mathcal{F}\ \forall F \in \mathcal{F}. \tag{4.9}$$

In terms of (4.9) we note two simple properties of operator (4.5). First of all

$$\mathbb{A}[M](F) \subset F\ \forall M \in \mathcal{F}\ \forall F \in \mathcal{P}(T \times \mathbb{R}^n). \tag{4.10}$$

According to (4.6), there is isotonic property:

$$\begin{aligned}\forall M_1 \in \mathcal{F}\,\forall F_1 \in \mathcal{P}(T \times \mathbb{R}^n)\ \forall M_2 \in \mathcal{F}\,\forall F_2 \in \mathcal{P}(T \times \mathbb{R}^n)\\ ((M_1 \subset M_2)\&(F_1 \subset F_2)) \Rightarrow (\mathbb{A}[M_1](F_1) \subset \mathbb{A}[M_2](F_2)).\end{aligned} \tag{4.11}$$

We note that there is property which is similar to sequential continuity (in terms of order) [26, Proposition 4.3]: if $(M_i)_{i\in\mathbb{N}} : \mathbb{N} \to \mathcal{F}$, $(F_i)_{i\in\mathbb{N}} : \mathbb{N} \to \mathcal{F}$, $M \in \mathcal{P}(T \times \mathbb{R}^n)$, $F \in \mathcal{P}(T \times \mathbb{R}^n)$ and besides of all that

$$((M_i)_{i\in\mathbb{N}} \downarrow M)\ \&\ ((F_i)_{i\in\mathbb{N}} \downarrow F), \tag{4.12}$$

then $M \in \mathcal{F}$, $F \in \mathcal{F}$ and

$$(\mathbb{A}[M_i](F_i))_{i\in\mathbb{N}} \downarrow \mathbb{A}[M](F). \tag{4.13}$$

In this property implication (4.12) $\Rightarrow$ (4.13) is most important. With (2.4), particularly we have following mappings

$$\varepsilon \longmapsto S_0(\mathbf{M}, \varepsilon) : [0, \infty[\to \mathcal{F}, \tag{4.14}$$

$$\varepsilon \longmapsto S_0(\mathbf{N}, \varepsilon) : [0, \infty[\to \mathcal{F}. \tag{4.15}$$

Thus, we have operators $\mathbb{A}[\mathbf{M}]$ and $\mathbb{A}[S_0(\mathbf{M}, \varepsilon)]$, where $\varepsilon \in [0, \infty[$.

Note that [26, Sect. 5] to each sets $M \in \mathcal{F}$ and $N \in \mathcal{F}$ we assign the sequence of sets $(\mathcal{W}_k(M, N))_{k \in \mathbb{N}_0} : \mathbb{N}_0 \to \mathcal{P}(T \times \mathbb{R}^n)$ and limit set

$$\mathcal{W}(M, N) = \bigcap_{k \in \mathbb{N}_0} \mathcal{W}_k(M, N) \in \mathcal{P}(T \times \mathbb{R}^n);$$

wherein

$$(\mathcal{W}_0(M, N) = N) \& (\mathcal{W}_{k+1}(M, N) = \mathbb{A}[M](\mathcal{W}_k(M, N)) \ \ \forall k \in \mathbb{N}_0). \tag{4.16}$$

Also $\mathcal{F} \subset \mathfrak{F}$, where $\mathfrak{F}$ is defined in [26, Sect. 4]; according to [26, (10.1)] with $M \in \mathcal{F}$ and $N \in \mathcal{F}$

$$(\mathcal{W}_k(M, N) \in \mathcal{F}|_N \ \ \forall k \in \mathbb{N}_0) \ \& \ (\mathcal{W}(M, N) \in \mathcal{F}|_N). \tag{4.17}$$

In our case (where $N \in \mathcal{F}$) the inclusion $\mathcal{F}|_N \subset \mathcal{F}$ holds; thus, regarding (4.17), follows, in particular, that $\forall M \in \mathcal{F} \ \ \forall N \in \mathcal{F}$

$$(\mathcal{W}_k(M, N) \in \mathcal{F} \ \ \forall k \in \mathbb{N}_0) \ \& \ (\mathcal{W}(M, N) \in \mathcal{F}). \tag{4.18}$$

Also, we note that according to (4.6), with $M \in \mathcal{F}$ and $F \in \mathcal{F}$

$$M \cap F \subset \mathbb{A}[M](F). \tag{4.19}$$

As a consequence of (4.16) and (4.18), we obtain

$$M \cap \mathcal{W}_k(M, N) \subset \mathcal{W}_{k+1}(M, N); \tag{4.20}$$

then (see (4.16), (4.20)), in particular, we have inclusion

$$M \cap N \subset \mathcal{W}_1(M, N). \tag{4.21}$$

Further, by induction (see (4.20)),

$$M \cap N \subset M \cap \mathcal{W}_k(M, N) \ \ \forall k \in \mathbb{N}. \tag{4.22}$$

Also, according to (4.16), $M \cap N \subset N = \mathcal{W}_0(M, N)$. As corollary (see (4.22)) we have

$$M \cap N \subset \mathcal{W}(M, N). \tag{4.23}$$

Properties (4.22) and (4.23) will hold with arbitrary $M \in \mathcal{F}$ and $N \in \mathcal{F}$. In particular, we have that

$$(\mathbf{M} \subset \mathcal{W}_k(\mathbf{M}, \mathbf{N}) \ \forall k \in \mathbb{N}_0) \ \& \ (\mathbf{M} \subset \mathcal{W}(\mathbf{M}, \mathbf{N})). \tag{4.24}$$

In regards to (4.24), we have

$$(\mathcal{W}_k(\mathbf{M}, \mathbf{N}) \neq \emptyset \ \forall k \in \mathbb{N}_0) \ \& \ (\mathcal{W}(\mathbf{M}, \mathbf{N}) \neq \emptyset), \tag{4.25}$$

where

$$\mathcal{W}(\mathbf{M}, \mathbf{N}) = \bigcap_{k \in \mathbb{N}_0} \mathcal{W}_k(\mathbf{M}, \mathbf{N}). \tag{4.26}$$

We recall that if $\varepsilon \in [0, \infty[$, then according to (4.12)–(4.16) we can define (see (4.18)) the sequence

$$(\mathcal{W}_k(S_0(\mathbf{M}, \varepsilon), S_0(\mathbf{N}, \varepsilon)))_{k \in \mathbb{N}_0} : \mathbb{N}_0 \to \mathcal{F}, \tag{4.27}$$

and limit set

$$\mathcal{W}(S_0(\mathbf{M}, \varepsilon), S_0(\mathbf{N}, \varepsilon)) \in \mathcal{F}; \tag{4.28}$$

wherein (see (4.16))

$$(\mathcal{W}_0(S_0(\mathbf{M}, \varepsilon), S_0(\mathbf{N}, \varepsilon)) = S_0(\mathbf{N}, \varepsilon)) \ \& \ (\mathcal{W}_{k+1}(S_0(\mathbf{M}, \varepsilon), S_0(\mathbf{N}, \varepsilon)) = \mathbb{A}[S_0(\mathbf{M}, \varepsilon)](\mathcal{W}_k(S_0(\mathbf{M}, \varepsilon), S_0(\mathbf{N}, \varepsilon))) \ \forall k \in \mathbb{N}_0), \tag{4.29}$$

moreover,

$$\mathcal{W}(S_0(\mathbf{M}, \varepsilon), S_0(\mathbf{N}, \varepsilon)) = \bigcap_{k \in \mathbb{N}_0} \mathcal{W}_k(S_0(\mathbf{M}, \varepsilon), S_0(\mathbf{N}, \varepsilon)). \tag{4.30}$$

Let's note that according to preliminaries of Sect. 1, related to (2.4) in considered case $\mathbf{M} \in \mathcal{F}$ with $\mathbf{N} \in \mathcal{F}$,

$$(S_0(\mathbf{M}, 0) = \mathbf{M}) \ \& \ (S_0(\mathbf{N}, 0) = \mathbf{N}). \tag{4.31}$$

Therefore, if $\varepsilon = 0$, then constructions (4.27)–(4.31) are reduced to (4.24)–(4.26). Hence, second property in (4.24) is more sufficient for considered case. Namely, $\mathbf{M} = \mathbf{M} \cap \mathbf{N}$ is nonempty closed (see (4.1)) subset of $\mathcal{W}(\mathbf{M}, \mathbf{N})$. Now, we will take in account a natural connection (4.24) and (4.28). For this we will consider following sets $S_0(\mathbf{M}, \varepsilon)$, $\varepsilon \in [0, \infty[$. Wherein (see (4.1)),

$$\bigcup_{\varepsilon \in [0, \infty[} S_0(\mathbf{M}, \varepsilon) = T \times \mathbb{R}^n. \tag{4.32}$$

As corollary, we have that

$$\bigcup_{\varepsilon\in[0,\infty[} (S_0(\mathbf{M}, \varepsilon) \cap S_0(\mathbf{N}, \varepsilon)) = T \times \mathbb{R}^n. \tag{4.33}$$

In regards to (4.14), (4.15) and (4.23), $\forall \varepsilon \in [0, \infty[$

$$S_0(\mathbf{M}, \varepsilon) = S_0(\mathbf{M}, \varepsilon) \cap S_0(\mathbf{N}, \varepsilon) \subset \mathcal{W}(S_0(\mathbf{M}, \varepsilon), S_0(\mathbf{N}, \varepsilon)). \tag{4.34}$$

Then, we have the equality

$$T \times \mathbb{R}^n = \bigcup_{\varepsilon\in[0,\infty[} \mathcal{W}(S_0(\mathbf{M}, \varepsilon), S_0(\mathbf{N}, \varepsilon)). \tag{4.35}$$

It means that $\forall (t, x) \in T \times \mathbb{R}^n\ \exists \varepsilon \in [0, \infty[\colon (t, x) \in \mathcal{W}(S_0(\mathbf{M}, \varepsilon), S_0(\mathbf{N}, \varepsilon))$. Thus, with $(t, x) \in T \times \mathbb{R}^n$

$$\Sigma_0(t, x) \triangleq \{\varepsilon \in [0, \infty[\ \mid (t, x) \in \mathcal{W}(S_0(\mathbf{M}, \varepsilon), S_0(\mathbf{N}, \varepsilon))\} \tag{4.36}$$

is nonempty subset of $[0, \infty[$, and therefore we have defined following

$$\varepsilon_0(t, x) \triangleq \inf(\Sigma_0(t, x)) \in [0, \infty[. \tag{4.37}$$

Proposition 1 *If* $(t_*, x_*) \in T \times \mathbb{R}^n$*, then* $\varepsilon_0(t_*, x_*) \in \Sigma_0(t_*, x_*)$.

Proof We fix position $(t_*, x_*) \in T \times \mathbb{R}^n$ and we have nonempty set $\Sigma_0(t_*, x_*)$, $\Sigma_0(t_*, x_*) \subset [0, \infty[$. Wherein, according to (4.37), $\varepsilon_0(t_*, x_*) = \inf(\Sigma_0(t_*, x_*)) \in [0, \infty[$. Besides,

$$\Sigma^{(k)} \triangleq \Sigma_0(t_*, x_*) \cap [\varepsilon_0(t_*, x_*), \varepsilon_0(t_*, x_*) + \frac{1}{k}[\neq \emptyset \quad \forall k \in \mathbb{N}.$$

Then $\Sigma^{(k)}$ is a nonempty subset of $[0, \infty[$ with $k \in \mathbb{N}$. Thus, according to countable axiom of choice we have

$$\prod_{k\in\mathbb{N}} \Sigma^{(k)} \neq \emptyset. \tag{4.38}$$

We choose a sequence $(\varepsilon_k)_{k\in\mathbb{N}} \in \prod_{k\in\mathbb{N}} \Sigma^{(k)}$; $(\varepsilon_k)_{k\in\mathbb{N}} : \mathbb{N} \to [0, \infty[$, with following property

$$\varepsilon_s \in \Sigma^{(s)} \quad \forall s \in \mathbb{N}. \tag{4.39}$$

In particular, with $s \in \mathbb{N}$ we have that

$$\varepsilon_0(t_*, x_*) \leqslant \varepsilon_s < \varepsilon_0(t_*, x_*) + \frac{1}{s} \quad \forall s \in \mathbb{N}. \tag{4.40}$$

Then $(\varepsilon_s)_{s\in\mathbb{N}} \to \varepsilon_0(t_*, x_*)$. And in regards to (4.36) we obtain, along with (4.39), that

$$(t_*, x_*) \in \mathcal{W}(S_0(\mathbf{M}, \varepsilon_s), S_0(\mathbf{N}, \varepsilon_s)) \ \forall s \in \mathbb{N}. \tag{4.41}$$

We introduce following numbers

$$\varsigma_k \triangleq \min_{1\leqslant s\leqslant k} \ \varepsilon_s \ \ \forall k \in \mathbb{N}. \tag{4.42}$$

It's correct to note that with $k \in \mathbb{N}$, $\varepsilon_0(t_*, x_*) \leqslant \varsigma_k < \varepsilon_0(t_*, x_*) + \frac{1}{k}$, since $\varsigma_k \leqslant \varepsilon_k < \varepsilon_0(t_*, x_*) + \frac{1}{k}$. Thus, we have the following

$$(\varsigma_k)_{k\in\mathbf{N}} \to \varepsilon_0(t_*, x_*). \tag{4.43}$$

From (4.41) and (4.42) follows that

$$(t_*, x_*) \in \mathcal{W}(S_0(\mathbf{M}, \varsigma_k), S_0(\mathbf{N}, \varsigma_k)) \ \forall k \in \mathbb{N}. \tag{4.44}$$

However, according to (4.42), we have $\varsigma_{k+1} \leqslant \varsigma_k \ \forall k \in \mathbb{N}$. It means that, with $k \in \mathbb{N}$ we obtain

$$(S_0(\mathbf{M}, \varsigma_{k+1}) \subset S_0(\mathbf{M}, \varsigma_k)) \ \& \ (S_0(\mathbf{N}, \varsigma_{k+1}) \subset S_0(\mathbf{N}, \varsigma_k)). \tag{4.45}$$

Therefore, we have acquired the following sets

$$\bigcap_{k\in\mathbb{N}} S_0(\mathbf{M}, \varsigma_k) \in \mathcal{F}, \ \bigcap_{k\in\mathbb{N}} S_0(\mathbf{N}, \varsigma_k) \in \mathcal{F}.$$

Wherein, for $k \in \mathbb{N}$

$$S_0(\mathbf{M}, \varepsilon_0(t_*, x_*)) \subset S_0(\mathbf{M}, \varsigma_k), \ S_0(\mathbf{N}, \varepsilon_0(t_*, x_*)) \subset S_0(\mathbf{N}, \varsigma_k).$$

By (4.42), (4.43), we obtain

$$(S_0(\mathbf{M}, \varsigma_k))_{k\in\mathbb{N}} \downarrow S_0(\mathbf{M}, \varepsilon_0(t_*, x_*)), \tag{4.46}$$

$$(S_0(\mathbf{N}, \varsigma_k))_{k\in\mathbb{N}} \downarrow S_0(\mathbf{N}, \varepsilon_0(t_*, x_*)). \tag{4.47}$$

Considering (4.27), (4.28), (4.46) and (4.47), we have that, according to [26, Corollary 5.2]

$$(\mathcal{W}(S_0(\mathbf{M}, \varsigma_k), S_0(\mathbf{N}, \varsigma_k)))_{k\in\mathbb{N}} \downarrow \mathcal{W}(S_0(\mathbf{M}, \varepsilon_0(t_*, x_*)), S_0(\mathbf{N}, \varepsilon_0(t_*, x_*))). \tag{4.48}$$

It means, in particular, that

$$\mathcal{W}(S_0(\mathbf{M}, \varepsilon_0(t_*, x_*)), S_0(\mathbf{N}, \varepsilon_0(t_*, x_*))) = \bigcap_{k \in \mathbb{N}} \mathcal{W}(S_0(\mathbf{M}, \varsigma_k), S_0(\mathbf{N}, \varsigma_k));$$

then, according to (4.44), we have

$$(t_*, x_*) \in \mathcal{W}(S_0(\mathbf{M}, \varepsilon_0(t_*, x_*)), S_0(\mathbf{N}, \varepsilon_0(t_*, x_*))). \tag{4.49}$$

From (4.36) and (4.49) follows $\varepsilon_0(t_*, x_*) \in \Sigma_0(t_*, x_*)$, thus, Proposition 1 is fully proven.

As a corollary, we note (see (4.36)) that

$$(t, x) \in \mathcal{W}(S_0(\mathbf{M}, \varepsilon_0(t, x)), S_0(\mathbf{N}, \varepsilon_0(t, x))) \ \forall (t, x) \in T \times \mathbb{R}^n. \tag{4.50}$$

Let us introduce another option for functional on trajectories of the process. If $t_* \in T, x(\cdot) \in C_n([t_*, \vartheta_0])$ and $\theta \in [t_*, \vartheta_0]$, then we assume that

$$\omega(t_*, x(\cdot), \theta) = \sup\,(\{\rho((\theta, x(\theta)), \mathbf{M});\ \max_{t_* \leqslant t \leqslant \theta} \rho((t, x(t)), \mathbf{N})\}), \tag{4.51}$$

$\omega(t_*, x(\cdot), \theta) \in [0, \infty[$. As a corollary, we define for $t_* \in T$ special payoff function

$$\gamma_{t_*} : C_n([t_*, \vartheta_0]) \to [0, \infty[, \tag{4.52}$$

by following conditions: $\forall\, x(\cdot) \in C_n([t_*, \vartheta_0])$.

$$\gamma_{t_*}(x(\cdot)) \triangleq \min_{\theta \in [t_*, \vartheta_0]} \omega(t_*, x(\cdot), \theta). \tag{4.53}$$

Minimum in (4.53) is attainable, since $\rho((\cdot\,, x(\cdot)), \mathbf{M})$ and $\rho((\cdot\,, x(\cdot)), \mathbf{N})$ - are uniformly continuous functions (see also (4.51)).

Proposition 2 *Let $t_* \in T,\ x(\cdot) \in C_n([t_*, \vartheta_0])$ and $\varepsilon_* \in [0, \infty[$ be given. Then, following two conditions are equivalent:*

1. $\exists \vartheta \in [t_*, \vartheta_0] :\ ((\vartheta, x(\vartheta)) \in S_0(\mathbf{M}, \varepsilon_*))\ \&\ ((t, x(t)) \in S_0(\mathbf{N}, \varepsilon_*)\ \ \forall t \in [t_*, \vartheta])$;
2. $\gamma_{t_*}(x(\cdot)) \leqslant \varepsilon_*$

Proof Let condition (1) holds. We choose $\vartheta_* \in [t_*, \vartheta_0]$ in a way, that

$$((\vartheta_*, x(\vartheta_*)) \in S_0(\mathbf{M}, \varepsilon_*))\ \&\ ((t, x(t)) \in S_0(\mathbf{N}, \varepsilon_*)\ \ \forall t \in [t_*, \vartheta_*]). \tag{4.54}$$

It means that

$$\rho((\vartheta_*, x(\vartheta_*)), \mathbf{M}) \leqslant \varepsilon_*. \tag{4.55}$$

Besides, from (4.54) follows inequality $\rho((t, x(t)), \mathbf{N}) \leqslant \varepsilon_*\ \ \forall t \in [t_*, \vartheta_*]$. Thus

$$\max_{t_* \leqslant t \leqslant \vartheta_*} \rho((t, x(t)), \mathbf{N}) \leqslant \varepsilon_*. \tag{4.56}$$

According to (4.51), (4.55), and (4.56)

$$\omega(t_*, x(\cdot), \vartheta_*) \leqslant \varepsilon_*. \tag{4.57}$$

However, in regards to (4.53), following holds $\gamma_{t_*}(x(\cdot)) \leqslant \omega(t_*, x(\cdot), \vartheta_*)$ therefore $\gamma_{t_*}(x(\cdot)) \leqslant \varepsilon_*$ (see (4.57)). Thus, we have 2). So, we have proven 1) $\Rightarrow$ 2). Now, we show the opposite. Indeed, let the following holds

$$\gamma_{t_*}(x(\cdot)) \leqslant \varepsilon_*. \tag{4.58}$$

This means, according to (4.53), that for some $\theta_* \in [t_*, \vartheta_0]$, following inequality holds

$$\omega(t_*, x(\cdot), \theta_*) \leqslant \varepsilon_*. \tag{4.59}$$

Wherein, with (4.51)

$$\omega(t_*, x(\cdot), \theta_*) = \sup\ (\{\rho((\theta_*, x(\theta_*)), \mathbf{M});\ \max_{t_* \leqslant t \leqslant \theta_*} \rho((t, x(t)), \mathbf{N})\}). \tag{4.60}$$

From (4.59) and (4.60) follows that

$$\rho((\theta_*, x(\theta_*)), \mathbf{M}) \leqslant \varepsilon_*, \tag{4.61}$$

and with it

$$\max_{t_* \leqslant t \leqslant \theta_*} \rho((t, x(t)), \mathbf{N}) \leqslant \varepsilon_*. \tag{4.62}$$

Also, from (2.2) and (4.61) we have that

$$((\theta_*, x(\theta_*)) \in S_0(\mathbf{M}, \varepsilon_*))\ \&\ (\max_{t_* \leqslant t \leqslant \theta_*} \rho((t, x(t)), \mathbf{N}) \leqslant \varepsilon_*). \tag{4.63}$$

However, according to (4.62), follows $\rho((t, x(t)), \mathbf{N}) \leqslant \varepsilon_*\ \ \forall t \in [t_*, \theta_*]$. Finally,

$$(t, x(t)) \in S_0(\mathbf{N}, \varepsilon_*)\ \ \forall t \in [t_*, \theta_*]. \tag{4.64}$$

Thus (see (4.63)), the implication holds: 2) $\Rightarrow$ 1); therefore, Proposition 2 is fully proven. □

We recall that $\mathbf{M} \subset \mathbf{N}$. If $\varepsilon \in [0, \infty[$, then $S_0(\mathbf{M}, \varepsilon) \subset S_0(\mathbf{N}, \varepsilon)$.

Hereinafter we consider multivalued non-anticipating strategies as admissible control procedures. We define considered strategies in terms of non-anticipating operators on control-measure spaces. For this, it is helpful to recall some of the properties of measurable spaces, used in paragraph 2. We should improve some constructions given there. For arbitrary $t \in T$, let's fix compactum $Z_t \triangleq [t, \vartheta_0] \times Q$

and σ-algebra $\mathcal{D}_t$ of Borel subsets of Z_t. Wherein, for moments of time $t_1 \in T$ and $t_2 \in [t_1, \vartheta_0]$ $\mathcal{D}_{t_2} = \mathcal{D}_{t_1}|_{Z_{t_2}} = \{D \in \mathcal{D}_{t_1} | D \subset Z_{t_2}\}$. Further, if $t \in T$, then following holds: $\Gamma \times Q \in \mathcal{D}_t \ \forall \Gamma \in \mathcal{T}_t$. Following those constructions, we consider the set of measures, which are similar to Borel mappings from $[t, \vartheta_0]$ to Q, namely: $\mathcal{E}_t \triangleq \{\nu \in (\sigma - \text{add})_+[\mathcal{D}_t] \mid \nu(\Gamma \times Q) = \lambda_t(\Gamma) \ \ \forall \Gamma \in \mathcal{T}_t\}$; we use $\mathcal{E}_t$ as set of generalized controls of player II. Moreover, we note that $D \underline{\times} P \triangleq \{(t, u, v) \in \Omega_t \mid (t, v) \in D\} \in \mathcal{C}_t \ \forall D \in \mathcal{D}_t$. Therefore, we have the following (see [33, (4.3)])

$$\Pi_t(\nu) \triangleq \{\eta \in \mathcal{H}_t \mid \eta(D \underline{\times} P) = \nu(D) \ \forall D \in \mathcal{D}_t\} \ \forall \nu \in \mathcal{E}_t. \qquad (4.65)$$

Thus, the family of programs [15, c. 162] for second player on a interval $[t, \vartheta_0]$ was introduced.

We also need additional operations for generalized controls, namely gluing and constriction. Let us recall another notation and define corresponding designation: if X and Y - nonempty sets, where $h \in Y^X$ and $\tilde{X} \in \mathcal{P}'(X)$, then $(h|\tilde{X}) \triangleq (h(x))_{x \in \tilde{X}} \in Y^{\tilde{X}}$. As X a family can be used and as h — set function.

If $t_1 \in T$ and $t_2 \in [t_1, \vartheta_0]$, then (see [33, sect. 4]) $\mathcal{C}_{t_2} = \mathcal{C}_{t_1} |_{\Omega_{t_2}} = \{H \in \mathcal{C}_{t_1} \mid H \subset \Omega_{t_2}\}$,

$$\mathcal{C}_{t_1}^{t_2} = \mathcal{C}_{t_1} |_{[t_1, t_2[\times P \times Q} = \{H \in \mathcal{C}_{t_1} \mid H \subset [t_1, t_2[\times P \times Q\},$$

$$\mathcal{D}_{t_1}^{t_2} = \mathcal{D}_{t_1} |_{[t_1, t_2[\times Q} = \{D \in \mathcal{D}_{t_1} \mid D \subset [t_1, t_2[\times Q\}.$$

In this case we have σ-algebras of sets. It is useful to note that (see [33, (4.4)])

$$\mathcal{H}_{t_2} = \{(\eta \mid \mathcal{C}_{t_2}) : \eta \in \mathcal{H}_{t_1}\}, \ \mathcal{E}_{t_2} = \{(\nu \mid \mathcal{D}_{t_2}) : \nu \in \mathcal{E}_{t_1}\}.$$

If $t_* \in T$, then we assume that $\tilde{A}_{t_*}$ is the set of all generalized multivalued non-anticipating strategies (see [33]) for first player on a line $[t_*, \vartheta_0]$:

$$\tilde{A}_{t_*} \triangleq \{\alpha \in \prod_{\nu \in \mathcal{E}_{t_*}} \mathcal{P}'(\Pi_{t_*}(\nu)) \mid \forall \nu_1 \in \mathcal{E}_{t_*} \ \forall \nu_2 \in \mathcal{E}_{t_*} \ \forall \theta \in [t_*, \vartheta_0] : ((\nu_1 \mid \mathcal{D}_{t_*}^{\theta})$$
$$= (\nu_2 \mid \mathcal{D}_{t_*}^{\theta})) \Rightarrow (\{(\eta \mid \mathcal{C}_{t_*}^{\theta}) : \eta \in \alpha(\nu_1)\} = \{(\eta \mid \mathcal{C}_{t_*}^{\theta}) : \eta \in \alpha(\nu_2)\})\}. \qquad (4.66)$$

If $\alpha \in \tilde{A}_{t_*}$ and $\nu \in \mathcal{E}_{t_*}$, then as $\alpha(\nu)$ we have nonempty subset of $\Pi_{t_*}(\nu)$; in particular, $\alpha(\nu) \subset \mathcal{H}_{t_*}$. Thus, we defined following union-set (see [33, (10.2)])

$$\tilde{\Pi}_{t_*}(\alpha) \triangleq \bigcup_{\nu \in \mathcal{E}_{t_*}} \alpha(\nu) \in \mathcal{P}'(\mathcal{H}_{t_*}), \qquad (4.67)$$

which is the set of all generalized controls-measures, generated by non-anticipating strategy α. From (4.36) and Proposition 1 we have following:

Proposition 3 *If* $(t_*, x_*) \in T \times \mathbb{R}^n$, *then*

$$\varepsilon_0(t_*, x_*) = \inf_{\alpha \in \tilde{A}_{t*}} \sup_{\eta \in \tilde{\Pi}_{t_*}(\alpha)} \gamma_{t_*}(\varphi(\cdot\,, t_*, x_*, \eta)),$$

wherein $\exists \tilde{\alpha}_* \in \tilde{A}_{t_*} : \varepsilon_0(t_*, x_*) = \sup_{\eta \in \tilde{\Pi}_{t_*}(\tilde{\alpha}_*)} \gamma_{t_*}(\varphi(\cdot\,, t_*, x_*, \eta)).$

Proof For this proof we use notation, similar to one in [33, Theorem 10.1]. Let us suppose that $\varepsilon_* \triangleq \varepsilon_0(t_*, x_*)$. In regards to [33, (10.20)], we introduce the following set

$$\begin{aligned} \mathcal{S}_{M_*,N_*}(t,x) \triangleq \{\eta \in \mathcal{H}_t \mid \exists \vartheta \in [t, \vartheta_0] : ((\vartheta, \varphi(\vartheta, t, x, \eta)) \in M_*) \\ \& \, ((\xi, \varphi(\xi, t, x, \eta)) \in N_* \; \forall \xi \in [t, \vartheta])\} \; \forall (t,x) \in N_*, \end{aligned} \tag{4.68}$$

where $M_* \triangleq S_0(\mathbf{M}, \varepsilon_*)$ and $N_* \triangleq S_0(\mathbf{N}, \varepsilon_*)$. In (4.68) we employ closedness property of N_*. According to statements provided in [25], for the set $W(M_*, N_*)$ used in [33], we have the equality $W(M_*, N_*) = \mathcal{W}(M_*, N_*)$ (see [26, Proposition 5.2] and [33, Theorem 10.1]). Therefore,

$$\mathcal{W}(M_*, N_*) = \{(t,x) \in N_* \mid \exists \alpha \in \tilde{A}_t : \tilde{\Pi}_t(\alpha) \subset \mathcal{S}_{M_*,N_*}(t,x)\} \tag{4.69}$$

In addition, by Proposition 1, $\varepsilon_* \in \Sigma_0(t_*, x_*)$. As a corollary, $(t_*, x_*) \in \mathcal{W}(M_*, N_*)$. In regards to (4.69), we select non-anticipating strategy $\alpha_* \in \tilde{A}_{t_*}$, such that $\tilde{\Pi}_{t_*}(\alpha_*) \subset \mathcal{S}_{M_*,N_*}(t,x)$. The last inclusion denotes that

$$\begin{aligned} \forall \eta \in \tilde{\Pi}_{t_*}(\alpha_*) \, \exists \vartheta \in [t_*, \vartheta_0] : ((\vartheta, \varphi(\vartheta, t_*, x_*, \eta)) \in M_*) \\ \& \, ((t, \varphi(t, t_*, x_*, \eta)) \in N_* \;\; \forall t \in [t_*, \vartheta]). \end{aligned} \tag{4.70}$$

By definition of M_* and N_*, we have that

$$\begin{aligned} \forall \eta \in \tilde{\Pi}_{t_*}(\alpha_*) \, \exists \vartheta \in [t_*, \vartheta_0] : (\rho((\vartheta, \varphi(\vartheta, t_*, x_*, \eta)), M_*) \leqslant \varepsilon_*) \\ \& \, ((\rho((t, \varphi(t, t_*, x_*, \eta)), N_*)) \leqslant \varepsilon_* \; \forall t \in [t_*, \vartheta]). \end{aligned} \tag{4.71}$$

According to closedness property of N_*, (4.70), (4.36) and (4.71), we obtain that $\forall \eta \in \tilde{\Pi}_{t_*}(\alpha_*) \, \exists \vartheta \in [t_*, \vartheta_0]$:

$$\omega(t_*, \varphi(\cdot\,, t_*, x_*, \eta), \vartheta) \leqslant \varepsilon_*.$$

As a corollary, by (4.50), $\gamma_{t_*}(\varphi(\cdot\,, t_*, x_*, \eta)) \leqslant \varepsilon_* \;\; \forall \eta \in \tilde{\Pi}_{t_*}(\alpha_*)$. As a result,

$$\sup_{\eta \in \tilde{\Pi}_{t_*}(\alpha_*)} \gamma_{t_*}(\varphi(\cdot\,, t_*, x_*, \eta)) \leqslant \varepsilon_*, \tag{4.72}$$

and, as a corollary,

$$\inf_{\alpha \in \tilde{A}_{t_*}} \sup_{\eta \in \tilde{\Pi}_{t_*}(\alpha_*)} \gamma_{t_*}(\varphi(\cdot\,, t_*, x_*, \eta)) \leqslant \varepsilon_*. \tag{4.73}$$

Further, let us select arbitrary $\alpha^* \in \tilde{A}_{t_*}$ and consider the set $\tilde{\Pi}_{t_*}(\alpha^*) \in \mathcal{P}'(\mathcal{H}_{t_*})$. With employment of the uniform boundedness condition, (4.51), and (4.53), we obtain that

$$\gamma_{t_*}(\varphi(\cdot\,, t_*, x_*, \eta)) \leqslant b_* \;\; \forall \eta \in \tilde{\Pi}_{t_*}(\alpha^*), \tag{4.74}$$

where $b_* \in [0, \infty[$. Let us also define

$$c^* \triangleq \sup_{\eta \in \tilde{\Pi}_{t_*}(\alpha^*)} \gamma_{t_*}(\varphi(\cdot\,, t_*, x_*, \eta)) \in [0, \infty[. \tag{4.75}$$

According to definitions above, $c^* \leqslant b_*$. Suppose that $M^* \triangleq S_0(\mathbf{M}, c^*)$ and $N^* \triangleq S_0(\mathbf{N}, c^*)$, where $M^* \in \mathcal{F}$ and $N^* \in \mathcal{F}$. By Proposition 2, (4.53), (4.75), and by definition of M^* and N^*

$$\forall \eta \in \tilde{\Pi}_{t_*}(\alpha^*)\, \exists \vartheta \in [t_*, \vartheta_0] : ((\vartheta, \varphi(\vartheta, t_*, x_*, \eta)) \in M^*) \\ \& \; ((t, \varphi(t, t_*, x_*, \eta)) \in N^* \;\; \forall t \in [t_*, \vartheta]). \tag{4.76}$$

Thus, from (4.76) and the fact that $\tilde{\Pi}_t(\alpha^*) \neq \emptyset$, follows $(t_*, x_*) \in N^*$. Let us introduce the set $\mathcal{S}_{M^*,N^*}(t_*, x_*)$, which is similar to one in (4.68), according to [33, (10.20)]. Then, $\tilde{\Pi}_t(\alpha^*) \subset \mathcal{S}_{M^*,N^*}(t_*, x_*)$ and $(t_*, x_*) \in \mathcal{W}(M^*, N^*)$. By definition of M^* and N^*, we obtain that $c^* \in \Sigma_0(t_*, x_*)$. As a corollary, $\varepsilon_* = \varepsilon_0(t_*, x_*) \leqslant c^*$. Since selection of α^* was arbitrary, we have shown that

$$\varepsilon_* \leqslant \inf_{\alpha \in \tilde{A}_{t_*}} \sup_{\eta \in \tilde{\Pi}_{t_*}(\alpha)} \gamma_{t_*}(\varphi(\cdot\,, t_*, x_*, \eta)). \tag{4.77}$$

Thus, by (4.73) and (4.77), we obtain the equality

$$\varepsilon_* = \inf_{\alpha \in \tilde{A}_{t_*}} \sup_{\eta \in \tilde{\Pi}_{t_*}(\alpha)} \gamma_{t_*}(\varphi(\cdot\,, t_*, x_*, \eta)). \tag{4.78}$$

Therefore,

$$\sup_{\eta \in \tilde{\Pi}_{t_*}(\alpha_*)} \gamma_{t_*}(\varphi(\cdot\,, t_*, x_*, \eta)) = \varepsilon_*$$

Our proposition is established. □

Therefore it is established that function $\varepsilon_0 : T \times \mathbb{R} \to [0, \infty[$, defined by (4.37), is equal to minimax of payoff function γ in terms of non-anticipating strategies for any fixed position.

5 Approximative Realization of $\varepsilon_0(t, x)$, $(t, x) \in T \times \mathbb{R}^n$

In this section, we should pay special attention to construction of function ε_0. In particular, we use sequence of functions $(\varepsilon_0^{(s)})_{s\in\mathbb{N}_0}$ constructed on each iteration with number s. Later in this section, we will explain how those functions relate in terms of iterations. Let us consider special kind of number sets. Each set is generated on respective stage of our iteration procedure. We denote them as follows

$$\Sigma_0^{(s)}(t, x) \triangleq \{\varepsilon \in [0, \infty[\,|\, (t, x) \in \mathcal{W}_s(S_0(\mathbf{M}, \varepsilon), S_0(\mathbf{N}, \varepsilon))\} \;\forall s \in \mathbb{N}_0 \;\forall (t, x) \in T \times \mathbb{R}^n. \tag{5.1}$$

Proposition 4 *If $s \in \mathbb{N}_0$ and $(t, x) \in T \times \mathbb{R}^n$, then $\Sigma_0(t, x) \subset \Sigma_0^{(s)}(t, x)$.*

Proof We fix $m \in \mathbb{N}_0$ and $(t_*, x_*) \in T \times \mathbb{R}^n$. Thus, we obtain following sets $\Sigma_0(t_*, x_*)$ and $\Sigma_0^{(m)}(t_*, x_*)$. Let $\varepsilon_* \in \Sigma_0(t_*, x_*)$. Then we have $\varepsilon_* \in [0, \infty[$ wherein $(t_*, x_*) \in \mathcal{W}(S_0(\mathbf{M}, \varepsilon_*), S_0(\mathbf{N}, \varepsilon_*))$, and according to (4.30) $\mathcal{W}(S_0(\mathbf{M}, \varepsilon_*), S_0(\mathbf{N}, \varepsilon_*)) \subset \mathcal{W}_m(S_0(\mathbf{M}, \varepsilon_*), S_0(\mathbf{N}, \varepsilon_*))$. Hereof we obtain, $(t_*, x_*) \in \mathcal{W}_m(S_0(\mathbf{M}, \varepsilon_*), S_0(\mathbf{N}, \varepsilon_*))$; thus $\varepsilon_* \in \Sigma_0^{(m)}(t_*, x_*)$. Since selection of ε_* was arbitrary, we have $\Sigma_0(t_*, x_*) \subset \Sigma_0^{(m)}(t_*, x_*)$. However, selection of m and position (t_*, x_*) was also arbitrary, thus Proposition 4 is established. □

From Propositions 1 and 4 we obtain that,

$$\varepsilon_0(t, x) \in \Sigma_0^{(s)}(t, x) \;\; \forall (t, x) \in T \times \mathbb{R}^n \;\; \forall s \in \mathbb{N}_0. \tag{5.2}$$

In particular, if $(t, x) \in T \times \mathbb{R}^n$ and $s \in \mathbb{N}_0$ as $\Sigma_0^{(s)}(t, x)$, we have non-empty subset of $[0, \infty[$, thus, $\inf(\Sigma_0^{(s)}(t, x)) \in [0, \infty[$ is defined.

Hereinafter, we assume that

$$\varepsilon_0^{(s)}(t, x) \triangleq \inf(\Sigma_0^{(s)}(t, x)) \;\; \forall (t, x) \in T \times \mathbb{R}^n \;\; \forall s \in \mathbb{N}_0. \tag{5.3}$$

Using (5.3) with each $s \in \mathbb{N}_0$, we can define function

$$\varepsilon_0^{(s)} : T \times \mathbb{R}^n \to [0, \infty[. \tag{5.4}$$

Also, from (5.3) and Proposition 4, we obtain that,

$$\varepsilon_0^{(s)}(t, x) \leqslant \varepsilon_0(t, x) \;\; \forall (t, x) \in T \times \mathbb{R}^n \;\; \forall s \in \mathbb{N}_0. \tag{5.5}$$

Let us designate point-wise order in the set of all functions from $T \times \mathbb{R}^n$ into $[0, \infty[$ by $\leqq$. Then from (5.5) we have

$$\varepsilon_0^{(s)} \leqq \varepsilon_0 \;\; \forall s \in \mathbb{N}_0. \tag{5.6}$$

Proposition 5 *If* $s \in \mathbb{N}_0$ *and* $(t_*, x_*) \in T \times \mathbb{R}^n$, *then* $\varepsilon_0^{(s)}(t_*, x_*) \in \Sigma_0^{(s)}(t_*, x_*)$.

Proof Fixate $s \in \mathbb{N}_0$ and $(t_*, x_*) \in T \times \mathbb{R}^n$. Following (5.1)

$$(t_*, x_*) \in \mathcal{W}_s(S_0(\mathbf{M}, \varepsilon), S_0(\mathbf{N}, \varepsilon)) \ \forall \varepsilon \in \Sigma_0^{(s)}(t_*, x_*). \tag{5.7}$$

Let us note that according to (5.3) $\varepsilon^* \triangleq \varepsilon_0^{(s)}(t_*, x_*) = \inf(\Sigma_0^{(s)}(t_*, x_*)) \in [0, \infty[$. Then, for some sequence $(\delta_k)_{k\in\mathbb{N}} : \mathbb{N} \to \Sigma_0^{(s)}(t_*, x_*)$ there is monotone convergence

$$(\delta_k)_{k\in\mathbb{N}} \downarrow \varepsilon^*, \tag{5.8}$$

ε^* is limit of sequence $(\delta_k)_{k\in\mathbb{N}}$ and $\delta_{j+1} \leqslant \delta_j \ \forall j \in \mathbb{N}$. According to (5.8), we obtain that

$$((S_0(\mathbf{M}, \delta_k))_{k\in\mathbb{N}} \downarrow S_0(\mathbf{M}, \varepsilon^*)) \ \& \ ((S_0(\mathbf{N}, \delta_k))_{k\in\mathbb{N}} \downarrow S_0(\mathbf{N}, \varepsilon^*)). \tag{5.9}$$

Following [26, Proposition 5.3], from (5.9) there is convergence

$$(\mathcal{W}_s(S_0(\mathbf{M}, \delta_k), S_0(\mathbf{N}, \delta_k)))_{k\in\mathbb{N}} \downarrow \mathcal{W}_s(S_0(\mathbf{M}, \varepsilon^*), S_0(\mathbf{N}, \varepsilon^*)). \tag{5.10}$$

However, according to (5.7) we have following property

$$(t_*, x_*) \in \mathcal{W}_s(S_0(\mathbf{M}, \delta_k), S_0(\mathbf{N}, \delta_k)) \ \forall k \in \mathbf{N}.$$

Thus, from (5.10) we obtain

$$(t_*, x_*) \in \mathcal{W}_s(S_0(\mathbf{M}, \varepsilon^*), S_0(\mathbf{N}, \varepsilon^*)). \tag{5.11}$$

From (5.1) and (5.11) we have property $\varepsilon^* \in \Sigma_0^{(s)}(t_*, x_*)$.

Also, from (5.5) we show, in particular, that exact upper bound is defined properly. That is $\sup(\{\varepsilon_0^{(s)}(t, x) : s \in \mathbb{N}_0\}) \in [0, \varepsilon_0(t, x)] \ \forall (t, x) \in T \times \mathbb{R}^n$.

Proposition 6 *If* $(t_*, x_*) \in T \times \mathbb{R}^n$, *then* $\varepsilon_0(t_*, x_*) = \sup(\{\varepsilon_0^{(s)}(t_*, x_*) : s \in \mathbb{N}_0\})$

Proof We assume that $\Xi \triangleq \{\varepsilon_0^{(s)}(t_*, x_*) : s \in \mathbb{N}_0\}$; thus $\Xi \in \mathcal{P}'([0, \varepsilon^*])$, where $\varepsilon^* \triangleq \varepsilon_0(t_*, x_*) \in [0, \infty[$. Hence, $\varepsilon_* \triangleq \sup(\Xi) \in [0, \varepsilon^*]$. Thereby, $0 \leqslant \varepsilon_* \leqslant \varepsilon^*$. Let us show that $\varepsilon_* = \varepsilon^*$. Indeed, assume by contradiction: $\varepsilon_* \neq \varepsilon^*$. Then we have $\varepsilon_* < \varepsilon^*$. At the same time, according to (4.37), $\varepsilon^* = \inf(\Sigma_0(t_*, x_*))$, where $\Sigma_0(t_*, x_*)$ - non-empty subset of $[0, \infty[$. Thus $\varepsilon_* \notin \Sigma_0(t_*, x_*)$, and following (4.36) we obtain that

$$(t_*, x_*) \notin \mathcal{W}(S_0(\mathbf{M}, \varepsilon_*), S_0(\mathbf{N}, \varepsilon_*)). \tag{5.12}$$

According to (4.30) and (5.12), for some $r \in \mathbb{N}_0$

$$(t_*, x_*) \notin \mathcal{W}_r(S_0(\mathbf{M}, \varepsilon_*), S_0(\mathbf{N}, \varepsilon_*)). \tag{5.13}$$

From (5.1) and (5.13) we have that

$$\varepsilon_* \notin \Sigma_0^{(r)}(t_*, x_*). \tag{5.14}$$

Wherein, following (5.5) $\varepsilon_0^{(r)}(t_*, x_*) \leqslant \varepsilon_0(t_*, x_*) = \varepsilon^*$. According to (5.13) let us recall following equality

$$\Sigma_0^{(r)}(t_*, x_*) = \{\varepsilon \in [0, \infty[\mid (t_*, x_*) \in \mathcal{W}_r(S_0(\mathbf{M}, \varepsilon), S_0(\mathbf{N}, \varepsilon))\}. \tag{5.15}$$

Wherein, in terms of Proposition 5, we obtain that $\varepsilon_0^{(r)}(t_*, x_*) \in \Sigma_0^{(r)}(t_*, x_*)$, also

$$\varepsilon_0^{(r)}(t_*, x_*) \leqslant \varepsilon \ \forall \varepsilon \in \Sigma_0^{(r)}(t_*, x_*). \tag{5.16}$$

By definition of Ξ following inequality holds

$$\varepsilon_0^{(r)}(t_*, x_*) \leqslant \varepsilon_*. \tag{5.17}$$

Considering properties of $\varepsilon_0^{(r)}(t_*, x_*)$, we obtain

$$(t_*, x_*) \in \mathcal{W}_r(S_0(\mathbf{M}, \varepsilon_0^{(r)}(t_*, x_*)), S_0(\mathbf{N}, \varepsilon_0^{(r)}(t_*, x_*))). \tag{5.18}$$

At the same time, according to (5.14) and (5.15), we have

$$(t_*, x_*) \notin \mathcal{W}_r(S_0(\mathbf{M}, \varepsilon_*), S_0(\mathbf{N}, \varepsilon_*)). \tag{5.19}$$

However, from (5.17) follows $(S_0(\mathbf{M}, \varepsilon_0^{(r)}(t_*, x_*)) \subset S_0(\mathbf{M}, \varepsilon_*))$ & $(S_0(\mathbf{N}, \varepsilon_0^{(r)}(t_*, x_*)) \subset S_0(\mathbf{N}, \varepsilon_*))$. As a corollary, we have

$$\mathcal{W}_r(S_0(\mathbf{M}, \varepsilon_0^{(r)}(t_*, x_*)), S_0(\mathbf{N}, \varepsilon_0^{(r)}(t_*, x_*))) \subset \mathcal{W}_r(S_0(\mathbf{M}, \varepsilon_*), S_0(\mathbf{N}, \varepsilon_*)). \tag{5.20}$$

Finally, according to (5.18) and (5.20), following inclusion holds $(t_*, x_*) \in \mathcal{W}_r(S_0(\mathbf{M}, \varepsilon_*), S_0(\mathbf{N}, \varepsilon_*))$, which contradicts (5.19). Thereby, we have shown that $\varepsilon_* = \varepsilon^*$. □

We consider another proposition which will establish relations between $\varepsilon_0^{(s)}(t_*, x_*)$ and $\varepsilon_0^{(s+1)}(t_*, x_*)$.

Proposition 7 *Let $(t_*, x_*) \in T \times \mathbb{R}^n$ and $s \in \mathbb{N}_0$. Then*

$$\varepsilon_0^{(s)}(t_*, x_*) \leqq \varepsilon_0^{(s+1)}(t_*, x_*). \tag{5.21}$$

Proof We fix position $(t_*, x_*) \in T \times \mathbb{R}^n$ and index $s \in \mathbb{N}_0$. By definition, $\varepsilon_*^{(s)} \triangleq \varepsilon_0^{(s)}(t_*, x_*) \in \Sigma_0^{(s)}(t_*, x_*)$; thus, according to (5.1), $(t_*, x_*) \in \mathcal{W}_s(S_0(\mathbf{M}, \varepsilon_*^{(s)}), S_0(\mathbf{N}, \varepsilon_*^{(s)}))$. Following (5.1) and Proposition 5, we obtain for $\varepsilon_*^{(s+1)} \triangleq \varepsilon_0^{(s+1)}(t_*, x_*)$:

$$(t_*, x_*) \in \mathcal{W}_{s+1}(S_0(\mathbf{M}, \varepsilon_*^{(s+1)}), S_0(\mathbf{N}, \varepsilon_*^{(s+1)})). \tag{5.22}$$

Wherein, according to (4.6) and (4.29), we have

$$\mathcal{W}_{s+1}(S_0(\mathbf{M}, \varepsilon_*^{(s+1)}), S_0(\mathbf{N}, \varepsilon_*^{(s+1)})) \subset \mathcal{W}_s(S_0(\mathbf{M}, \varepsilon_*^{(s+1)}), S_0(\mathbf{N}, \varepsilon_*^{(s+1)})). \quad (5.23)$$

Hence, position (t_*, x_*) belongs to set - iteration with index s, namely

$$(t_*, x_*) \in \mathcal{W}_s(S_0(\mathbf{M}, \varepsilon_*^{(s+1)}), S_0(\mathbf{N}, \varepsilon_*^{(s+1)})). \quad (5.24)$$

Thereby, according to (5.3), we obtain $\varepsilon_*^{(s+1)} \in \Sigma_0^{(s)}(t_*, x_*)$, so $\varepsilon_*^{(s)} \leqslant \varepsilon_*^{(s+1)}$. Thus, Proposition 7 is proved. □

Finally, we have shown following property:

$$\varepsilon_0^{(s)} \leqq \varepsilon_0^{(s+1)} \;\; \forall s \in \mathbb{N}_0. \quad (5.25)$$

6 Conversion Operator for Functions Defined on Space of Positions

In this section we construct program operator, which will define for $s \in \mathbb{N}_0$ conversion from $\varepsilon_0^{(s)}$ to $\varepsilon_0^{(s+1)}$ without using sets of type (4.27). To achieve this, we will use special type of construction, which is similar to one introduced in [34]. The modification of program iterations method described in [34] corresponds to differential game with non-fixed moment of termination. In regards with constructions of program iterations method, we note [35–37]. Also, let us introduce new notations according to those in [33, 34, 38].

First of all, we introduce the function $\psi : T \times \mathbb{R}^n \to [0, \infty[$ by condition:

$$\psi(t, x) \triangleq \rho((t, x), \mathbf{M}) \;\; \forall (t, x) \in T \times \mathbb{R}^n. \quad (6.1)$$

According to the properties of the distance function from a point to a nonempty set $\mathbf{M}$ we have that $\psi \in C(T \times \mathbb{R}^n)$, where $C(T \times \mathbb{R}^n)$ - set of all continuous real-valued functions on $T \times \mathbb{R}^n$. We note that

$$\psi^{-1}([0, c]) \in \mathcal{F} \;\; \forall c \in [0, \infty[. \quad (6.2)$$

Thus, according to (6.2) and due to non-negativity of ψ, we have lower semi-continuity property.

Subsequently, for every non-empty set H as $\mathfrak{R}_+(H)$ we denote the set of all non-negative real-valued functions on H.

In terms of (6.2) we define, according to [34, §2], the following set

$$\mathfrak{M} \triangleq \{g \in \mathfrak{R}_+(T \times \mathbb{R}^n) | \ g^{-1}([0, c]) \in \mathcal{F} \ \forall c \in [0, \infty[\}. \tag{6.3}$$

Besides, we require set

$$\mathfrak{M}_\psi \triangleq \{g \in \mathfrak{M} | \ g \leqq \psi\}; \tag{6.4}$$

where $\psi \in \mathfrak{M}_\psi$. Thus, (6.4) is non-empty subset of $\mathfrak{M}$.

Proposition 8 *If* $(t, x) \in T \times \mathbb{R}^n$ *and* $s \in \mathbb{N}_0$, *then* $\Sigma_0^{(s)}(t, x) = [\varepsilon_0^{(s)}(t, x), \infty[$.

Proof Let us fix $(t_*, x_*) \in T \times \mathbb{R}^n$ and $m \in \mathbb{N}_0$. Then, by (5.1) we obtain

$$\Sigma_0^{(m)}(t_*, x_*) = \{\varepsilon \in [0, \infty[\ | \ (t_*, x_*) \in \mathcal{W}_m(S_0(\mathbf{M}, \varepsilon), S_0(\mathbf{N}, \varepsilon))\}. \tag{6.5}$$

According to (5.3) and (5.5), we have that

$$\Sigma_0^{(m)}(t_*, x_*) \subset [\varepsilon_0^{(m)}(t_*, x_*), \infty[. \tag{6.6}$$

Let $\varepsilon_* \in [\varepsilon_0^{(m)}(t_*, x_*), \infty[$. Then we obtain following inclusions

$$(S_0(\mathbf{M}, \varepsilon_0^{(m)}(t_*, x_*)) \subset S_0(\mathbf{M}, \varepsilon_*)) \ \& \ (S_0(\mathbf{N}, \varepsilon_0^{(m)}(t_*, x_*)) \subset S_0(\mathbf{N}, \varepsilon_*)). \tag{6.7}$$

From (4.10), (4.26), and (6.7), we obtain that

$$\mathcal{W}_m(S_0(\mathbf{M}, \varepsilon_0^{(m)}(t_*, x_*)), S_0(\mathbf{N}, \varepsilon_0^{(m)}(t_*, x_*))) \subset \mathcal{W}_m(S_0(\mathbf{M}, \varepsilon_*), S_0(\mathbf{N}, \varepsilon_*)), \tag{6.8}$$

where, by Proposition 5, $(t_*, x_*) \in \mathcal{W}_m(S_0(\mathbf{M}, \varepsilon_0^{(m)}(t_*, x_*)), S_0(\mathbf{N}, \varepsilon_0^{(m)}(t_*, x_*)))$. Using (6.8), we have inclusion

$$(t_*, x_*) \in \mathcal{W}_m(S_0(\mathbf{M}, \varepsilon_*), S_0(\mathbf{N}, \varepsilon_*)). \tag{6.9}$$

By (6.5) and (6.9) we obtain that $\varepsilon_* \in \Sigma_0^{(m)}(t_*, x_*)$. As a corollary, the following inclusion

$$[\varepsilon_0^{(m)}(t_*, x_*), \infty[\subset \Sigma_0^{(m)}(t_*, x_*) \tag{6.10}$$

is established. From (6.6) and (6.10), we obtain equality $\Sigma_0^{(m)}(t_*, x_*) = [\varepsilon_0^{(m)}(t_*, x_*), \infty[$. Since selection of position (t_*, x_*) and index m was arbitrary, the required statement is established. □

Proposition 9 *If* $b \in [0, \infty[$ *and* $s \in \mathbb{N}_0$, *then* $(\varepsilon_0^{(s)})^{-1}([0, b]) = \mathcal{W}_s(S_0(\mathbf{M}, b), S_0(\mathbf{N}, b))$.

Proof Let $(t_*, x_*) \in \mathcal{W}_s(S_0(\mathbf{M}, b), S_0(\mathbf{N}, b))$. According to (5.1), $b \in \Sigma_0^{(s)}(t_*, x_*)$, then by condition (5.3) we have $\varepsilon_0^{(s)}(t_*, x_*) \leqslant b$. Thus, $(t_*, x_*) \in (\varepsilon_0^{(s)})^{-1}([0, b])$ (see (5.1)). Therefore,

$$\mathcal{W}_s(S_0(\mathbf{M}, b), S_0(\mathbf{N}, b)) \subset (\varepsilon_0^{(s)})^{-1}([0, b]). \tag{6.11}$$

Conversely, let $(t^*, x^*) \in (\varepsilon_0^{(s)})^{-1}([0, b])$, i.e. $(t^*, x^*) \in T \times \mathbb{R}^n$ and $\varepsilon_0^{(s)}(t^*, x^*) \leqslant b$. According to Proposition 8, $\Sigma_0^{(s)}(t^*, x^*) = [\varepsilon_0^{(s)}(t^*, x^*), \infty[$; thus $b \in \Sigma_0^{(s)}(t^*, x^*)$ and, hence (see (5.1)) $(t^*, x^*) \in \mathcal{W}_s(S_0(\mathbf{M}, b), S_0(\mathbf{N}, b))$, which completes the verification of the opposite to (6.11) inclusion. □

Proposition 10 *If* $b \in [0, \infty[$, *then* $(\varepsilon_0)^{-1}([0, b]) = \mathcal{W}(S_0(\mathbf{M}, b), S_0(\mathbf{N}, b))$.

The proof is similar to one in Proposition 9.

By (4.27) and Proposition 9, we have that $(\varepsilon_0^{(s)})^{-1}([0, b]) \in \mathcal{F}\ \forall b \in [0, \infty[\ \forall s \in \mathbb{N}_0$. However, from (6.3) follows that

$$\varepsilon_0^{(s)} \in \mathfrak{M}\ \forall s \in \mathbb{N}_0. \tag{6.12}$$

In a similar way, by (4.28) and Proposition 10 we have that $(\varepsilon_0)^{-1}([0, b]) \in \mathcal{F}\ \forall b \in [0, \infty[$. In regards to (6.3), we obtain

$$\varepsilon_0 \in \mathfrak{M}. \tag{6.13}$$

Proposition 11 *If* $(t_*, x_*) \in T \times \mathbb{R}^n$, *then* $\varepsilon_0(t_*, x_*) \leqslant \psi(t_*, x_*)$.

Proof Considering inclusion $\mathbf{M} \subset \mathbf{N}$, we obtain $\rho((t_*, x_*), \mathbf{N}) \leqslant \rho((t_*, x_*), \mathbf{M}) = \psi(t_*, x_*)$. Thus, $(t_*, x_*) \in S_0(\mathbf{M}, \psi(t_*, x_*)) \cap S_0(\mathbf{N}, \psi(t_*, x_*))$ (see (2.2), (2.4)). However,

$$S_0(\mathbf{M}, \psi(t_*, x_*)) \subset S_0(\mathbf{N}, \psi(t_*, x_*)). \tag{6.14}$$

By (4.34), (4.36), and (6.14), we get $\psi(t_*, x_*) \in \Sigma_0(t_*, x_*)$. In regards to (4.37), we obtain the required inequality $\varepsilon_0(t_*, x_*) \leqslant \psi(t_*, x_*)$. □

By Proposition 11, we have that $\varepsilon_0 \leqq \psi$. However, by (6.4) and (6.13), we have following inclusion

$$\varepsilon_0 \in \mathfrak{M}_\psi. \tag{6.15}$$

On the other hand, considering (5.5) and Proposition 11, we get $\varepsilon_0^{(s)}(t, x) \leqslant \psi(t, x)$ $\forall s \in \mathbb{N}_0\ \forall (t, x) \in T \times \mathbb{R}^n$. Thus, $\varepsilon_0^{(s)} \leqq \psi\ \ \forall s \in \mathbb{N}_0$. Finally, by (6.4) and (6.12),

$$\varepsilon_0^{(s)} \in \mathfrak{M}_\psi\ \ \forall s \in \mathbb{N}_0. \tag{6.16}$$

Proposition 12 *If* $(t_*, x_*) \in T \times \mathbb{R}^n$, *then* $\exists c \in [0, \infty[$:

$$\psi(t, x(t)) \leqslant c\ \ \forall v \in Q\ \forall x(\cdot) \in \mathcal{X}_\pi(t_*, x_*, v)\ \forall t \in [t_*, \vartheta_0].$$

This proposition is an immediate consequence of the uniform boundedness condition and triangle inequality.

If $g \in \mathfrak{M}_\psi$ and $(t_*, x_*) \in T \times \mathbb{R}^n$, then (since $g \leqq \psi$) we obtain by Proposition 12, that for some $c \in [0, \infty[$

$$\{g(t, x(t)) : t \in [t_*, \vartheta]\} \in \mathcal{P}'([0, c]) \ \forall v \in Q \ \forall x(\cdot) \in \mathcal{X}_\pi(t_*, x_*, v) \ \forall \vartheta \in [t_*, \vartheta_0]. \tag{6.17}$$

Following property (6.17), we have useful corollary:

$$\sup_{t \in [t_*, \vartheta]} g(t, x(t)) \leqslant c \ \forall v \in Q \ \forall x(\cdot) \in \mathcal{X}_\pi(t_*, x_*, v) \ \forall \vartheta \in [t_*, \vartheta_0]. \tag{6.18}$$

By (6.18) and Proposition 12, with $g \in \mathfrak{M}_\psi$ and $(t_*, x_*) \in T \times \mathbb{R}^n$, we obtain that $\exists \tilde{c} \in [0, \infty[$:

$$\sup(\{ \sup_{t \in [t_*, \bar{\vartheta}]} g(t, \bar{x}(t)); \psi(\bar{\vartheta}, \bar{x}(\bar{\vartheta}))\}) \in [0, \tilde{c}]$$
$$\forall v \in Q \ \forall \bar{x}(\cdot) \in \mathcal{X}_\pi(t_*, x_*, v) \ \forall \bar{\vartheta} \in [t_*, \vartheta_0].$$

Then for $g \in \mathfrak{M}_\psi$ and $(t_*, x_*) \in T \times \mathbb{R}^n$, we obtain following

$$\sup_{v \in Q} \inf_{(x(\cdot), \vartheta) \in \mathcal{X}_\pi(t_*, x_*, v) \times [t_*, \vartheta_0]} \sup(\{ \sup_{t \in [t_*, \vartheta]} g(t, x(t)); \psi(\vartheta, x(\vartheta))\}) \in [0, \infty[. \tag{6.19}$$

Considering (6.19), we define the operator $\Gamma : \mathfrak{M}_\psi \to \mathfrak{R}_+[T \times \mathbb{R}^n]$ as follows:

$$\Gamma(g)(t_*, x_*) \triangleq \sup_{v \in Q} \inf_{(x(\cdot), \vartheta) \in \mathcal{X}_\pi(t_*, x_*, v) \times [t_*, \vartheta_0]} \sup(\{ \sup_{t \in [t_*, \vartheta]} g(t, x(t));$$
$$\psi(\vartheta, x(\vartheta))\}) \ \forall g \in \mathfrak{M}_\psi \ \forall (t_*, x_*) \in T \times \mathbb{R}^n.$$

If $g \in \mathfrak{M}_\psi$, $(t_*, x_*) \in T \times \mathbb{R}^n$ and $v \in Q$, then by $\mathfrak{h}[g; t_*; x_*; v]$ we denote the functional

$$(x(\cdot), \vartheta) \longmapsto \sup(\{ \sup_{t \in [t_*, \vartheta]} g(t, x(t)); \psi(\vartheta, x(\vartheta))\}) : \mathcal{X}_\pi(t_*, x_*, v) \times [t_*, \vartheta_0] \to [0, \infty[; \tag{6.20}$$

also, we consider following Lebesgue sets: for $b \in [0, \infty[$

$$\mathcal{Y}_b[g; t_*; x_*; v] \triangleq \{(x(\cdot), \vartheta) \in \mathcal{X}_\pi(t_*, x_*, v) \times [t_*, \vartheta_0] | \ \mathfrak{h}[g; t_*; x_*; v](x(\cdot), \vartheta) \leqslant b\}.$$

Proposition 13 *Let $g \in \mathfrak{M}_\psi$, $(t_*, x_*) \in T \times \mathbb{R}^n$, $v \in Q$ and $b \in [0, \infty[$, then the set $\mathcal{Y}_b \triangleq \mathcal{Y}_b[g; t_*; x_*; v]$ is closed in $\mathcal{X}_\pi(t_*, x_*, v) \times [t_*, \vartheta_0]$ with topology, which is the product of the uniform convergence topology $\mathcal{X}_\pi(t_*, x_*, v)$ and usual $|\cdot|$-topology of $[t_*, \vartheta_0]$.*

Proof We fix g, position (t_*, x_*), control v and number b according to our definitions. Following metrizability of the topology-multiplication, we restrict ourselves to proving sequential closure property. Therefore, we select and fix following sequence $(z_k^*)_{k\in\mathbb{N}} : \mathbb{N} \to \mathcal{Y}_b$ with $z^* \in \mathcal{X}_\pi(t_*, x_*, v) \times [t_*, \vartheta_0]$, where $(z_k^*)_{k\in\mathbb{N}} \to z^*$. Let us denote $x_s^*(\cdot) \triangleq pr_1(z_s^*)$ and $\vartheta_s^* \triangleq pr_2(z_s^*)$, where $pr_1(z_s^*)$ and $pr_2(z_s^*)$ - first and second elements of ordered pair (of sequence) z_s^* respectively, $s \in \mathbb{N}$. Then

$$((x_s^*(\cdot))_{s\in\mathbb{N}} \rightrightarrows x^*(\cdot)) \;\&\; ((\vartheta_s^*)_{s\in\mathbb{N}} \to \vartheta^*), \tag{6.21}$$

where $x^*(\cdot) \triangleq pr_1(z^*)$ and $\vartheta^* \triangleq pr_2(z^*)$. For the sake of brewity, we agree to reduce our notations in a following way

$$(\mathfrak{h}(z_s^*) \triangleq \mathfrak{h}[g; t_*; x_*; v](z_s^*) \;\; \forall s \in \mathbb{N}) \;\&\; (\mathfrak{h}(z^*) \triangleq \mathfrak{h}[g; t_*; x_*; v](z^*)). \tag{6.22}$$

It's clear that $\mathfrak{h}(z_s^*) = \mathfrak{h}(x_s^*(\cdot), \vartheta_s^*)$, $s \in \mathbb{N}$, and $\mathfrak{h}(z^*) = \mathfrak{h}(x^*(\cdot), \vartheta^*)$. Then $\mathfrak{h}(z_k^*) \leqslant b \;\; \forall k \in \mathbb{N}$. Let us show that $\mathfrak{h}(z^*) \leqslant b$. By contradiction, let $\mathfrak{h}(z^*) > b$. Then (see (6.22))

$$(\sup_{t\in[t_*,\vartheta^*]} g(t, x^*(t)) > b) \vee (\psi(\vartheta^*, x^*(\vartheta^*)) > b). \tag{6.23}$$

Let us note that ψ is continuous function; thus, according to (6.21), we have following convergence

$$(\psi(\vartheta_k^*, x_k^*(\vartheta_k^*)))_{k\in\mathbb{N}} \to \psi(\vartheta^*, x^*(\vartheta^*)). \tag{6.24}$$

At the same time, with $k \in \mathbb{N}$, we obtain $\psi(\vartheta_k^*, x_k^*(\vartheta_k^*)) \leqslant \mathfrak{h}(z_k^*) \leqslant b$. Therefore (see (6.24)) $\psi(\vartheta^*, x^*(\vartheta^*)) \leqslant b$; thus, in regards to (6.23)

$$\sup_{t\in[t_*,\vartheta^*]} g(t, x^*(t)) > b. \tag{6.25}$$

However, by selection of z_s^*, $s \in \mathbb{N}$, we obtain this property:

$$\sup_{t\in[t_*,\vartheta_k^*]} g(t, x_k^*(t)) \leqslant \mathfrak{h}(z_k^*) \leqslant b \;\; \forall k \in \mathbb{N}. \tag{6.26}$$

From (6.25) we have for some $t^* \in [t_*, \vartheta^*]$

$$b < g(t^*, x^*(t^*)). \tag{6.27}$$

Let $t_k^* \triangleq \inf(\{\vartheta_k^*; t^*\}) \;\; \forall k \in \mathbb{N}$. Considering (6.21), we can easily show that $(t_k^*)_{k\in\mathbb{N}} \to t^*$. As a corollary, we obtain the convergence

$$(\rho((t_k^*, x_k^*(t_k^*)), (t^*, x^*(t^*))))_{k\in\mathbb{N}} \to 0. \tag{6.28}$$

Thus, according to (6.26), we have the following $g(t_k^*, x_k^*(t_k^*)) \leqslant b \ \forall k \in \mathbb{N}$. Hence, $(t_k^*, x_k^*(t_k^*)) \in g^{-1}([0, b]) \ \forall k \in \mathbb{N}$. Since $g \in \mathfrak{M}_\psi$, we obtain $g^{-1}([0, b]) \in \mathcal{F}$; then by (6.28), we have inclusion $(t^*, x^*(t^*)) \in g^{-1}([0, b])$. So, we shown that $g(t^*, x^*(t^*)) \leqslant b$, which contradicts (6.27).

Therefore, $\mathfrak{h}(z^*) \leqslant b$. Taking into account that $(z_k^*)_{k \in \mathbb{N}}$ and selection of z^* was arbitrary, we have obtained that $\mathcal{Y}_b$ is closed in considered topology of $\mathcal{X}_\pi(t_*, x_*, v) \times [t_*, \vartheta_0]$. □

Corollary 1 *If* $g \in \mathfrak{M}_\psi$, $(t_*, x_*) \in T \times \mathbb{R}^n$ *and* $v \in Q$, *then*

$$\exists\, (\bar{x}(\cdot), \bar{\vartheta}) \in \mathcal{X}_\pi(t_*, x_*, v) \times [t_*, \vartheta_0]:$$
$$\mathfrak{h}[g; t_*; x_*; v](\bar{x}(\cdot), \bar{\vartheta}) = \inf_{(x(\cdot),\vartheta) \in \mathcal{X}_\pi(t_*, x_*, v) \times [t_*, \vartheta_0]} \mathfrak{h}[g; t_*; x_*; v](x(\cdot), \vartheta).$$

Proof We fix $g \in \mathfrak{M}_\psi$, $(t_*, x_*) \in T \times \mathbb{R}^n$ and $v \in Q$. For the sake of brewity, we assume that $\mathfrak{h} \triangleq \mathfrak{h}[g; t_*; x_*; v]$. Then

$$\mathfrak{a} \triangleq \inf_{(x(\cdot),\vartheta) \in \mathcal{X}_\pi(t_*, x_*, v) \times [t_*, \vartheta_0]} \mathfrak{h}(x(\cdot), \vartheta) \in [0, \infty[. \tag{6.29}$$

Let $\mathcal{Y}_b \triangleq \mathfrak{h}^{-1}([0, b]) \ \forall b \in [0, \infty[$. Also, by Proposition 13, we obtain that each of sets $\mathcal{Y}_b$, $b \in [0, \infty[$, is closed in metrizable compactum $\mathcal{X}_\pi(t_*, x_*, v) \times [t_*, \vartheta_0]$ with topology, which is a product of metrizable topologies $\mathcal{X}_\pi(t_*, x_*, v)$ and $[t_*, \vartheta_0]$. With $b_1 \in [0, \infty[$, $b_2 \in [0, \infty[$, we have $(b_1 \leqslant b_2) \Rightarrow (\mathcal{Y}_{b_1} \subset \mathcal{Y}_{b_2})$. Therefore $\{\mathcal{Y}_b : b \in]\mathfrak{a}, \infty[\}$ is centered family of closed subsets of compactum. As a corollary, by definition of $\mathfrak{a}$, we obtain:

$$\bigcap_{a \in]\mathfrak{a}, \infty[} \mathcal{Y}_a \neq \emptyset. \tag{6.30}$$

Let $(x^*(\cdot), \vartheta^*) \in \bigcap_{a \in]\mathfrak{a}, \infty[} \mathcal{Y}_a$. Then $(x^*(\cdot), \vartheta^*) \in \mathcal{X}_\pi(t_*, x_*, v) \times [t_*, \vartheta_0]$ and $\mathfrak{h}(x^*(\cdot), \vartheta^*) \leqslant a \ \forall a \in]\mathfrak{a}, \infty[$. Hence, $\mathfrak{h}(x^*(\cdot), \vartheta^*) \leqslant \mathfrak{a}$, which conclude the proof, since according to (6.29) $\mathfrak{h}(x^*(\cdot), \vartheta^*) \geqslant \mathfrak{a}$. □

By definition of operator Γ and Corollary 1, we obtain that

$$\Gamma(g)(t_*, x_*) = \sup_{v \in Q} \min_{(x(\cdot),\vartheta) \in \mathcal{X}_\pi(t_*, x_*, v) \times [t_*, \vartheta_0]} \mathfrak{h}[g; t_*; x_*; v](x(\cdot), \vartheta)$$
$$\forall g \in \mathfrak{M}_\psi \ \forall (t_*, x_*) \in T \times \mathbb{R}^n. \tag{6.31}$$

Theorem 1 *If* $k \in \mathbb{N}_0$, *then* $\varepsilon_0^{(k+1)} = \Gamma(\varepsilon_0^{(k)})$.

Proof We recall that $\varepsilon_0^{(k)} \in \mathfrak{M}_\psi$, $\varepsilon_0^{(k+1)} \in \mathfrak{M}_\psi$, and also $\Gamma(\varepsilon_0^{(k)}) : T \times \mathbb{R}^n \to [0, \infty[$. We consider position $(t_*, x_*) \in T \times \mathbb{R}^n$. Let us compare $(\varepsilon_0^{(k+1)}(t_*, x_*) \in [0, \infty[)$ & $(\Gamma(\varepsilon_0^{(k)})(t_*, x_*) \in [0, \infty[)$. Here, according to (6.31), we have

$$\Gamma(\varepsilon_0^{(k)})(t_*, x_*) = \sup_{v \in Q} \min_{(x(\cdot),\vartheta) \in \mathcal{X}_\pi(t_*, x_*, v) \times [t_*, \vartheta_0]} \mathfrak{h}[\varepsilon_0^{(k)}; t_*; x_*; v](x(\cdot), \vartheta). \tag{6.32}$$

Further in this proof, we assume that

$$(a_* \triangleq \varepsilon_0^{(k+1)}(t_*, x_*)) \ \& \ (b_* \triangleq \Gamma(\varepsilon_0^{(k)})(t_*, x_*)); \tag{6.33}$$

thus, we obtain two non-negative numbers, where $a_* \in \Sigma_0^{(k+1)}(t_*, x_*)$. Therefore by (5.1)

$$(t_*, x_*) \in \mathcal{W}_{k+1}(S_0(\mathbf{M}, a_*), S_0(\mathbf{N}, a_*)). \tag{6.34}$$

Then $(t_*, x_*) \in \mathbb{A}[S_0(\mathbf{M}, a_*)](\mathcal{W}_k(S_0(\mathbf{M}, a_*), S_0(\mathbf{N}, a_*)))$.

It means that $(t_*, x_*) \in \mathcal{W}_k(S_0(\mathbf{M}, a_*), S_0(\mathbf{N}, a_*))$ and $\forall v \in Q \ \exists (x(\cdot), \vartheta) \in \mathcal{X}_\pi(t_*, x_*, v) \times [t_*, \vartheta_0]$:

$$((\vartheta, x(\vartheta)) \in S_0(\mathbf{M}, a_*)) \ \& \ ((t, x(t)) \in \mathcal{W}_k(S_0(\mathbf{M}, a_*), S_0(\mathbf{N}, a_*)) \ \ \forall t \in [t_*, \vartheta]). \tag{6.35}$$

Let $\bar{v} \in Q$. Moreover, let $\bar{x}(\cdot) \in \mathcal{X}_\pi(t_*, x_*, \bar{v})$ with $\bar{\vartheta} \in [t_*, \vartheta_0]$ such that

$$((\bar{\vartheta}, \bar{x}(\bar{\vartheta})) \in S_0(\mathbf{M}, a_*)) \ \& \ ((t, \bar{x}(t)) \in \mathcal{W}_k(S_0(\mathbf{M}, a_*), S_0(\mathbf{N}, a_*)) \ \ \forall t \in [t_*, \bar{\vartheta}]). \tag{6.36}$$

Then (see (4.1), (5.1), (5.3), (6.36)) we obtain following properties

$$(\psi(\bar{\vartheta}, \bar{x}(\bar{\vartheta})) \leqslant a_*) \ \& \ (\varepsilon_0^{(k)}(t, \bar{x}(t)) \leqslant a_* \ \ \forall t \in [t_*, \bar{\vartheta}]). \tag{6.37}$$

In regards to (6.20) and (6.37), we obtain that $\mathfrak{h}[\varepsilon_0^{(k)}; t_*; x_*; \bar{v}](\bar{x}(\cdot), \bar{\vartheta}) \leqslant a_*$. Moreover, we have inequality

$$\min_{(x(\cdot), \vartheta) \in \mathcal{X}_\pi(t_*, x_*, \bar{v}) \times [t_*, \vartheta_0]} \mathfrak{h}[\varepsilon_0^{(k)}; t_*; x_*; \bar{v}](x(\cdot), \vartheta) \leqslant a_*. \tag{6.38}$$

Since selection of $\bar{v}$ was arbitrary, we have shown (see (6.32), (6.38)) that

$$b_* = \Gamma(\varepsilon_0^{(k)})(t_*, x_*) \leqslant a_*. \tag{6.39}$$

Let us verify the validity of the opposite inequality. Consider arbitrary control $\hat{v} \in Q$. In regards to (6.32), (6.33), and also Corollary 1, we obtain, for some $\hat{x}(\cdot) \in \mathcal{X}_\pi(t_*, x_*, \hat{v})$ and $\hat{\vartheta} \in [t_*, \vartheta_0]$, following inequality $\mathfrak{h}[\varepsilon_0^{(k)}; t_*; x_*; \hat{v}](\hat{x}(\cdot), \hat{\vartheta}) \leqslant b_*$. Thus, $\psi(\hat{\vartheta}, \hat{x}(\hat{\vartheta})) \leqslant b_*$ and $\varepsilon_0^{(k)}(t, \hat{x}(t)) \leqslant b_* \ \ \forall t \in [t_*, \hat{\vartheta}]$. Then according to (6.1)

$$(\hat{\vartheta}, \hat{x}(\hat{\vartheta})) \in S_0(\mathbf{M}, b_*). \tag{6.40}$$

On the other hand, according to (5.1) and Proposition 8, we obtain for $t \in [t_*, \hat{\vartheta}]$, that $b_* \in \Sigma_0^{(k)}(t, \hat{x}(t))$; thus $(t, \hat{x}(t)) \in \mathcal{W}_k(S_0(\mathbf{M}, b_*), S_0(\mathbf{N}, b_*))$. Since selection of $\hat{v} \in Q$ was arbitrary, we have shown that

$$\forall v \in Q\ \exists (x(\cdot), \vartheta) \in \mathcal{X}_\pi(t_*, x_*, v) \times [t_*, \vartheta_0] : ((\vartheta, x(\vartheta)) \in S_0(\mathbf{M}, b_*))\ \& \\ ((t, x(t)) \in \mathcal{W}_k(S_0(\mathbf{M}, b_*), S_0(\mathbf{N}, b_*))\ \forall t \in [t_*, \vartheta]). \quad (6.41)$$

Since $(t_*, x_*) \in \mathcal{W}_k(S_0(\mathbf{M}, b_*), S_0(\mathbf{N}, b_*))$, following inclusion (see (4.6), (4.29)) holds

$$(t_*, x_*) \in \mathcal{W}_{k+1}(S_0(\mathbf{M}, b_*), S_0(\mathbf{N}, b_*)). \quad (6.42)$$

Thus, according to (5.1) and (6.42), we obtain that $b_* \in \Sigma_0^{(k+1)}(t_*, x_*)$; therefore we have shown that $a_* \leqslant b_*$. Hence (see (6.39)), $a_* = b_*$. In regards to definitions of a_* and b_*, we have the following $\varepsilon_0^{(k+1)}(t_*, x_*) = \Gamma(\varepsilon_0^{(k)})(t_*, x_*)$. Since selection of position (t_*, x_*) was arbitrary, then $\varepsilon_0^{(k+1)} = \Gamma(\varepsilon_0^{(k)})$. Thus, Theorem 1 is fully proven. □

7 Fixed Point Property of Operator Γ

In this section we consider an important property of operator Γ, namely its fixed point. But first let us denote some additional propositions.

Proposition 14 *Following condition holds* $\varepsilon_0^{(0)} = \rho(\cdot, \mathbf{N})$.

Proof Let us note that

$$\mathcal{W}_0(S_0(\mathbf{M}, \varepsilon), S_0(\mathbf{N}, \varepsilon)) = S_0(\mathbf{N}, \varepsilon)\ \forall \varepsilon \in [0, \infty[. \quad (7.1)$$

Next, according to (5.1) and (7.1), we obtain the following

$$\Sigma_0^{(0)}(t, x) = \{\varepsilon \in [0, \infty[\ |\ (t, x) \in S_0(\mathbf{N}, \varepsilon)\}\ \forall (t, x) \in T \times \mathbb{R}^n. \quad (7.2)$$

Wherein, in regards to (5.3), we have the following

$$\varepsilon_0^{(0)}(t, x) = \inf(\Sigma_0^{(0)}(t, x))\ \forall (t, x) \in T \times \mathbb{R}^n; \quad (7.3)$$

$$\varepsilon_0^{(0)}(t, x) \in \Sigma_0^{(0)}(t, x)\ \forall (t, x) \in T \times \mathbb{R}^n. \quad (7.4)$$

Next, according to Proposition 8, we have

$$\Sigma_0^{(0)}(t, x) = [\varepsilon_0^{(0)}(t, x), \infty[\ \forall (t, x) \in T \times \mathbb{R}^n. \quad (7.5)$$

Let us note that

$$S_0(\mathbf{N}, \varepsilon) \triangleq \{(t, x) \in T \times \mathbb{R}^n\ |\ \rho((t, x), \mathbf{N}) \leqslant \varepsilon\}\ \forall \varepsilon \in [0, \infty[. \quad (7.6)$$

Wherein $\rho(\cdot\,, \mathbf{N}) = (\rho((t, x), \mathbf{N}))_{(t,x)\in T\times\mathbb{R}^n} \in \mathfrak{R}_+[T \times \mathbb{R}^n]$. and $\varepsilon_0^{(0)} \in \mathfrak{R}_+[T \times \mathbb{R}^n]$. Let $(t_*, x_*) \in T \times \mathbb{R}^n$. Now, we compare $\rho((t_*, x_*), \mathbf{N})$ and $\varepsilon_0^{(0)}(t_*, x_*)$. By (7.4), we have that $\varepsilon_* \triangleq \varepsilon_0^{(0)}(t_*, x_*) \in \Sigma_0^{(0)}(t_*, x_*)$; thus, in regards to (7.2) we obtain

$$(t_*, x_*) \in S_0(\mathbf{N}, \varepsilon_*); \tag{7.7}$$

therefore, according to (7.6) and (7.7), we have the following

$$\rho((t_*, x_*), \mathbf{N}) \leqslant \varepsilon_*. \tag{7.8}$$

On the other hand, $\rho((t_*, x_*), \mathbf{N}) \in [0, \infty[$ and also

$$S_0(\mathbf{N}, \rho((t_*, x_*), \mathbf{N})) = \{(t, x) \in T \times \mathbb{R}^n \mid \rho((t, x), \mathbf{N}) \leqslant \rho((t_*, x_*), \mathbf{N})\},$$

which implies that $(t_*, x_*) \in S_0(\mathbf{N}, \rho((t_*, x_*), \mathbf{N}))$; thus, according to (7.2) and (7.5)

$$\rho((t_*, x_*), \mathbf{N}) \in \Sigma_0^{(0)}(t_*, x_*). \tag{7.9}$$

However, in regards to (7.5), $\Sigma_0^{(0)}(t_*, x_*) = [\varepsilon_0^{(0)}(t_*, x_*), \infty[$; then, according to (7.9) we obtain

$$\varepsilon_* = \varepsilon_0^{(0)}(t_*, x_*) \leqslant \rho((t_*, x_*), \mathbf{N}). \tag{7.10}$$

Therefore, following (7.8), (7.10) and the fact, that selection of (t_*, x_*) was arbitrary, we have $\varepsilon_0^{(0)} = \rho(\cdot, \mathbf{N})$, which concludes our proof. □

Proposition 15 *If $g \in \mathfrak{M}_\psi$, then $g \leqq \Gamma(g)$.*

Proof We fix $g \in \mathfrak{M}_\psi$ and position $(t_*, x_*) \in T \times \mathbb{R}^n$. Consider with $v \in Q$ function $\mathfrak{h}_*^{(v)} \triangleq \mathfrak{h}[g; t_*; x_*; v]$. We define it as follows:

$$\mathfrak{h}_*^{(v)}(x(\cdot), \vartheta) = \sup(\{\sup_{t\in[t_*,\vartheta]} g(t, x(t)); \psi(\vartheta, x(\vartheta))\})$$
$$\forall x(\cdot) \in \mathcal{X}_\pi(t_*, x_*, v)\ \forall \vartheta \in [t_*, \vartheta_0]. \tag{7.11}$$

Let us note, that following equality holds

$$\Gamma(g)(t_*, x_*) = \sup_{v\in Q}\ \min_{(x(\cdot),\vartheta)\in\mathcal{X}_\pi(t_*,x_*,v)\times[t_*,\vartheta_0]} \mathfrak{h}_*^{(v)}(x(\cdot), \vartheta). \tag{7.12}$$

Also, let us note that $Q \neq \emptyset$. Next, we select and fix control $v_0 \in Q$; thus, we obtain, according to (7.11)

$$\mathfrak{h}_*^{(v_0)}(x(\cdot), \vartheta) = \sup(\{\sup_{t\in[t_*,\vartheta]} g(t, x(t)); \psi(\vartheta, x(\vartheta))\})$$
$$\forall x(\cdot) \in \mathcal{X}_\pi(t_*, x_*, v_0)\ \forall \vartheta \in [t_*, \vartheta_0]. \tag{7.13}$$

We note that according to (7.12), following holds

$$\min_{(x(\cdot),\vartheta)\in\mathcal{X}_\pi(t_*,x_*,v_0)\times[t_*,\vartheta_0]} \mathfrak{h}_*^{(v_0)}(x(\cdot),\vartheta) \leqslant \Gamma(g)(t_*,x_*). \qquad (7.14)$$

Let us take a closer look on left-hand side of (7.14). Let $\bar{x}(\cdot) \in \mathcal{X}_\pi(t_*, x_*, v_0)$ and $\bar{\vartheta} \in [t_*, \vartheta_0]$; then

$$\mathfrak{h}_*^{(v_0)}(\bar{x}(\cdot),\bar{\vartheta}) = \sup(\{\sup_{t\in[t_*,\bar{\vartheta}]} g(t,\bar{x}(t)); \psi(\bar{\vartheta},\bar{x}(\bar{\vartheta}))\})$$
$$\geqslant \sup(\{g(t_*,\bar{x}(t_*)); \psi(\bar{\vartheta},\bar{x}(\bar{\vartheta}))\}) \geqslant g(t_*,\bar{x}(t_*)),$$

where $\bar{x}(t_*) = x_*$. We obtain the following $\mathfrak{h}_*^{(v_0)}(\bar{x}(\cdot),\bar{\vartheta}) \geqslant g(t_*, x_*)$. Since selection of trajectory $\bar{x}(\cdot)$ and moment of time $\bar{\vartheta}$ was arbitrary, we have

$$\mathfrak{h}_*^{(v_0)}(x(\cdot),\vartheta) \geqslant g(t_*,x_*) \ \forall (x(\cdot),\vartheta) \in \mathcal{X}_\pi(t_*,x_*,v_0)\times[t_*,\vartheta_0]. \qquad (7.15)$$

Hence,

$$\min_{(x(\cdot),\vartheta)\in\mathcal{X}_\pi(t_*,x_*,v_0)\times[t_*,\vartheta_0]} \mathfrak{h}_*^{(v_0)}(x(\cdot),\vartheta) \geqslant g(t_*,x_*),$$

wherein, considering (7.14), we obtain $g(t_*, x_*) \leqslant \Gamma(g)(t_*, x_*)$. Finally, since selection of (t_*, x_*) was arbitrary, we have shown that $g \leqq \Gamma(g)$. □

Before we formulate the main theorem of the fixed point of the operator Γ, let us note one useful proposition, namely isotonic property of Γ.

Proposition 16 *If $g_1 \in \mathfrak{M}_\psi$ and $g_2 \in \mathfrak{M}_\psi$, then following implication holds $(g_1 \leqq g_2) \Rightarrow (\Gamma(g_1) \leqq \Gamma(g_2))$.*

Proof of this proposition can be easily obtained from given definitions (see (6.20) and (6.31)).

Theorem 2 *Function ε_0 is the fixed point of operator Γ: $\varepsilon_0 = \Gamma(\varepsilon_0)$.*

Proof According to Propositions 15 and (6.15), we obtain that $\varepsilon_0 \leqq \Gamma(\varepsilon_0)$. We consider arbitrary position $(t_*, x_*) \in T \times \mathbb{R}^n$. Let $a_* \triangleq \varepsilon_0(t_*, x_*)$, $b_* \triangleq \Gamma(\varepsilon_0)(t_*, x_*)$, $\mathfrak{h}_*^{(v)} \triangleq \mathfrak{h}_*^{(v)}[\varepsilon_0; t_*; x_*; v]\ \forall v \in Q$. Then

$$b_* = \sup_{v\in Q} \min_{(x(\cdot),\vartheta)\in\mathcal{X}_\pi(t_*,x_*,v)\times[t_*,\vartheta_0]} \mathfrak{h}_*^{(v)}(x(\cdot),\vartheta). \qquad (7.16)$$

On the other hand, according to Proposition 1, $a_* \in \Sigma_0(t_*, x_*)$; thus, considering (4.36), we get $(t_*, x_*) \in \mathcal{W}(S_0(\mathbf{M}, a_*), S_0(\mathbf{N}, a_*))$. However, in regards to (see [26, Proposition 5.2]), we obtain $(t_*, x_*) \in \mathbb{A}[S_0(\mathbf{M}, a_*)](\mathcal{W}(S_0(\mathbf{M}, a_*),\ S_0(\mathbf{N}, a_*)))$. Hence,

$$\forall v \in Q\ \exists (x(\cdot), \vartheta) \in \mathcal{X}_\pi(t_*, x_*, v) \times [t_*, \vartheta_0] : ((\vartheta, x(\vartheta)) \in S_0(\mathbf{M}, a_*))\ \&\ ((t, x(t)) \in \mathcal{W}(S_0(\mathbf{M}, a_*), S_0(\mathbf{N}, a_*))\ \ \forall t \in [t_*, \vartheta]).$$

Thus, by (2.2), (2.4), (6.1), (4.36) and (4.37) we obtain that

$$\forall v \in Q\ \exists (x(\cdot), \vartheta) \in \mathcal{X}_\pi(t_*, x_*, v) \times [t_*, \vartheta_0] : (\psi(\vartheta, x(\vartheta)) \leqslant a_*)\ \&\ (\varepsilon_0(t, x(t)) \leqslant a_*\ \forall t \in [t_*, \vartheta]).$$

In other words, $\forall v \in Q\ \exists (x(\cdot), \vartheta) \in \mathcal{X}_\pi(t_*, x_*, v) \times [t_*, \vartheta_0] : \mathfrak{h}_*^{(v)}(x(\cdot), \vartheta) \leqslant a_*$. As a corollary, $\forall v \in Q$

$$\min_{(x(\cdot),\vartheta) \in \mathcal{X}_\pi(t_*, x_*, v) \times [t_*, \vartheta_0]} \mathfrak{h}_*^{(v)}(x(\cdot), \vartheta) \leqslant a_*.$$

But in this case (see (6.31)) $\Gamma(\varepsilon_0)(t_*, x_*) \leqslant a_*$, i.e. $b_* \leqslant a_*$; however, as shown before, $a_* \leqslant b_*$. Hence, $a_* = b_*$. Since selection of our position was arbitrary, we obtain $\varepsilon_0 = \Gamma(\varepsilon_0)$. Theorem 2 is proved. □

Let us consider the set of all fixed points of operator Γ, namely $\mathfrak{M}_\psi^{(\Gamma)} \triangleq \{g \in \mathfrak{M}_\psi \mid g = \Gamma(g)\}$. In regards to (6.15) and Theorem 2, we obtain that $\varepsilon_0 \in \mathfrak{M}_\psi^{(\Gamma)}$. Let $\tilde{\mathfrak{M}}_\psi^{(\Gamma)} \triangleq \{g \in \mathfrak{M}_\psi^{(\Gamma)} \mid \varepsilon_0^{(0)} \leqq g\} = \{g \in \mathfrak{M}_\psi^{(\Gamma)} \mid \rho(\cdot, \mathbf{N}) \leqq g\}$. According to Proposition 6, we obtain that $\varepsilon_0^{(0)} \leqq \varepsilon_0$. Then, following our definitions, $\varepsilon_0 \in \tilde{\mathfrak{M}}_\psi^{(\Gamma)}$.

Theorem 3 *Function ε_0 is the smallest element of $\tilde{\mathfrak{M}}_\psi^{(\Gamma)}$, in other words, $\varepsilon_0 \in \tilde{\mathfrak{M}}_\psi^{(\Gamma)}$ and $\varepsilon_0 \leqq \mathbf{g}\ \forall \mathbf{g} \in \tilde{\mathfrak{M}}_\psi^{(\Gamma)}$.*

Proof Let us select element $\mathbf{g} \in \tilde{\mathfrak{M}}_\psi^{(\Gamma)}$. By construction, $\mathbf{g} \in \mathfrak{M}_\psi^{(\Gamma)}$ and $\varepsilon_0^{(0)} \leqq \mathbf{g}$. Hence, $r = 0$ belongs to $\mathfrak{N} \triangleq \{s \in \mathbb{N}_0 \mid \varepsilon_0^{(s)} \leqq \mathbf{g}\}$. Next, let us assume that $r \in \mathfrak{N}$. By induction, we show that $r + 1$ also belongs to $\mathfrak{N}$. According to (6.16), $\varepsilon_0^{(r)} \in \mathfrak{M}_\psi$, wherein $\varepsilon_0^{(r)} \leqq \mathbf{g}$. In regards to Theorem 1, we obtain $(\Gamma(\varepsilon_0^{(r)}) \leqq \mathbf{g}) \Rightarrow (\varepsilon_0^{(r+1)} \leqq \mathbf{g})$. Indeed, $\Gamma(\mathbf{g}) = \mathbf{g}$, by selection of $\mathbf{g}$ and Proposition 16, $\Gamma(\varepsilon_0^{(r)}) \leqq \mathbf{g}$; thus, $\varepsilon_0^{(r+1)} = \Gamma(\varepsilon_0^{(r)})$. Then $\varepsilon_0^{(r+1)} \leqq \mathbf{g}$ and $r + 1 \in \mathfrak{N}$. So, $k + 1 \in \mathfrak{N}\ \forall k \in \mathfrak{N}$. Thus, $\mathfrak{N} = \mathbb{N}_0$, i.e. $\varepsilon_0^{(s)} \leqq \mathbf{g}\ \forall s \in \mathbb{N}_0$. Finally, according to Proposition 6, we obtain $\varepsilon_0(t, x) \leqslant \mathbf{g}(t, x)\ \ \forall (t, x) \in T \times \mathbb{R}^n$ or $\varepsilon_0 \leqq \mathbf{g}$. Since selection of $\mathbf{g}$ was arbitrary, theorem is fully proved. □

Let us consider one special case of our setting. Namely, let

$$(\mathbf{M} \triangleq T \times \mathcal{M})\ \&\ (\mathbf{N} \triangleq T \times \mathcal{N}), \tag{7.17}$$

where $\mathcal{M}$ and $\mathcal{N}$ are closed non-empty sets in $\mathbb{R}^n$ with usual topology generated by euclidean norm $\|\cdot\|$. Moreover, let $\mathcal{M} \subset \mathcal{N}$. If $H \in \mathcal{P}'(\mathbb{R}^n)$ and $x \in \mathbb{R}^n$, then we introduce

$$(\|\cdot\| - \inf)[x; H] \triangleq \inf(\{\|x - h\| : h \in H\}) \in [0, \infty[. \tag{7.18}$$

We obtain the following property: for $H \in \mathcal{P}'(\mathbb{R}^n)$ and $x_* \in \mathbb{R}^n$,

$$\rho((t, x_*), T \times H) = (\|\cdot\| - \inf)[x_*; H] \ \forall t \in T. \tag{7.19}$$

Indeed, we fix $t_* \in T$. Then,

$$\rho((t_*, x_*), T \times H) = \inf(\{\rho((t_*, x_*), (t, h)) : (t, h) \in T \times H\}).$$

In addition, for $h \in H$, we obtain that $(t, h) \in T \times H$, where $t \in T$, and

$$\|x_* - h\| \leqslant \rho((t_*, x_*), (t, h)). \tag{7.20}$$

Also, $(t_*, h) \in T \times H$ and $\rho((t_*, x_*), (t_*, h)) = \|x_* - h\|$. Wherein, $\rho((t_*, x_*), T \times H) \leqslant \rho((t_*, x_*), (t_*, h))$. As a corollary,

$$\rho((t_*, x_*), T \times H) \leqslant \|x_* - h\|. \tag{7.21}$$

Since the selection of h was arbitrary, by (7.18)

$$\rho((t_*, x_*), T \times H) \leqslant (\|\cdot\| - \inf)[x_*; H]. \tag{7.22}$$

On the other hand, from (7.20), we have that

$$(\|\cdot\| - \inf)[x_*; H] \leqslant \rho((t_*, x_*), (t, \tilde{h})) \ \forall t \in T \ \forall \tilde{h} \in H.$$

Then $(\|\cdot\| - \inf)[x_*; H] \leqslant \rho((t_*, x_*), T \times H)$, and by (7.22), we obtain $\rho((t_*, x_*), T \times H) = (\|\cdot\| - \inf)[x_*; H]$. Since the selection of t_* was arbitrary, the property (7.19) is established. So, by (7.17) and (7.19), $\forall (t, x) \in T \times \mathbb{R}^n$

$$(\rho((t, x), \mathbf{M}) = (\|\cdot\| - \inf)[x; \mathcal{M}]) \ \& \ (\rho((t, x), \mathbf{N}) = (\|\cdot\| - \inf)[x; \mathcal{N}]).$$

Therefore, we obtain that

$$\omega(t, x(\cdot), \theta) = \sup(\{(\|\cdot\| - \inf)[x(\theta); \mathcal{M}]; \max_{t \leqslant \xi \leqslant \theta} (\|\cdot\| - \inf)[x(\xi); \mathcal{N}]\}) \\ \forall t \in T \ \forall x(\cdot) \in C_n([t, \vartheta_0]) \ \forall \theta \in [t, \vartheta_0].$$

As a corollary, in considered case (7.17), we obtain

$$\gamma_t(x(\cdot)) = \min_{\theta \in [t, \vartheta_0]} \sup(\{(\|\cdot\| - \inf)[x(\theta); \mathcal{M}]; \max_{t \leqslant \xi \leqslant \theta} (\|\cdot\| - \inf)[x(\xi); \mathcal{N}]\}) \\ \forall t \in T \ \forall x(\cdot) \in C_n([t, \vartheta_0]).$$

Thus, we have natural payoff.

8 Conclusion

In this paper the relaxation of the non-linear zero-sum pursuit-evasion differential game was considered. The optimal strategy of player I was constructed. Namely, for special payoff, defining the quality of pursuit for player I, the minimax was found using special iterative procedure. It was shown that minimax function is the fixed point of unique operator, which implements the procedure.

The problems of finding new ways to relax this setting of pursuit-evasion differential game as well as construction of optimal strategy for the player II are still open and will be considered in future work.

References

1. Isaacs, R.: Differential Games. Wiley, New York (1965)
2. Pontryagin, L.S.: Linear differential games of pursuit. Mat. Sb. (N.S.) **112**(154):**3**(7), 307–330 (1980)
3. Krasovskii, N.N.: Igrovye zadachi o vstreche dvizhenii. [Game problems on the encounter of motions]. Moscow, Nauka Publ., 1970, 420 p. (in Russian)
4. Pshenichnii, B.N.: The structure of differential games. Sov. Math. Dokl. **10**, 70–72
5. Kuntsevich, V.M., Kuntsevich, A.V.: Optimal Control over the approach of conflicting moving objects under uncertainty. Cybern. Syst. Anal. **38**, 230–237 (2002). https://doi.org/10.1023/A:1016395429276
6. Kuntsevich, V.M.: Control Under Uncertainty: Guaranteed Results in Control and Identification Problems. Naukova Dumka, Kyiv (2006). (in Russian)
7. Kuntsevich, V.M., Gubarev, V.F., Kondratenko, Y.P., Lebedev, D.V., Lysenko, V.P. (eds.): Control Systems: Theory and Applications. Series in Automation, Control and Robotics. River Publishers (2018)
8. Krasovskii, N.N., Subbotin, A.I.: An alternative for the game problem of convergence. J. Appl. Math. Mech. **34**(6), 948–965 (1970). https://doi.org/10.1016/0021-8928(70)90158-9
9. Krasovsky, N.N., Subbotin, A.I.: Game-Theoretical Control Problems. Springer, New York (1988)
10. Kryazhimskii, A.V.: On the theory of positional differential games of approach-evasion. Sov. Math. Dokl. **19**(2), 408–412 (1978)
11. Roxin, E.: Axiomatic approach in differential games. J. Optim. Theory Appl. **3**, 153 (1969). https://doi.org/10.1007/BF00929440
12. Elliott, R.J., Kalton, N.J.: Values in differential games. Bull. Am. Math. Soc. **78**(3), 427–431 (1972)
13. Chentsov, A.G.: On a game problem of guidance with information memory. Sov. Math. Dokl. **17**, 411–414 (1976)
14. Chentsov, A.G.: On a game problem of converging at a given instant of time. Math. USSR-Sb. **28**(3), 353–376 (1976). https://doi.org/10.1070/SM1976v028n03ABEH001657
15. Subbotin, A.I., Chentsov, A.G.: Optimizatsiya garantii v zadachakh upravleniya [Optimization of guarantee in control problems]. Moscow, Nauka Publ., 1981, 288 p. (in Russian)
16. Ushakov, V.N., Ershov, A.A.: On the solution of control problems with fixed terminal time. Vestn. Udmurt. Univ. Mat. Mekh. Kompyut. Nauki **26**(4), 543–564 (2016) (in Russian). https://doi.org/10.20537/vm160409
17. Ushakov, V.N., Matviychuk A.R.: Towards solution of control problems of nonlinear systems on a finite time interval. Izv. IMI UdGU (46) no. 2, 202–215 (2015). (in Russian)

18. Ushakov, V.N., Ukhobotov, V.I., Ushakov, A.V., Parshikov, G.V.: On solving approach problems for control systems. Proc. Steklov Inst. Math. **291**(1), 263–278 (2015). https://doi.org/10.1134/S0081543815080210
19. Chikrii, A.A.: Conflict-Controlled Processes, p. 424. Springer Science and Busines Media, Boston, London, Dordrecht (2013)
20. Krasovskii, N.N.: A differential game of approach and evasion. I. Cybernetics **11**(2), 189–203 (1973)
21. Krasovskii, N.N.: A differential game of approach and evasion. II. Cybernetics **11**(3), 376–394 (1973)
22. Chentsov, A.G.: The structure of a certain game-theoretic approach problem. Sov. Math. Dokl. **16**(5), 1404–1408 (1975)
23. Chistyakov, S.V.: On solutions for game problems of pursuit. Prikl. Mat. Mekh. **41**(5), 825–832 (1977). (in Russian)
24. Ukhobotov, V.I.: Construction of a stable bridge for a class of linear games. J. Appl. Math. Mech. **41**(2), 350–354 (1977). https://doi.org/10.1016/0021-8928(77)90021-1
25. Chentsov, A.G.: Metod programmnykh iteratsii dlya differentsialnoi igry sblizheniya-ukloneniya [The method of program iterations for a differential approach-evasion game]. Sverdlovsk, 1979. Available from VINITI, no. 1933–79, 102 p
26. Chentsov, A.G.: Stability iterations and an evasion problem with a constraint on the number of switchings. Trudy Inst. Mat. i Mekh. UrO RAN **23**(2), 285–302 (2017) (in Russian). https://doi.org/10.21538/0134-4889-2017-23-2-285-302
27. Chentsov, A.G.: O zadache upravleniya s ogranichennym chislom pereklyuchenii [On a control problem with a bounded number of switchings]. Sverdlovsk, 1987. Available from VINITI, no. 4942-B87, 44 p (in Russian)
28. Chentsov, A.G.: O differentsialnykh igrakh s ogranicheniem na chislo korrektsii, 2 [On differential games with restriction on the number of corrections, 2]. Sverdlovsk, 1980. Available from VINITI, no. 5406-80, 55 p. (in Russian)
29. Chentsov, A.G., Khachay, D.M.: Relaxation of the pursuit-evasion differential game and iterative methods. Trudy Inst. Mat. i Mekh. UrO RAN **24**(4), 246–269 (in Russian). https://doi.org/10.21538/0134-4889-2018-24-4-246-269
30. Neveu, J.: Mathematical foundations of the calculus of probability. San Francisco: Holden-Day, 1965, 223 p. Translated to Russian under the title Matematicheskie osnovy teorii veroyatnostei. Moscow: Mir Publ., 1969, 309 p
31. Billingsley, P.: Convergence of probability measures. N-Y: Wiley, 1968, 253 p. ISBN: 0471072427. Translated to Russian under the title Skhodimost veroyatnostnykh mer. Moscow: Nauka Publ., 1977, 352 p
32. Dieudonne, J.: Foundations of modern analysis. N Y: Acad. Press Inc., 1960, 361 p. Translated to Russian under the title Osnovy sovremennogo analiza. Moscow: Mir Publ., 1964, 430 p
33. Chentsov, A.G.: The program iteration method in a game problem of guidance. Proc. Steklov Inst. Math. **297**(suppl. 1), 43–61 (2017). https://doi.org/10.1134/S0081543817050066
34. Chentsov, A.G.: On the game problem of convergence at a given moment of time. Math. USSR-Izv. **12**(2), 426–437 (1978). https://doi.org/10.1070/IM1978v012n02ABEH001985
35. Nikitin, F.F., Chistyakov, S.V.: On zero-sum two-person differential games of infinite duration. Vestn. St. Petersbg. Univ. Math. **37**(3), 28–32 (2004)
36. Chistyakov, S.V.: Programmed iterations and universal ε-optimal strategies in a positional differential game. Sov. Math. Dokl. **44**(1), 354–357 (1992)
37. Chistyakov, S.V.: On functional equations in games of encounter at a prescribed instant. J. Appl. Math. Mech. **46**(5), 704–706 (1982). https://doi.org/10.1016/0021-8928(82)90023-5
38. Chentsov, A.G.: Iterations of stability and the evasion problem with a constraint on the number of switchings of the formed control. Izv. IMI UdGU **49**, 17–54 (2017) (in Russian). https://doi.org/10.20537/2226-3594-2017-49-02

Part II
Artificial Intelligence and Soft Computing in Control and Decision Making Systems

Fuzzy Real-Time Multi-objective Optimization of a Prosthesis Test Robot Control System

Yuriy P. Kondratenko, Poya Khalaf, Hanz Richter and Dan Simon

Abstract This paper investigates the fuzzy real-time multi-objective optimization of a combined test robot/transfemoral prosthesis system with three degrees of freedom. Impedance control parameters are optimized with respect to the two objectives of ground force and vertical hip position tracking. Control parameters are first optimized off-line with an evolutionary algorithm at various values of walking speed, surface friction, and surface stiffness. These control parameters comprise a gait library of Pareto-optimal solutions for various walking scenarios. The user-preferred Pareto point for each walking scenario can be selected either by expert decision makers or by using an automated selection mechanism, such as the point that is the minimum distance to the ideal point. Then, given a walking scenario that has not yet been optimized, a fuzzy logic system is used to interpolate in real time among control parameters. This approach enables automated real-time multi-objective optimization. Simulation results confirm the effectiveness of the proposed approach.

1 Introduction

Prosthesis research is a multi-disciplinary area that has grown significantly over the past couple of decades due to the increasing number of amputees [1]. The CYBER-LEGs Alpha-Prototype prosthesis [2] produces a natural walking gait by reproducing the joint torques and kinematics of a non-amputee, and uses mechanical energy regeneration to transfer the negative work done by the knee to the ankle to assist ankle push-off. Independent control of constrained motion and force trajectories [3] can be used in robotic prostheses with redundant sensory and actuation channels. Prostheses have also been controlled with approaches based on the Stewart platform [4]

Y. P. Kondratenko (✉)
Intelligent Information Systems Department, Petro Mohyla Black Sea National University, Mykolaiv 54003, Ukraine
e-mail: yuriy.kondratenko@chmnu.edu.ua

P. Khalaf · H. Richter · D. Simon
Washkewicz College of Engineering, Cleveland State University, Cleveland, OH 44115, USA

Y. P. Kondratenko et al. (eds.), *Advanced Control Techniques in Complex Engineering Systems: Theory and Applications*, Studies in Systems, Decision and Control 203, https://doi.org/10.1007/978-3-030-21927-7_8

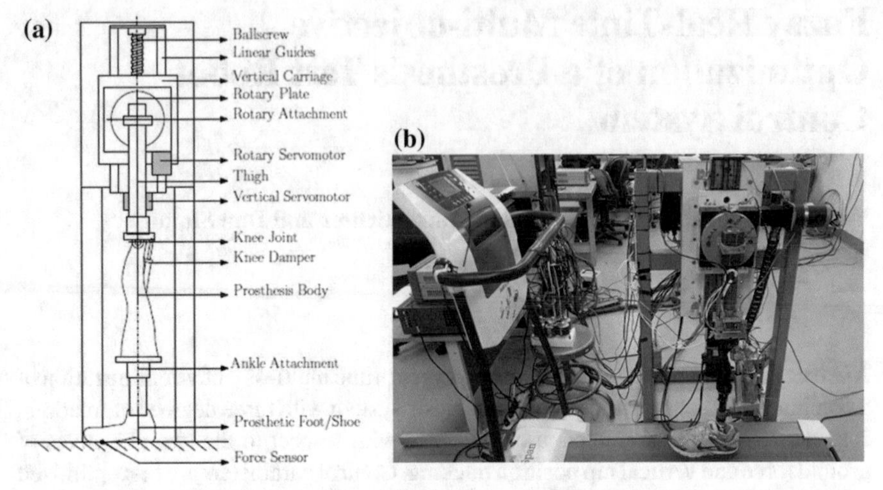

Fig. 1 The Cleveland State University prosthesis test robot (with a prosthesis): **a** kinematic structure, and **b** photograph of the test robot and a transfemoral prosthesis walking on a treadmill

and with the use of pneumatic actuators [5]. Myoelectric-driven finite state control has been used for powered ankle-foot prostheses [6] to modulate both impedance and power output during stance. Myoelectric control has also been used for assistive robotics and rehabilitation [7].

Prosthesis design and control parameters have been optimized with evolutionary algorithms [8–11], fuzzy logic [12, 13], adaptive and optimal control [14, 15], and neural networks [4, 11]. Clinical tests show that the biomechanical energy cost for above-knee amputee walking is much higher than it is for able-bodied individuals [15], which indicate the need for continued prosthesis research.

Much of the above research has been facilitated by gait studies and human motion analysis for able-bodied individuals [16–18]. In this paper, prosthesis design and optimization is enabled with a prosthesis test robot (Fig. 1) that allows prosthesis testing without human subjects [15, 19–22]. Vertical hip displacement and thigh angle motion profiles are applied to the test robot, and any transfemoral prosthesis can be connected to the robot for test and evaluation of the prosthesis or the control algorithms. A treadmill is used as the walking surface. In addition to tracking the hip and thigh motion trajectories, the test robot control system can regulate the contact force between the treadmill and the prosthesis.

Most human motion and control optimization research has focused on a single objective, such as user effort [23] or human kinetics [24]. However, human gait is inherently multi-objective [25], thus motivating recent research into multi-objective prosthesis control optimization [26, 27]. Multi-objective optimization (MOO) algorithms have been used in previous research to optimize this test robot's impedance controller [20]. However, those results were obtained for only one walking scenario—a given walking speed, a given treadmill belt stiffness, and a given coefficient of friction of the treadmill. In practice, a change in the walking scenario has a significant effect on the optimal control parameters. For example, it has been shown that optimal control parameter values strongly depend on walking speed [28].

The effect of walking surface stiffness and friction on optimal control parameter values will be shown later in this paper, and provides an intermediate but important contribution to the literature. In this paper, the two objectives that are optimized are vertical ground force tracking, and vertical position tracking, both measured at the hip. Both objectives are important because they significantly affect amputee health, as discussed in the following section.

When walking on a treadmill, walking surface stiffness and friction are constant. However, the purpose of the test robot is to prototype, test, and tune prostheses for amputee use outside the lab, which is an environment where walking surface stiffness and friction can significantly change over a short period of time. The purpose of this research is to develop and verify a fuzzy multi-objective impedance control optimization algorithm with the test robot, with the goal of eventual implementation outside the lab with amputee subjects.

The goal of this research is fourfold: (1) extend the simulation model of the prosthesis test robot to a set of walking scenarios greater than those that have yet been implemented, including different walking surface conditions; (2) develop a fuzzy classification system to identify the user-preferred weight vector of the two objectives of vertical hip force tracking and hip trajectory tracking; (3) develop a gait library that includes optimized control parameters for several combinations of walking scenarios; and (4) integrate the above results into a strategy to optimize the test robot's impedance controller in real time for previously non-optimized walking scenarios.

Note that this is the first time that these goals have been reported in the literature. That is: (1) the prosthesis test robot simulation has not yet been extended as in this paper; (2) fuzzy classification of user-preferred weight vectors in multi-objective optimization has not yet been reported as in this paper; (3) gait libraries for prosthetic walking have not been reported before now; and (4) real-time multi-objective optimization of prosthesis control has not yet been reported.

The research of this paper is important for robotic testing of prostheses, because it shows how the balance between force tracking and position tracking can be specified. The results are also important for real-time multi-objective optimization in general, because they show how Pareto-optimal points can be automatically selected in real time without user intervention.

The robot-prosthesis system in this paper is a rough approximation to human gait. Typical prosthesis development research begins with simulation [29], proceeds to the use of a prosthesis test robot [20], proceeds from there to the use of able-bodied subjects with bent-knee adapters [30], and concludes with amputee trials [31]. This paper focuses on the test robot aspect of prosthesis development.

Section 2 reviews the model of the prosthesis test robot, its control system, and the MOO algorithm for the optimization of impedance control parameters. Section 3 analyzes the factors that affect the desired trajectories and the desired user-specified MOO weight vector, which balances the two objectives of hip force tracking and hip position tracking. Section 3 also provides MOO simulation results (Pareto fronts) for 27 walking scenarios, which include all possible combinations of three different walking speeds, three different walking surface stiffness values, and three different walking surface friction values; these simulation results are denoted as the *gait library*. Section 4 presents the fuzzy system that identifies the user-preferred weight vector for a previously non-optimized walking scenario; the fuzzy system is based on the gait library developed in the previous section. Section 5 presents the algorithm that integrates the fuzzy logic-defined user-preferred weight vector, the gait library, and the final determination of the optimal impedance control parameters. Section 6 provides an example of the proposed algorithm and presents simulation results. Section 7 summarizes the paper and suggests some directions for future research.

2 Multi-objective Optimization of Prosthesis Test Robot Impedance Parameters

The test robot's optimal impedance control parameters depend on the desired relation between two objectives, denoted Q_1 and Q_2 [20]. Q_1 is the vertical ground force tracking of able-bodied reference data, and Q_2 is the vertical position tracking of able-bodied reference data, both measured at the hip. Both objectives are important because they have a significant impact on ancillary health issues for amputees. Amputees often need to compensate for prosthesis shortcomings with unnatural hip and thigh motions that lead to health problems such as back pain and arthritis [32, 33]. Symmetric gait reduces the occurrence of ancillary health issues, and so we develop a method in this paper to emulate able-bodied gait during amputee use. The solution of this MOO problem is based on the following approach.

Step 1. Formulate the mathematical model of the test robot/prosthesis system as a three-link robot with two actuated joints (hip and thigh, which are actuated by DC servo-motors) and one unactuated joint (knee). We assume here that the prosthesis is passive so the knee joint is unactuated. The foot/ground interaction model includes two contact points between the treadmill and the prosthesis (toe and heel).

Step 2. Obtain desired force F_d and motion (q_{1d}, q_{2d}) trajectories based on gait lab measurements of able-bodied walking. Reference data for this paper were collected in the Human Motion and Control Laboratory at Cleveland State University at normal walking speeds [34].

Step 3. Design two decoupled single-input-single-output controllers for the separate control of vertical hip motion and angular thigh motion. Vertical hip motion is controlled with an impedance controller that provides control signal U_1, and angular thigh motion is controlled with a sliding mode controller that provides control signal U_2 [20]:

$$U_1 = \frac{1}{K_1}\left\{\begin{array}{l}\frac{M_1}{M_m}\left[M_m\ddot{q}_{1d} + K_f(F - F_d) - C_m(\dot{q}_1 - \dot{q}_{1d}) + K_m(q_1 - q_{1d})\right]\\ + f_1 + g_1 - F_{int}\end{array}\right\}; \quad (1)$$

$$U_2 = \frac{J_2}{K_2}\left[(\ddot{q}_{2d} - \lambda\dot{q}_{2d}) + \left(\frac{b_2}{J_2} - \lambda\right)\dot{q}_2 + \eta\text{sign}(s)\right], \quad (2)$$

where q_1, q_{1d} are the hip's vertical displacement and desired (reference) value; q_2, q_{2d} are the thigh's angular position and desired (reference) value; M_1 is the inertia of the test robot/prosthesis system; J_2 is the inertia of the thigh angle load and motor; f_1 is the friction coefficient of the vertical hip motion; g_1 is force due to gravity; F_{int} is the vertical hip component of the ground reaction force F as measured at the hip joint; F_d is the desired (reference) force trajectory; K_1 is a constant that includes a combination of servo amplifier gain, motor torque constant, and rotary/linear motion transformation at the vertical hip joint; K_2 is a constant that includes a combination of servo amplifier gain and motor torque constant at the angular thigh joint; M_m, C_m, K_m are the desired inertia, damping, and stiffness parameters, respectively; K_f is a force tracking gain; b_2 is a viscous damping coefficient; s is a sliding function for the sliding mode controller; and $\lambda, \eta (\lambda > 0, \eta > 0)$ are sliding mode control constants. We treat the parameters M_m, C_m, K_m, K_f as independent variables that will be tuned with an MOO algorithm. Examples of reference trajectories are shown in Fig. 2 for different walking scenarios. In Fig. 2a, the vertical hip position is measured relative to the walking surface; in Fig. 2b, the thigh angle is measured relative to the vertical, with a positive angle in the direction of clockwise rotation as shown in Fig. 1a.

Step 4. Define the vertical hip force objective Q_1 and the vertical hip motion objective Q_2:

$$Q_1 = \sqrt{\frac{1}{N}\sum_{i=1}^{N}(F_i - F_{id})^2}; \quad (3)$$

$$Q_2 = \sqrt{\frac{1}{N}\sum_{i=1}^{N}(q_{1i} - q_{1id})^2}, \quad (4)$$

where N is the number of discrete time steps in the simulation or experiment.

Step 5. Solve the MOO problem to find optimal impedance control parameters M_m, C_m, K_m, K_f and calculate the resulting force and motion profiles (F_i, q_{1i}, q_{2i}) using the simulation of the test robot/prosthesis system. Denote the Pareto-optimal set and its solutions as $P(D) = \{D_1, D_2, \ldots, D_i, \ldots, D_m\}$. Any MOO algorithm can be used for this step: a genetic algorithm, biogeography based optimization, invasive weed optimization, swarm optimization, ant colony optimization, bee colony algorithm, and so on [8, 9, 11]. Most of these algorithms perform similarly to each other. The computational effort of evolutionary MOO algorithms is almost entirely comprised of function evaluation, or experiments, and so all of these algorithms have almost the same computational requirements. In this paper (in the next section) we will use invasive weed optimization. It will be shown in the next section that

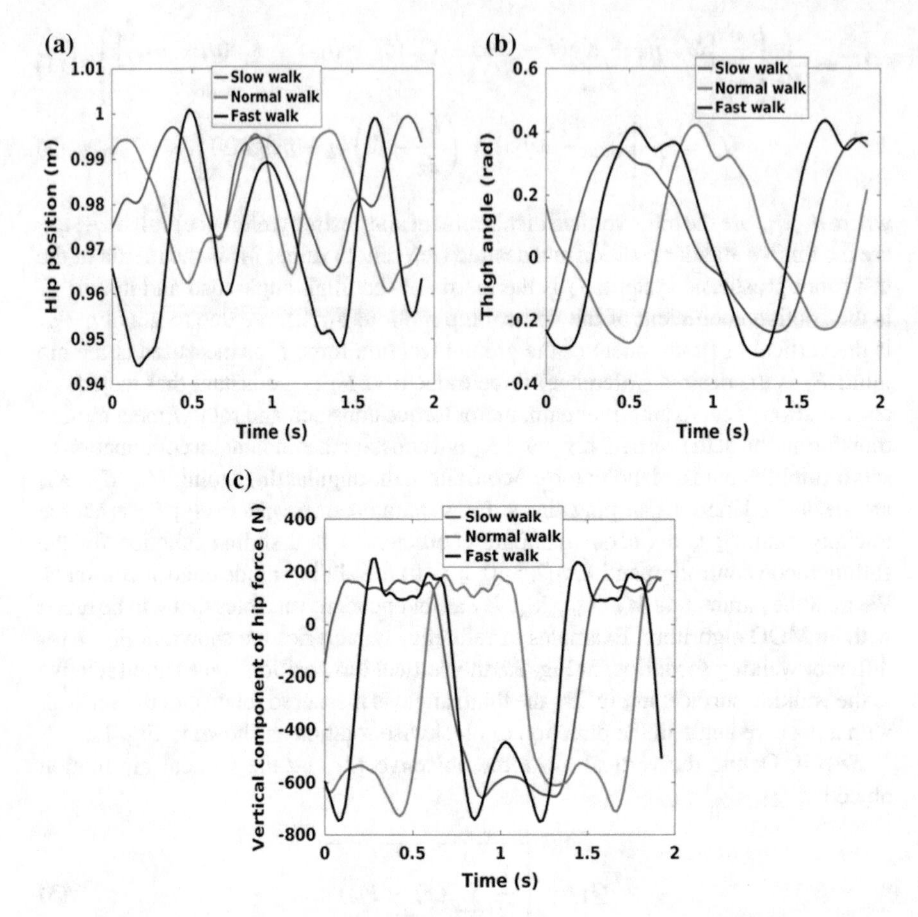

Fig. 2 Desired trajectories for: **a** vertical hip motion, **b** angular thigh motion, and **c** vertical component of ground force measured at the hip

objectives Q_1 and Q_2 conflict with each other [8, 20]. Eventually a specific Pareto-optimal solution D^*, $D^* \subset P(D)$ will need to be selected from the Pareto set, and this will depend on either a human expert's subjective evaluation, or will be based on the minimum distance to the ideal point. In this paper, as explained in the following section, the Pareto-optimal solution is chosen based on a minimum distance, thus avoiding the need of kernel function-based selection.

Step 6. To simplify the selection of the best solution $D^* \subset P(D)$ use a pseudo-weight approach to transform the nonlinear image of the Pareto front $P(D)$ (Fig. 3) to a linear function in the space of pseudo-weight components (w_1, w_2) [8, 20]:

$$w_i = \frac{(Q_i\max - Q_i)/\left(Q_i\max - Q_i\min\right)}{\sum_{j=1}^{2}\left(Q_j\max - Q_j\right)/\left(Q_j\max - Q_j\min\right)}, \quad (i = 1, 2), \tag{5}$$

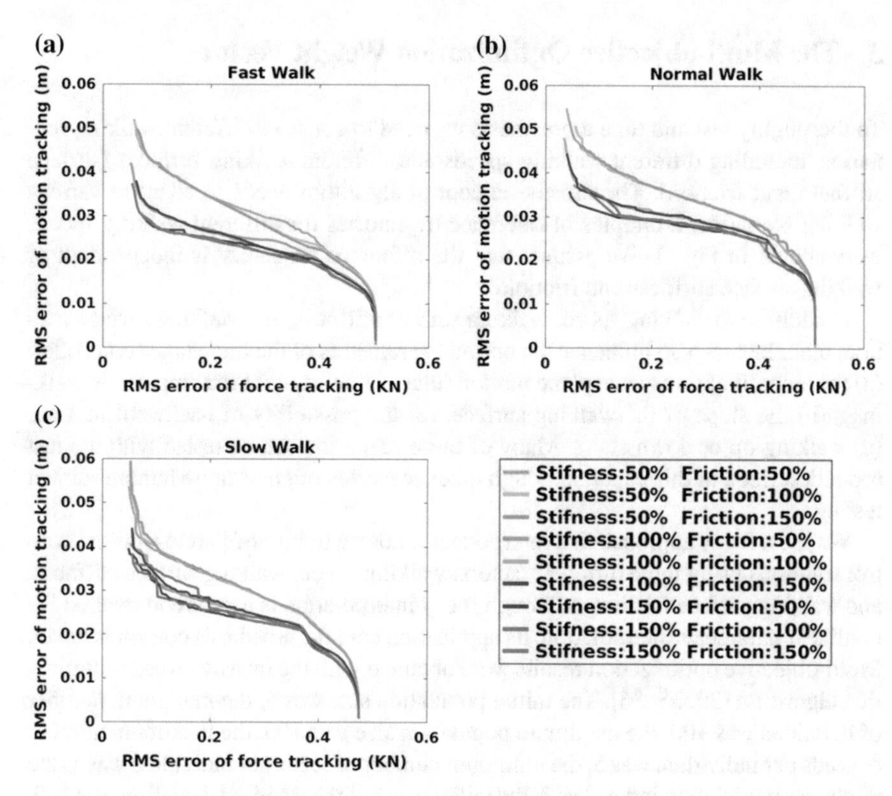

Fig. 3 Pareto fronts in the criteria space (Q_1, Q_2) for different values of walking surface stiffness, walking surface friction, and walking speed: **a** fast walking speed, **b** normal walking speed, **c** slow walking speed

where $Q_{i\max}$, $Q_{i\min}$ are the maximum and minimum values of $Q_i \in \{Q_1, Q_2\}$, and $w_2 = 1 - w_1$.

This pseudo-weight approach will help the decision-maker select the best solution $D^* \subset P(D)$ from the Pareto set. This approach avoids the problem of subjectively choosing weight coefficients as in the case of convolution methods, but we still need to subjectively choose the user-preferred pseudo-weight vector $\left(w_1^*, w_2^*\right)$. In other words, in this section we have discussed how to obtain a gait library with Pareto-optimal MOO control solutions; in Sect. 5 we propose a fuzzy logic-based method for determining the weight $\left(w_1^*, w_2^*\right)$ and thus selecting a user-preferred solution from the gait library.

3 The Multi-objective Optimization Weight Vector

To thoroughly test and tune a prosthesis we need to consider different walking scenarios, including different walking speeds and different walking surfaces (surface stiffness and friction). The prosthesis control algorithm needs to adapt to various walking scenarios. Examples of reference trajectories for different walking speeds were shown in Fig. 2. We assume that the reference trajectory is independent of walking surface stiffness and friction.

In addition to walking speed, walking surface stiffness, and walking surface friction, other factors also influence the optimal parameters of the impedance controller: (a) the amount of transverse plane motion (that is, how straight the amputee is walking); (b) the slope of the walking surface; (c) the possibility of backward motion; (d) walking up or down stairs. Many of these scenarios can be tested with the test robot described in this paper, although other scenarios might require human subject testing.

We use a fuzzy approach that interpolates among a library of Pareto-optimal control solutions using three different factors: walking speed, walking surface stiffness, and walking surface friction. Although fuzzy interpolation is a common method for nonlinear parameter interpolation, its application here for prosthesis control is novel. Multi-objective optimization results were obtained with the invasive weed optimization algorithm [20, 35, 36]. The initial population size was 5, the maximum number of iterations was 100, the maximum population size was 100, the maximum number of seeds per individual was 5, the minimum number of seeds per individual was 1, the nonlinear modulation index was 3, the initial value of the standard deviation was 100, and the final value of the standard deviation was 1. These values were chosen based on the literature, our previous experience with IWO, and trial and error. However, we note that the performance of IWO is relatively robust to IWO parameter values.

Figure 3 shows the resulting Pareto fronts and demonstrates the influence of the walking scenario on the optimal impedance control parameters. In Fig. 3 the walking surface friction coefficient γ was selected from the set $\gamma \in \{0.5, 1.0, 1.5\}$ and the walking surface stiffness parameter k_2 (N/m) was selected from the set $k_2 \in \{1850, 3700, 5550\}$, which are denoted as $\{50\%, 100\%, 150\%\}$ in Fig. 3. The walking speed was selected from the set $\{1.25\,\text{m/s}, 1.0\,\text{m/s}, 0.75\,\text{m/s}\}$. These values were selected based on extremes that are encountered in daily walking.

4 Fuzzy Identification of the Multi-objective Weight Vector

Past research on the use of fuzzy logic to determine preference values [37, 38] motivates our development of a novel fuzzy logic-based approach for Pareto front weight determination. We consider a Mamdani-type fuzzy logic system [39–44] that consists of three input signals $\{x_1, x_2, x_3\}$ and one output signal w_1^*.

The first input signal, x_1, is *walking speed* and is characterized by three linguistic terms (LTs): **Slow**, **Normal**, **Fast**. The second input signal, x_2, is *walking surface stiffness* and is characterized by three LTs: **Low**, **Medium**, **High**. The third input signal, x_3, is *walking surface friction* and is characterized by three LTs: **Low**, **Medium**, **High**. The numerical values and the normalized ranges for the input signals are selected as follows:

$$\begin{aligned} &\text{Walking Speed}\, x_1 \in [0.75\,\text{m/s}, 1.25\,\text{m/s}] \propto [0, 2.0]; \\ &\text{Walking Surface Stiffness}\, x_2 \in [1850\,\text{N/m}, 5550\,\text{N/m}] \propto [0, 2.0]; \\ &\text{Walking Surface Friction Coefficient}\, x_i \in [0.5, 1.5] \propto [0, 2.0]. \end{aligned} \quad (6)$$

The output signal, w_1, corresponds to the desired value of the user-preferred weight that corresponds to hip force; that is, the importance of hip force tracking relative to vertical hip motion tracking. w_1 is characterized by five LTs: **L (low)**, **LM (low-medium)**, **M (medium)**, **MH (medium-high)**, **H (high)**. The rule base of the fuzzy system consists of 27 fuzzy rules as shown in Table 1. The rule base was developed using the Pareto data of Fig. 3 and the preferred solutions were calculated based on the minimal Euclidean distance to the ideal point [45].

The relationships between the inputs $\{x_1, x_2, x_3\}$ and output w_1, shown in Table 1, can be visualized by the three output surfaces in Fig. 4.

The dependence of the weights on the environmental parameters clearly varies depending on other parameter values. Some of the areas in the plots are flat, showing

Table 1 Fuzzy rule base of the multi-objective weight-selection system. w_1 is the importance of hip force tracking relative to hip motion tracking

Rule number	Antecedents			Consequent
	x_1 (speed)	x_2 (stiffness)	x_3 (friction)	w_1 (MOO weight)
1	Fast	Low	Low	Low-medium
2	Fast	Low	Medium	Low
3	Fast	Low	High	Low
4	Fast	Medium	Low	Low-medium
5	Fast	Medium	Medium	Medium
6	Fast	Medium	High	Low-medium
7	Fast	High	Low	High
8	Fast	High	Medium	High
9	Fast	High	High	Low-medium
10	Normal	Low	Low	Low-medium
11	Normal	Low	Medium	Low-medium
12	Normal	Low	High	Medium-high
13	Normal	Medium	Low	Medium-high

(continued)

Table 1 (continued)

Rule number	Antecedents			Consequent
	x_1 (speed)	x_2 (stiffness)	x_3 (friction)	w_1 (MOO weight)
14	Normal	Medium	Medium	Medium
15	Normal	Medium	High	Medium
16	Normal	High	Low	Medium-high
17	Normal	High	Medium	Medium-high
18	Normal	High	High	Medium-high
19	Slow	Low	Low	Low-medium
20	Slow	Low	Medium	Low-medium
21	Slow	Low	High	Low-medium
22	Slow	Medium	Low	Low-medium
23	Slow	Medium	Medium	Low-medium
24	Slow	Medium	High	Low-medium
25	Slow	High	Low	Low-medium
26	Slow	High	Medium	Low
27	Slow	High	High	Medium

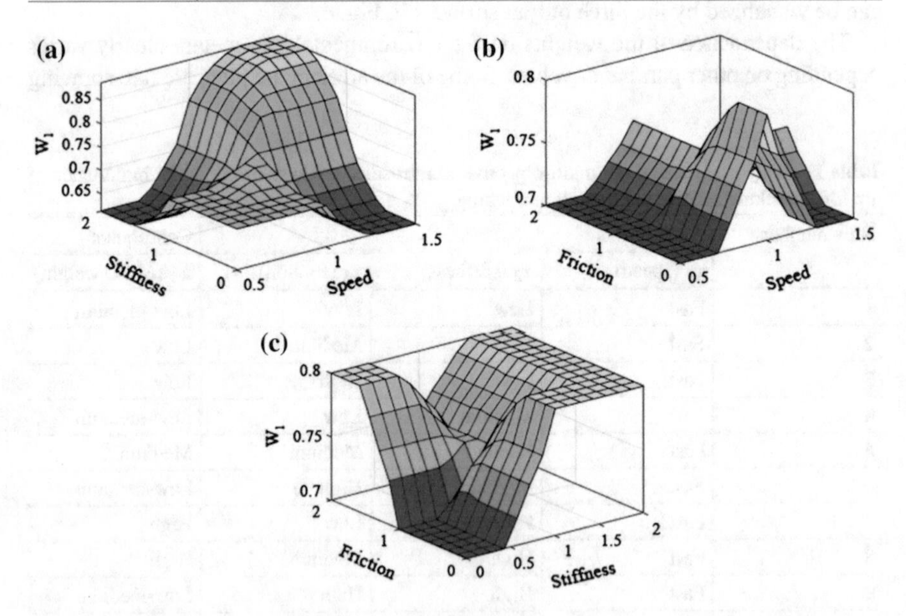

Fig. 4 Surfaces of the fuzzy multi-objective weight determination system: **a** friction $x_3 = Medium$; **b** stiffness $x_2 = Medium$; **c** speed $x_1 = Medium$

Table 2 Illustrative results of the effect of walking speed, surface stiffness, and surface friction on w_1, which is the relative weight of the hip force tracking objective

	Antecedents			Consequent
	x_1 (speed)	x_2 (stiffness)	x_3 (friction)	w_1 (MOO weight)
1	0.65	1.31	1.27	0.67
2	0.92	1.31	1.27	0.71
3	1.18	1.31	1.27	0.81
4	1.18	0.98	1.27	0.72
5	1.18	0.59	1.27	0.67
6	1.18	1.22	0.59	0.81
7	1.18	1.22	0.952	0.82
8	1.18	1.22	1.48	0.75

little sensitivity of the weights to parameter values. Other areas of the plots have a steep slope, showing high sensitivity.

Some illustrative fuzzy inference results are shown in Table 2. The table shows that the fuzzy inference system: (a) increases w_1 (the weight of the hip force objective function) when walking speed increases (rows 1–3); (b) increases w_1 when walking surface stiffness increases (rows 3–5); and (c) has an unpredictable effect on w_1 when walking surface friction changes (rows 6–8).

5 Fuzzy Multi-objective Optimization of Impedance Parameters

This section exploits the MOO approach and the fuzzy weight identification approach of the previous sections to optimize the impedance parameters of the prosthesis test robot without any human input or decision making. We start with a gait library, derived in Sect. 2, which includes multiple volumes of gait information. Each volume in the gait library is associated with a specific walking speed (Sp_i), walking surface stiffness (St_j), and walking surface friction (Fr_k):

$$\{x_1, x_2, x_3\} = \{Sp_i, St_j, Fr_k\}, \tag{7}$$

where $i \in \{1, \ldots, I\}$, and I is the number of different walking speeds; $j \in \{1, \ldots, J\}$, and J is the number of different values of walking surface stiffness; and $k \in \{1, \ldots, K\}$, and K is the number of different values of walking surface friction. In our example, $I = J = K = 3$, so there are 27 volumes in the gait library. Each volume in the gait library includes the following three components.

1. A set of reference trajectories derived from human motion data. The reference trajectories include the reference hip force trajectory, the reference hip motion

Table 3 Objective values, weight w_1, and Pareto-optimal impedance parameters for walking scenario $Sp = 1.25$, $St = 0.5$, $Fr = 1.0$. Note that $w_2 = 1 - w_1$. For the sake of conciseness, the table shows only a portion of the Pareto set

Q_1	Q_2	w_1	C_m	C_f	K_m	M_m
0.0578	0.0514	1.0000	8.7654	0.7390	804.7	0.0094
…	…	…	…	…	…	…
0.1999	0.0341	0.6618	116.2331	0.2976	1532.6	0.1847
0.2054	0.0337	0.6532	102.8675	0.2772	1525.1	1.0421
0.2100	0.0335	0.6462	81.1119	0.2667	1577.6	1.3729
0.2164	0.0330	0.6353	83.9620	0.2562	1578.9	1.2675
…	…	…	…	…	…	…
0.5025	0.0017	0.0000	401.9973	0.0000	1460.0	0.0809

trajectory, and the reference thigh angle trajectory, each for three different walking speeds (see Fig. 2). The reference trajectories are independent of walking surface stiffness and friction.

2. The Pareto front (Q_1, Q_2) (Fig. 3), the weight vector (w_1, w_2), and the Pareto set (Table 3).
3. The functional dependence of the optimal impedance parameters $\{C_m, K_f, K_m, M_m\}$ on the weight w_1 (Table 3):

$$C_m = f_1(w_1);\ K_f = f_2(w_1);\ K_m = f_3(w_1);\ M_m = f_4(w_1). \tag{8}$$

This functional dependence is based on the numerical data in Table 3.

For each volume we define the notation $Mask\{i, j, k\}$ as a shorthand indicator [46] for the volume that corresponds to the i-th walking speed value, the j-th walking surface stiffness value, and the k-th walking surface friction value. The number of volumes in the gait library is equal to $I \times J \times K$. For example, in Sect. 4 we considered $I = 3$, $J = 3$, and $K = 3$, in which case the library consists of 27 volumes.

The optimal impedance control parameters for a given walking scenario are determined by the following five steps.

Step 1. Measure or estimate the walking scenario parameters to form the set of input signals $\{x_1, x_2, x_3\} = \{\text{walking speed, surface stiffness, surface friction}\}$.

Step 2. Use the input signals and the fuzzy logic system of Sect. 4 to calculate the desired weight value w_1^*.

Step 3. Find the volume in the gait library that is nearest to the current walking scenario:

$$i = \begin{cases} 1, \text{ if } \max\left(\mu^{Sp}_{Slow}(x_1), \mu^{Sp}_{Normal}(x_1), \mu^{Sp}_{Fast}(x_1)\right) = \mu^{Sp}_{Fast}(x_1) \\ 2, \text{ if } \max\left(\mu^{Sp}_{Slow}(x_1), \mu^{Sp}_{Normal}(x_1), \mu^{Sp}_{Fast}(x_1)\right) = \mu^{Sp}_{Normal}(x_1); \\ 3, \text{ if } \max\left(\mu^{Sp}_{Slow}(x_1), \mu^{Sp}_{Normal}(x_1), \mu^{Sp}_{Fast}(x_1)\right) = \mu^{Sp}_{Slow}(x_1) \end{cases} \tag{9}$$

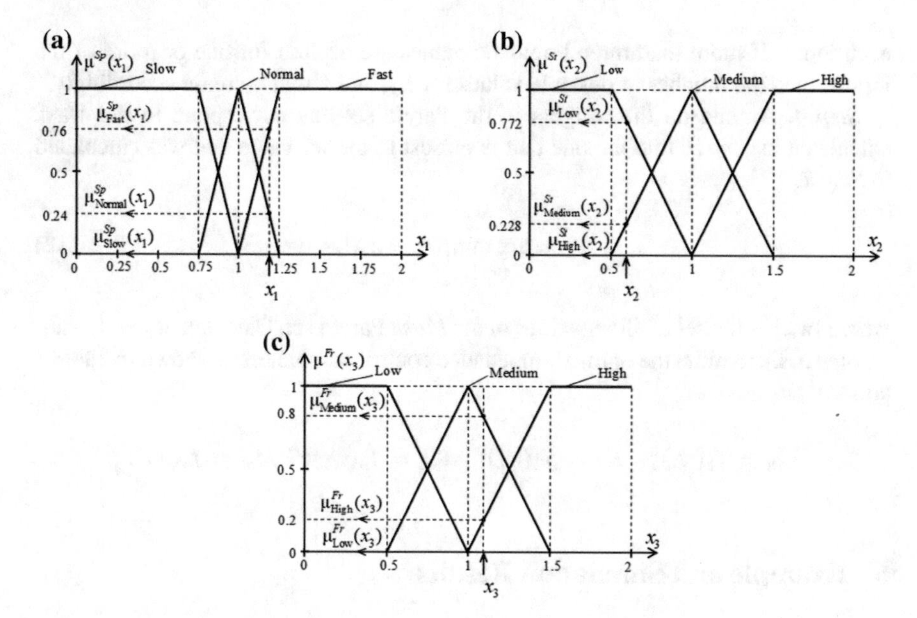

Fig. 5 Linguistic terms and fuzzy sets of the input signals x_1 (walking speed), x_2 (walking surface stiffness), and x_3 (walking surface friction)

$$j = \begin{cases} 1, \text{ if } \max\left(\mu^{St}_{Low}(x_2), \mu^{St}_{Medium}(x_2), \mu^{St}_{High}(x_2)\right) = \mu^{St}_{Low}(x_2) \\ 2, \text{ if } \max\left(\mu^{St}_{Low}(x_2), \mu^{St}_{Medium}(x_2), \mu^{St}_{High}(x_2)\right) = \mu^{St}_{Medium}(x_2); \\ 3, \text{ if } \max\left(\mu^{St}_{Low}(x_2), \mu^{St}_{Medium}(x_2), \mu^{St}_{High}(x_2)\right) = \mu^{St}_{High}(x_2) \end{cases} \quad (10)$$

$$k = \begin{cases} 1, \text{ if } \max\left(\mu^{Fr}_{Low}(x_3), \mu^{Fr}_{Medium}(x_3), \mu^{Fr}_{High}(x_3)\right) = \mu^{Fr}_{Low}(x_3) \\ 2, \text{ if } \max\left(\mu^{Fr}_{Low}(x_3), \mu^{Fr}_{Medium}(x_3), \mu^{Fr}_{High}(x_3)\right) = \mu^{Fr}_{Medium}(x_3); \\ 3, \text{ if } \max\left(\mu^{Fr}_{Low}(x_3), \mu^{Fr}_{Medium}(x_3), \mu^{Fr}_{High}(x_3)\right) = \mu^{Fr}_{High}(x_3) \end{cases} \quad (11)$$

where the values of membership functions $\mu^{Sp}_{Slow}(x_1)$, $\mu^{Sp}_{Normal}(x_1)$, $\mu^{Sp}_{Fast}(x_1)$, $\mu^{St}_{Low}(x_2)$, $\mu^{St}_{Medium}(x_2)$, $\mu^{St}_{High}(x_2)$, $\mu^{Fr}_{Low}(x_3)$, $\mu^{Fr}_{Medium}(x_3)$, and $\mu^{Fr}_{High}(x_3)$ are determined from the fuzzy membership functions of Fig. 5. These membership functions are chosen somewhat arbitrarily in this paper; we could try other membership functions, other shapes, or even optimize the fuzzy membership functions. However, those questions are peripheral to the main point of this paper, which is the use of real-time fuzzy interpolation for robot control adaptation.

As an example, $Mask\{i, j, k\}= Mask\{1, 3, 2\}$ corresponds to the scenario {*fast walking speed, high walking surface stiffness, medium walking surface friction*}. The fuzzification of these three inputs can be realized using direct models of triangular linguistic terms [47–51]. Here we assume that we have three linguistic values for each input. If more than three linguistic values are desired for one or more of the inputs, then the number of possible values for i, j, and k will increase accordingly.

Step 4. Among all the weights in the Pareto set that correspond to the *Mask* calculated in *Step 3*, find the one that is closest to the w_1^* value that was calculated in *Step 2*:

$$w_1 = \arg\min_{\{w_m\}} \left| w_1^* - w_m \right|, \tag{12}$$

where $\{w_m\}$ is the set of all w_1 values in the *Mask* Pareto set. Calculate $w_2 = 1 - w_1$.

Step 5. Determine the optimal impedance control parameters as shown in Table 3 and (8):

$$C_m = f_1(w_1);\quad K_f = f_2(w_1);\quad K_m = f_3(w_1);\quad M_m = f_4(w_1).$$

6 Example and Simulation Results

Here we illustrate the approach of the preceding section with a step-by-step example and some simulation results.

Step 1. Suppose that the normalized numerical values of walking speed, walking surface stiffness, and walking surface friction values are $x_1 = 1.19$, $x_2 = 0.614$, $x_3 = 1.1$ (see Eq. 6).

Step 2. Suppose that the fuzzy system of Sect. 4 processes the input data {1.19, 0.614, 1.1} to obtain the weight value $w_1^* = 0.651$. This means that for the given walking scenario, the desired human gait will be emulated by the test robot with a relative tracking weight of 0.651 for the hip force trajectory and 0.349 for the hip position trajectory.

Step 3. Suppose that the membership functions grades for the given inputs $x_1 = 1.19$, $x_2 = 0.614$, and $x_3 = 1.1$ are calculated from Fig. 5 as follows:

$$\mu_{Slow}^{Sp}(x_1) = 0.00,\ \mu_{Normal}^{Sp}(x_1) = 0.24,\ \mu_{Fast}^{Sp}(x_1) = 0.76;$$
$$\mu_{Low}^{St}(x_2) = 0.772,\ \mu_{Medium}^{St}(x_2) = 0.228,\ \mu_{High}^{St}(x_2) = 0.00;$$
$$\mu_{Low}^{Fr}(x_3) = 0.00,\ \mu_{Medium}^{Fr}(x_3) = 0.80,\ \mu_{High}^{Fr}(x_3) = 0.20.$$

Suppose that the gait library volume that is closest to this walking scenario is calculated as

$$\max(0.00, 0.24, 0.76) = 0.76 = \mu_{Fast}^{Sp}(x_1) \Rightarrow x_1 = 1.19;$$
$$\max(0.772, 0.228, 0.00) = 0.772 = \mu_{Low}^{St}(x_2) \Rightarrow x_2 = 0.614;$$
$$\max(0.00, 0.80, 0.20) = 0.20 = \mu_{Medium}^{Fr}(x_3) \Rightarrow x_3 = 1.1.$$

This means that $Mask\{1, 1, 2\}$ is the nearest volume in the gait library to the given walking scenario, which corresponds to {*fast walking speed, low walking surface stiffness, medium walking surface friction*}.

Step 4. Suppose that the $Mask\{1, 1, 2\}$ volume in the gait library includes the following Pareto set, which is an example of the data in Table 3:

$$\{w_1\} = \begin{pmatrix} 1.0000 \\ 0.9890 \\ 0.9340 \\ \vdots \\ 0.6675 \\ 0.6618 \\ 0.6532 \\ 0.6462 \\ 0.6353 \\ \vdots \\ 0.0002 \\ 0.0000 \end{pmatrix}. \tag{13}$$

Recall that *Step 2* above computed the weight value $w_1^* = 0.651$. The closest weight value in (13) to w_1^* is 0.6532. We therefore set $w_1 = 0.6532$ and $w_2 = 1 - w_1 = 0.3468$.

Step 5. Use w_1 as calculated in *Step 4* and the *Mask* volume of Pareto-optimal solutions (Table 3 and Eq. 8) to determine the impedance control parameters $\{C_m, K_f, K_m, M_m\}$. For this particular walking scenario, these parameters are found from Table 3 as follows:

$$C_m = 102.8675,\ K_f = 0.2772,\ K_m = 1525.1,\ M_m = 1.0421.$$

Figure 6 shows simulation results for the three degree-of-freedom prosthesis test robot using these impedance control parameters with fast walking speed, low walking surface stiffness, and medium walking surface friction. An examination of the figure confirms the intuitive expectation that hip force tracking is better than hip displacement tracking, which agrees with their relative weights in the Pareto solution (65% and 35% respectively). The results show that hip force tracking has an RMS error of 206 N, which is about of 43% of the hip's peak load, and vertical hip displacement has an RMS error of 3.4 cm, which is about of 89% of the hip's peak displacement. The tracking of the angular thigh position by the sliding mode controller (Fig. 6c) shows that it is practically identical to the reference trajectory.

Discussion—In step 1 above, we need to find normalized values of walking speed, walking surface stiffness, and walking surface friction. These values should be inferred in real time by the supervisory control algorithm. This implies that minima and maxima need to be defined a priori. If the ranges of these values are too large

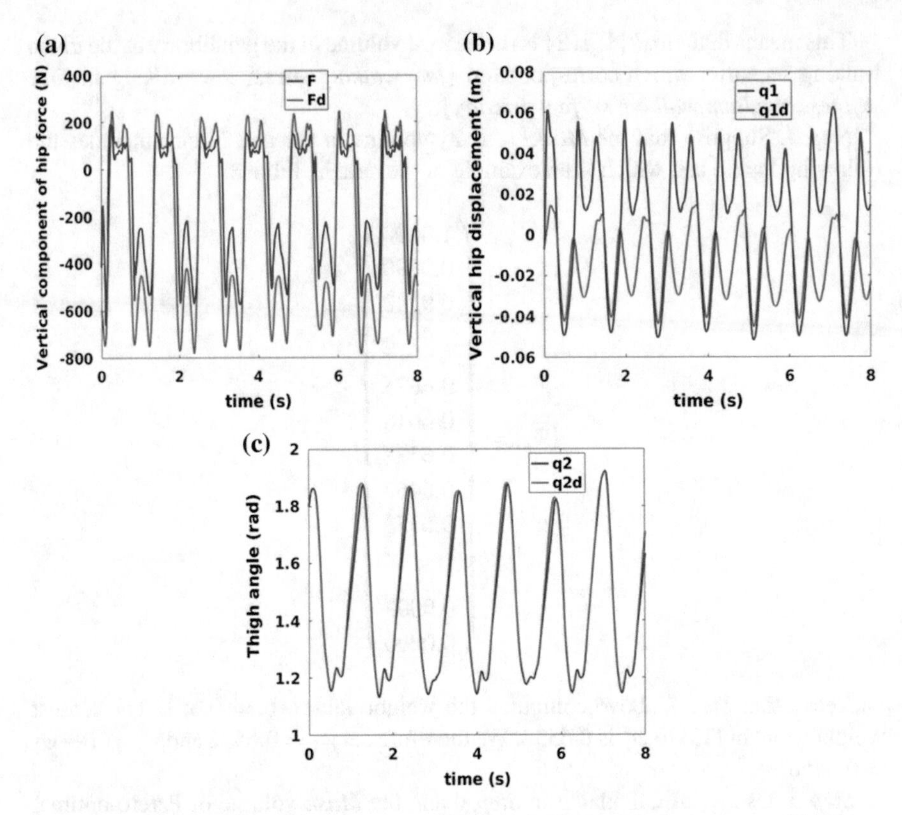

Fig. 6 Simulation results for walking scenario $Mask\{1, 1, 2\}$: **a** vertical component of hip force, **b** vertical hip displacement, and **c** thigh angle

(or too small), then multiple values of speed, stiffness, and friction will be mapped into similar (or the same) normalized values, which will result in a loss of control resolution.

In step 2 above, the fuzzy system processes the input data to obtain the MOO weight value. The correct operation of this step depends on the suitability of using the minimum Euclidean distance to the ideal point as the desired weight vector. If this approach is not suitable, then expert training can be used in Sect. 4 to find suitable weight vectors.

In step 3 above, membership grades are calculated from Fig. 5, which illustrates triangular membership functions. Other membership function shapes (trapezoidal, Gaussian, etc.) could also be used (or optimized) in this research.

In step 4 above, we find the closest value in the gait library to the weight calculated in step 3. However, instead of taking values directly from the gait library, we could instead interpolate between gait library data; that is, we could instead directly use the weight value from step 3.

In step 5 above, we calculate impedance control parameters from Table 3 and Eq. 8. However, if we instead directly use the weight value as suggested in the above paragraph, then we could interpolate between impedance control parameters rather than directly using parameter values from Table 3.

7 Conclusions

Prosthesis test robot control is a multi-objective optimization problem. This paper considered the two objectives of tracking able-bodied trajectories of hip force and hip position, and these objectives depend on the parameters of an impedance control framework. This multi-objective optimization problem was solved with invasive weed optimization, an evolutionary algorithm. The Pareto front point that is closest to the ideal point (zero tracking error for both hip force and hip trajectory) is used in this paper as the optimal Pareto point. The optimal Pareto point is characterized by a pseudo-weight that quantifies the importance of hip force tracking relative to hip trajectory tracking.

In addition to the complication of multiple objectives, this paper has shown that the objective function values and the optimal pseudo-weight depend on the walking scenario, which in this paper includes walking speed, walking surface stiffness, and walking surface friction. This results in $I \times J \times K$ Pareto fronts and pseudo-weights, where I, J, and K are the number of different walking speeds, surface stiffness values, and surface friction values, respectively. The resultant mapping from the walking scenario parameters to the optimal pseudo-weight is then fuzzified to obtain a fuzzy mapping from walking scenario parameters to fuzzy pseudo-weight.

Then, any new walking scenario is passed through the fuzzy mapping and defuzzified to obtain a crisp pseudo-weight. The walking scenario is projected onto the nearest Pareto set based on its similarity to previously optimized walking scenarios. The defuzzified pseudo-weight is rounded to the nearest pseudo-weight in the Pareto set, which defines the impedance parameters to use in the controller for the new walking scenario. Simulation results confirm that the proposed method is effective for impedance control optimization in the case of walking scenarios that have not been explicitly optimized during training. Results show that a combination of hip force trajectory tracking, and hip position trajectory tracking, are balanced in way that is consistent with user preferences.

The method presented in this paper specifically applies to prosthesis test robot impedance control optimization, but it could be applied to any multi-objective optimization problem whose performance depends on parameters that can vary from one experiment to the next. When parameters vary in multi-objective optimization problems, their Pareto fronts also vary. The fuzzy logic approach presented here is an effective way of interpolating between Pareto fronts and selecting a multi-objective pseudo-weight that accurately reflects user preferences.

Future research could extend in several directions: expanding the number of objectives or independent variables in the prosthesis test robot gait library to explore the scalability of the proposed method with higher dimensions; reducing the fuzzy rule base [43, 52–54] to achieve better real-time performance; optimizing the fuzzy membership function parameters [55–58] using classical or evolutionary optimization algorithms [8, 9, 59–61]; testing the proposed method with an extended set of simulations as well as experimental tests; and extending the proposed method to other multi-objective optimization applications [38, 44, 62, 63].

Acknowledgements The authors thank the Fulbright Scholar Program (USA) for supporting Prof. Y. P. Kondratenko with a Fulbright scholarship and for making it possible for this team to conduct research together in the USA. This research was partially supported by National Science Foundation Grant 1344954.

References

1. Meng, W., Liu, Q., Zhou, Z., Ai, Q., Sheng, B., Xie, S.: Recent development of mechanisms and control strategies for robot-assisted lower limb rehabilitation. Mechatronics **31**, 132–145 (2015)
2. Flynn, L., Geeroms, J., Jimenez-Fabian, R., Vanderborght, B., Vitiello, N., Lefeber, D.: Ankle-knee prosthesis with active ankle and energy transfer: development of the CYBERLEGs Alpha-Prosthesis. Robot. Auton. Syst. **73**, 4–15 (2015)
3. Hemami, H., Dariush, B.: Control of constraint forces and trajectories in a rich sensory and actuation environment. Math. Biosci. **228**, 171–184 (2010)
4. Bevilacqua, V., Dotoli, M., Foglia, M.M., Acciani, M., Tattoli, G., Valori, M.: Artificial neural networks for feedback control of human elbow hydraulic prosthesis. Neurocomputing **137**, 3–11 (2014)
5. Zheng, H., Shen, X.: Design and control of a pneumatically actuated transtibial prosthesis. J. Bionic Eng. **12**, 217–226 (2015)
6. Au, S., Berniker, M., Herr, H.: Powered ankle-foot prosthesis to assist level-ground and stair-descent gaits. Neural Netw. **21**, 654–666 (2008)
7. Oskoei, M.A., Hu, H.: Myoelectric control systems—a survey. Biomed. Signal Process. **2**, 275–294 (2007)
8. Deb, K.: Multi-objective Optimization Using Evolutionary Algorithms, vol. 16. Wiley (2001)
9. Simon, D.: Evolutionary Optimization Algorithms: Biologically Inspired and Population-Based Approaches to Computer Intelligence. Wiley (2013)
10. Thiele, L., Miettinen, K., Korhonen, P., Molina, J.: A preference-based evolutionary algorithm for multi-objective optimization. Evol. Comput. **17**(3), 411–436 (2009)
11. Thomas, G., Wilmot, T., Szatmary, S., Simon, D., Smith, W.: Evolutionary optimization of artificial neural networks for prosthetic knee control. In: Igelnik, B., et al. (eds.) Efficiency and Scalability Methods for Computational Intellect, Chapter 7, pp. 142–161 (2013)
12. Khushaba, R.N., Al-Jumaily, A., Al-Ani, A.: Evolutionary fuzzy discriminant analysis feature projection technique in myoelectric control. Pattern Recognit. Lett. **30**, 699–707 (2009)
13. Zadeh, L.A.: Fuzzy sets. Inform. Control **8**, 338–353 (1965)
14. Amin, A.T.M., Rahim, A.H.A., Low, C.Y.: Adaptive controller algorithm for 2-DOF humanoid arm. Procedia Technol. **15**, 765–774 (2014)
15. Popovic, D., Ogustoreli, M.N., Stein, R.B.: Optimal control for an above-knee prosthesis with two degrees of freedom. J. Biomech. **28**(1), 89–98 (1995)
16. Simon, S.R.: Quantification of human motion: gait analysis—benefits and limitations to its application to clinical problems. J. Biomech. **37**, 1869–1880 (2004)
17. Wang, L., Hu, W., Tan, T.: Recent developments in human motion analysis. Pattern Recognit. **36**, 585–601 (2003)

18. Zhou, H., Hu, H.: Human motion tracking for rehabilitation—a survey. Biomed. Signal Process. **3**, 1–18 (2008)
19. Davis, R., Richter, H., Simon, D., van den Bogert, A.: Evolutionary ground reaction force optimisation of a prosthesis leg testing robot. In: American Control Conference (ACC). IEEE, Portland, Oregon, USA, 4–6 June 2014 (2014)
20. Khalaf, P., Richter, H., van den Bogert, A.J., Simon, D.: Multi-objective optimization of impedance parameters in a prosthesis test robot. In: Proceedings of ASME Dynamic Systems and Control Conference, Columbus, Ohio, USA, 28–30 October 2015 (2015)
21. Richter, H., Simon, D.: Robust tracking control of the prosthesis test robot. J. Dyn. Syst. T ASME **136**(3), 031011 (2014)
22. Richter, H., Simon, D., Smith, W., Samorezov, S.: Dynamic modeling, parameter estimation and control of a leg prosthesis test robot. Appl. Math. Model. **39**, 559–573 (2015)
23. Felt, W., Selinger, J., Donelan, J., Remy, C.: Body-in-the-loop: optimizing device parameters using measures of instantaneous energetic cost. PLoS ONE **10**(8), e0135342 (2015)
24. Richter, H., Hui, X., van den Bogert, A., Simon, D.: Semiactive virtual control of a hydraulic prosthetic knee. In: IEEE Conference on Control Applications, Buenos Aires, Argentina, 19–22 September 2016 (2016)
25. Ries, A., Novacheck, T., Schwartz, M.: The efficacy of ankle-foot orthoses on improving the gait of children with diplegic cerebral palsy. Mult. Outcome Anal. Phys. Med. Rehabil. **7**(9), 922–929 (2015)
26. Khademi, G., Mohammadi, H., Richter, H., Simon, D.: Optimal mixed tracking/impedance control with application to transfemoral prostheses with energy regeneration. IEEE Trans. Biomed. Eng. **65**(4), 894–910 (2018)
27. Kondratenko, Y., Khademi, G., Azimi, V., Ebeigbe, D., Abdelhady, M., Fakoorian, S.A., Barto, T., Roshanineshat, A.Y., Atamanyuk, I., Simon, D.: Robotics and Prosthetics at Cleveland State University: modern information, communication, and modeling technologies. In: Ginige, A., et al. (eds.) Information and Communication Technologies in Education, Research, and Industrial Applications, ICTERI 2016. Communications in Computer and Information Science, vol. 783, pp. 133–155. Springer, Cham (2017)
28. Hansen, H., Childress, D.S., Miff, S.C., Gard, S.A., Mesplay, K.P.: The human ankle during walking: implications for design of biomimetic ankle prosthesis. J. Biomech. **37**, 1467–1474 (2004)
29. Fakoorian, S., Azimi, V., Moosavi, M., Richter, H., Simon, D.: Ground reaction force estimation in prosthetic legs with nonlinear Kalman filtering methods. ASME J. Dyn. Syst. Meas. Control **139**(11), DS-16-1583 (2017)
30. Azimi, V., Shu, T., Zhao, H., Ambrose, E., Ames, A., Simon, D.: Robust control of a powered transfemoral prosthesis device with experimental verification. In: IEEE American Control Conference (ACC), Seattle, WA, USA, 24–26 May 2017 (2017)
31. Khalaf, P., Warner, H., Hardin, E., Richter, H., Simon, D.: Development and experimental validation of an energy regenerative prosthetic knee controller and prototype. In: ASME Dynamics Systems and Control Conference, Atlanta, Georgia, USA, 30 September–3 October 2018 (2018)
32. Nolan, L., Wit, A., Dudziñski, K., Lees, A., Lake, M., Wychowañski, M.: Adjustments in gait symmetry with walking speed in trans-femoral and trans-tibial amputees. Gait Posture **17**(2), 142–151 (2003)
33. Tura, A., Raggi, M., Rocchi, L., Cutti, A.G., Chiari, L.: Gait symmetry and regularity in transfemoral amputees assessed by trunk accelerations. J. Neuroeng. Rehabil. **7**(1), 4 (2010)
34. Moore, J., Hnat, S., van den Bogert, A.: An elaborate data set on human gait and the effect of mechanical perturbations. PeerJ **3**, e918 (2015)
35. Basak, A., Pal, S., Das, S., Abraham, A., Snasel, V.: A modified invasive weed optimization algorithm for time-modulated linear antenna array synthesis. In: IEEE Congress on Evolutionary Computation (CEC), Barcelona, Spain, 18–23 July 2010 (2010)
36. Jafari, S., Khalaf, P., Montazeri-Gh, M.: Multi-objective meta heuristic optimization algorithm with multi criteria decision making strategy for aero-engine controller design. Int. J. Aerosp. Sci. **3**(1), 6–17 (2014)

37. Teodorovic, D., Pavkovich, G.: The fuzzy set theory approach to the vehicle routing problem when demand at nodes is uncertain. Fuzzy Set Syst. **82**, 307–317 (1996)
38. Werners, B., Kondratenko, Y.: Alternative fuzzy approaches for efficiently solving the capacitated vehicle routing problem in conditions of uncertain demands. In: Berger-Vachon, C., et al. (eds.) Complex Systems: Solutions and Challenges in Economics, Management and Engineering, Studies in Systems, Decision and Control, vol. 125, pp. 521–543. Springer, Berlin, Heidelberg (2018)
39. Kondratenko, Y.P., Encheva, S.B., Sidenko, E.V.: Synthesis of intelligent decision support systems for transport logistic. In: 6th IEEE International Conference on Intelligent Data Acquisition and Advanced Computing Systems: Technology and Applications (IDAACS), Prague, Czech Republic, 15–17 September 2011 (2011)
40. Kondratenko, Y.P., Al Zubi, E.Y.M.: The optimisation approach for increasing efficiency of digital fuzzy controllers. In: Annals of DAAAM and Proceedings, Vienna, Austria, January 2009 (2009)
41. Mamdani, E.H.: Application of fuzzy algorithm for control of a simple dynamic plant. Proc. Inst. Electr. Eng. **121**(12), 1585–1588 (1974)
42. Simon, D.: Training fuzzy systems with the extended Kalman filter. Fuzzy Set Syst. **132**, 189–199 (2002)
43. Kondratenko, Y., Simon, D.: Structural and parametric optimization of the fuzzy control and decision making systems. In: Zadeh, L.A., et al. (eds.) Resent Development and New Direction in Soft Computing Foundations and Applications. Studies in Fuzziness and Soft Computing, vol. 361, pp. 273–289. Springer, Cham (2018)
44. Kuntsevich, V.M., Gubarev, V.F., Kondratenko, Y.P., Lebedev, D.V., Lysenko, V.P. (eds.): Control Systems: Theory and Applications, Series in Automation, Control and Robotics. River Publishers, Gistrup, Delft (2018)
45. Sinha, A., Korhonen, P., Wallenius, J., Deb, K.: An interactive evolutionary multi-objective optimization algorithm with a limited number of decision maker calls. Eur. J. Oper. Res. **233**(3), 674–688 (2014)
46. Fleiss, J.L., Levin, B., Paik, M.C.: Statistical Methods for Rates and Proportions. Wiley (2013)
47. Kaufmann, A., Gupta, M.: Introduction to Fuzzy Arithmetic: Theory and Applications. Van Nostrand Reinhold Company, New York (1985)
48. Kondratenko, Y.P., Kondratenko, N.Y.: Soft computing analytic models for increasing efficiency of fuzzy information processing in decision support systems. In: Hudson, R. (ed.) Decision Making: Processes, Behavioral Influences and Role in Business Management, pp. 41–78. Nova Science Publishers, New York (2015)
49. Kondratenko, Y., Kondratenko, V.: Soft computing algorithm for arithmetic multiplication of fuzzy sets based on universal analytic models. In: Ermolayev, V., et al. (eds.) 10th International Conference on Information and Communication Technologies in Education, Research, and Industrial Applications, ICTERI 2014, pp. 49–77. Springer, Cham (2014)
50. Piegat, A.: Fuzzy Modeling and Control, vol. 69. Physica (2013)
51. Simon, D.: Sum normal optimization of fuzzy membership functions. Int. J. Uncertain. Fuzz. **10**, 363–384 (2002)
52. Kondratenko, Y.P., Klymenko, L.P., Al Zu'bi, E.Y.M.: Structural optimization of fuzzy systems' rules base and aggregation models. Kybernetes **42**(5), 831–843 (2013)
53. Pedrycz, W., Li, K., Reformat, M.: Evolutionary reduction of fuzzy rule-based models. In: Tamir, D., et al. (eds.) Fifty Years of Fuzzy Logic and Its Applications, pp. 459–481. Springer, Cham (2015)
54. Simon, D.: Design and rule base reduction of a fuzzy filter for the estimation of motor currents. Int. J. Approx. Reason. **25**, 145–167 (2000)
55. Kondratenko, Y.P., Altameem, T.A., Al Zubi, E.Y.M.: The optimisation of digital controllers for fuzzy systems design. Adv. Model. Anal. **A47**, 19–29 (2010)
56. Lodwick, W.A., Kacprzyk, J. (eds.): Fuzzy Optimization: Resent Advances and Applications, vol. 254. Springer, Heidelberg (2010)

57. Merigo, J.M., Gil-Lafuente, A.M., Yager, R.R.: An overview of fuzzy research with bibliometric indicators. Appl. Soft Comput. **27**, 420–433 (2015)
58. Simon, D.: $H\infty$ estimation for fuzzy membership function optimization. Int. J. Approx. Reason. **40**, 224–242 (2005)
59. Jamshidi, M., Kreinovich, V., Kacprzyk, J. (eds.): Advance trends in soft computing. In: Proceedings WCSC. Springer, Heidelberg (2013)
60. Tamir, D.E., Rishe, N.D., Kandel, A. (eds.): Fifty Years of Fuzzy Logic and Its Applications, vol. 326. Springer, Cham (2015)
61. Zadeh, L.A., Abbasov, A.M., Yager, R.R., Shahbazova, S.N., Reformat, M.Z. (eds.): Recent Developments and New Directions in Soft Computing, vol. 317. Springer, Berlin (2014)
62. Kuntsevich, V.M.: Control Under Uncertainty: Guaranteed Results in Control and Identification Problems. Naukova Dumka, Kyiv (2006) (in Russian)
63. Solesvik, M., Kondratenko, Y., Kondratenko, G., Sidenko, I., Kharchenko, V., Boyarchuk, A.: Fuzzy decision support systems in marine practice. In: IEEE International Conference on Fuzzy Systems (FUZZ-IEEE), Naples, Italy, 9–12 July 2017 (2017)

Bio-inspired Optimization of Type-2 Fuzzy Controllers in Autonomous Mobile Robot Navigation

Oscar Castillo

Abstract In this paper we perform a comparison of using type-2 fuzzy logic in two different bio-inspired methods: Ant Colony Optimization (ACO) and Gravitational Search Algorithm (GSA). Each of these methods is enhanced with a methodology for dynamic parameter adaptation using interval type-2 fuzzy logic, where based on some metrics about the algorithm, like the percentage of iterations elapsed or the diversity of the population, we aim at controlling its behavior and therefore control its abilities to perform a global or a local search. To test these methods two benchmark control problems were used in which a fuzzy controller is optimized to minimize the error in the simulation with nonlinear complex plants.

Keywords Interval type-2 fuzzy logic · Ant colony optimization · Gravitational search algorithm · Dynamic parameter adaptation

1 Introduction

Bio-inspired optimization algorithms can be applied to most combinatorial and continuous optimization problems, but different problems may need different parameter values, in order to obtain better results. There exist in the literature, several methods aiming at modeling better the behavior of these algorithms by adapting some of their parameters [1, 2], introducing different parameters in the equations of the algorithms [3], performing a hybridization with other algorithm [4], and using fuzzy logic [5–11]. In this paper a methodology for parameter adaptation using an interval type-2 fuzzy system is presented, where on each method a better model of the behavior is used in order to obtain better quality results. The proposed methodology has been previously successfully applied to different bio-inspired optimization methods like BCO (Bee Colony Optimization) in [12], CSA (Cuckoo Search Algorithm) in [13], PSO (Particle Swarm optimization) in [5, 7], ACO (Ant Colony Optimization) in [6,

O. Castillo (✉)
Tijuana Institute of Technology, Calzada Tecnologico s/n, Tomas Aquino, 22379 Tijuana, Mexico
e-mail: ocastillo@tectijuana.mx

Y. P. Kondratenko et al. (eds.), *Advanced Control Techniques in Complex Engineering Systems: Theory and Applications*, Studies in Systems, Decision and Control 203, https://doi.org/10.1007/978-3-030-21927-7_9

8], GSA (Gravitational Search Algorithm) in [9, 11], DE (Differential Evolution) in [14], HSA (Harmony Search Algorithm) in [15], BA (Bat Algorithm) in [16] and in FA (Firefly Algorithm) in [17].

The algorithms used in this particular work are: ACO (Ant Colony Optimization) from [8] and GSA (Gravitational Search Algorithm) from [9], each one with dynamic parameter adaptation using an interval type-2 fuzzy system. Fuzzy logic proposed by Zadeh in [18–20] help us to model a complex problem, with the use of membership functions and fuzzy rules, with the knowledge of a problem from an expert, fuzzy logic can bring tools to create a model and attack a complex problem. As related work in control, we can mention the interesting papers by Kuntsevich in [21, 22].

The contribution of this paper is the comparison between the bio-inspired methods which use an interval type-2 fuzzy system for dynamic parameter adaptation, in the optimization of fuzzy controllers for nonlinear complex plants. In particular, the problem of autonomous mobile robot navigation is considered. The adaptation of parameters with fuzzy logic helps to perform a better design of the fuzzy controllers, based on the results which are better than the original algorithms.

2 Bio-inspired Optimization Methods

ACO is a bio-inspired algorithm based on swarm intelligence emerged from ants, proposed by Dorigo in [23], where each individual helps each other to find the best route from their nest to a food source. Artificial ants represent the solutions to a particular problem, where each ant is a tour and each node is a dimension or a component of the problem. Biological ants use pheromone trails to communicate to other ants which path is the best and the artificial ant tries to mimic that behavior in the algorithm.

Artificial ants use probability to select the next node using Eq. 1, where with this equation calculate the probability of an ant k to select the node j from node i.

$$P_{ij}^{k} = \frac{[\tau_{ij}]^{\alpha}[\eta_{ij}]^{\beta}}{\sum_{l} \in N_i^k [\tau_{il}]^{\alpha}[\eta_{il}]^{\beta}}, \quad \text{if } j \in N_i^k \tag{1}$$

The components of Eq. 1 are: P^k is the probability of an ant k to select the node j from node i, τ_{ij} represents the pheromone in the arc that joins the nodes i and j and η_{ij} represents the visibility from node i to node j, with the condition that node j must be in the neighborhood of node i. Also like in nature the pheromone trail evaporates over time, and the ACO algorithm uses Eq. 2 to simulate the evaporation of pheromone in the trails.

$$\tau_{ij} \leftarrow (1-\rho)\tau_{ij}, \quad \forall (i,j) \in L \tag{2}$$

The components of Eq. 2 are: τ_{ij} representing the pheromone trail in the arc that joins the nodes i and j, ρ represents the percentage of evaporation of pheromone, and this equation is applied to all arcs in the graph L.

There are more equations for ACO, but these two equations are the most important in the dynamics of the algorithm, also these equations contain the parameters used to model a better behavior of the algorithm using an interval type-2 fuzzy system.

GSA originally proposed by Rashedi in [24], and is a population based algorithm that uses laws of physics to update its individuals, more particularly uses the Newtonian law of gravity and the second motion law. In this algorithm each individual is considered as an agent, where each one represent a solution to a problem and each agent has its own mass and can move to another agent. The mass of an agent is given by the fitness function, agents with bigger mass are better. Each agent applies some gravitational force to all other agents, and is calculated using Eq. 3.

$$F_{ij}^{d}(t) = G(t)\frac{M_{pi}(t) \times M_{aj}(t)}{R_{ij}(t) + \varepsilon}(x_{j}^{d}(t) - x_{i}^{d}(t)) \tag{3}$$

The components of Eq. 3 are: F_{ij}^{d} is the gravity force between agents i and j, G is the gravitational constant, M_{pi} is the mass of agent i or passive mass, and M_{aj} is the mass of agent j or active mass, R_{ij} is the distance between agents i and j, ε is an small number used to avoid division by zero, x_{j}^{d} is the position of agent j and x_{i}^{d} is the position of agent j.

The gravitational force is used to calculate the acceleration of the agent using Eq. 4.

$$a_{i}^{d}(t) = \frac{F_{i}^{d}(t)}{M_{ii}(t)} \tag{4}$$

The components of Eq. 4 are: a_{i}^{d} is the acceleration force of agent i, F_{i}^{d} is the gravitational force of agent i, and M_{ii} is the inertial mass of agent i.

In GSA the gravitational constant G from Eq. 3, unlike in real life here it can be variable and is given by Eq. 5.

$$G(t) = G_{0}^{-\alpha t/T} \tag{5}$$

The components of Eq. 5 are: G is the gravitational constant, G_{0} is the initial gravitational constant, α is a parameter defined by the user of GSA and is used to control the change in the gravitational constant, t is the actual iteration and T is the total number of iterations. To control the elitism GSA uses Eq. 6 to allow only the best agents to apply their force to other agents, and in initial iterations all the agents apply their force but Kbest will decrease over time until only a few agents are allowed to apply their force.

$$F_{i}^{d}(t) = \sum_{j \in Kbest, j \neq 1} rand_{i} F_{ij}^{d}(t) \tag{6}$$

The components of Eq. 6 are: F_i^d is the new gravity force of agent *i*, *Kbest* is the number of agents allowed to apply their force, sorted by their fitness the best *Kbest* agent can apply their force to all other agents, in this equation *j* is the number of dimension of agent *i*.

3 Methodology for Parameter Adaptation

The optimization methods involved in this comparison have dynamic parameter adaptation using interval type-2 fuzzy systems, and these adaptations are described in detail for ACO in [8] and for GSA in [9]. The way in which this adaptation of parameters was performed is as follows: first a metric about the performance of the algorithms needs to be created, in this case the metrics are a percentage of iteration elapsed described by Eq. 7 and the diversity of individuals described by Eq. 8, then after the metrics are defined we need to select the best parameters to be dynamically adjusted, and this was done based on experimentation with different levels of all the parameters of each optimization method.

$$Iteration = \frac{Current\,Iteration}{Maximum\,of\,Iterations} \tag{7}$$

The components of Eq. 7 are: *Iteration* is a percentage of the elapsed iterations, *current iteration* is the number of elapsed iterations, and *maximum of iterations* is the total number iterations set for the optimization algorithm to find the best possible solution.

$$Diversity\,(S(t)) = \frac{1}{n_s}\sum_{i=1}^{n_s}\sqrt{\sum_{j=1}^{n_x}\left(x_{ij}(t) - \bar{x}_j(t)\right)^2} \tag{8}$$

The components of Eq. 8 are: *Diversity*(*S*) is a degree of dispersion of the population *S*, n_s is the number of individuals in the population *S*, n_x is the number of dimensions in each individual from the population, x_{ij} is the *j* dimension of the individual *i*, *tested* x_j is the *j* dimension of the best individual in the population. After the metrics are defined and the parameters selected, a fuzzy system is created to adjust just one parameter, and with this obtain a fuzzy rule set to control this parameter, and for all the parameters we need to do the same, and at the end only one fuzzy system will be created to control all the parameters at the same time combining all the created fuzzy systems. The proposed methodology for parameter adaptation is illustrated in Fig. 1, where it contains the optimization method, which has an interval type-2 fuzzy system for parameter adaptation.

Figure 1 illustrates the general scheme for parameter adaptation, in which the bio-inspired optimization algorithm is evaluated by the metrics and these are used as inputs for the interval type-2 fuzzy system, which will adapt some parameters

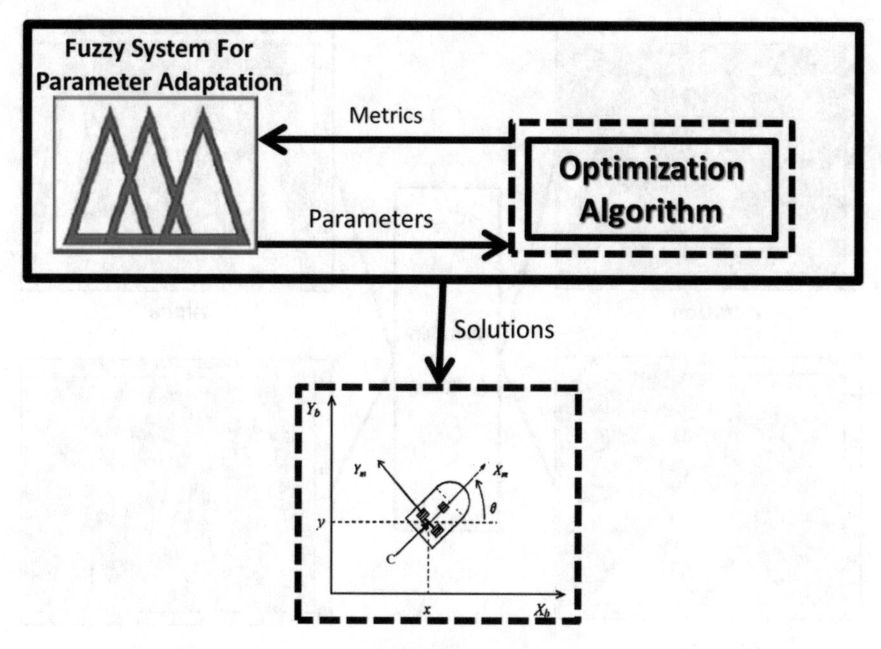

Fig. 1 General scheme of the proposal for parameter adaptation

of the optimization algorithm based on the metrics and the fuzzy rules. Then this method with parameter adaptation will provide the parameters or solutions for a problem, in this case the parameters for the fuzzy system used for control. The final interval type-2 fuzzy systems for each optimization method are illustrated in Fig. 2 and Fig. 3 respectively, for ACO and GSA correspondingly. Each of these fuzzy systems has iteration and diversity as inputs, with a range from 0 to 1 using the Eqs. 7 and 8 correspondingly to each input, and two outputs but these differs from each optimization method because each one has its own parameters to be dynamically adjusted.

The interval type-2 fuzzy system from Fig. 2 has two inputs and two outputs, the inputs are granulated into three type-2 triangular membership functions and the outputs into five type-2 triangular membership functions, and nine rules, in this case the parameters to be dynamically adjusted over the iterations are α *(alpha)* and ρ *(rho)* from Eq. 1 and Eq. 2 respectively, both with a range from 0 to 1.

The interval type-2 fuzzy system from Fig. 3 has *iteration* and *diversity* as inputs with three type-2 triangular membership functions and two outputs, which are the parameters to be adjusted in this case, α *(alpha)* with a range from 0 to 100 and *Kbest* from 0 to 1, each output is granulated into five type-2 triangular membership functions with a fuzzy rule set of nine rules. The parameters α *(alpha)* and *Kbest* are from Eq. 5 and Eq. 6 respectively.

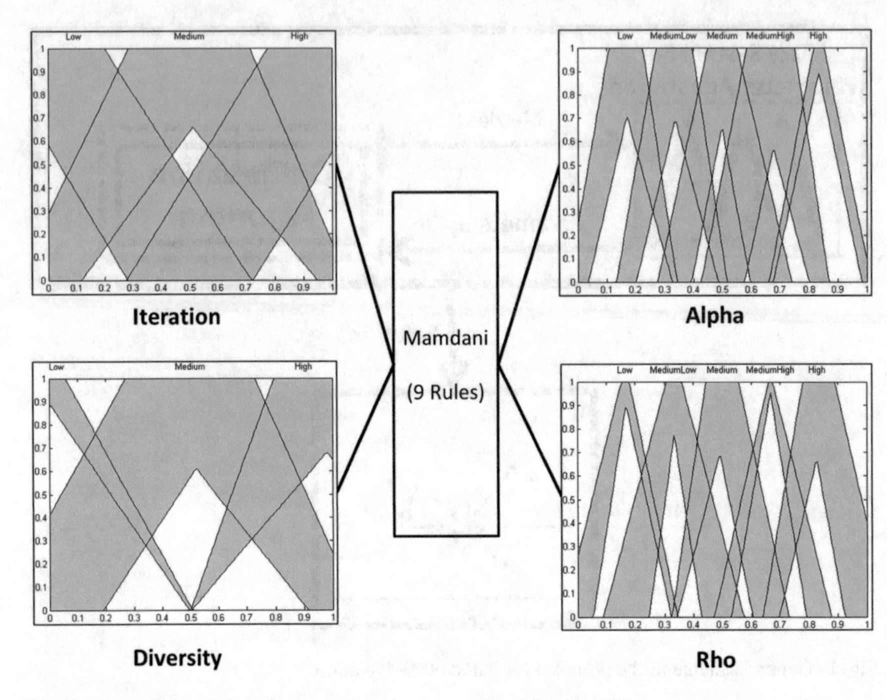

Fig. 2 Interval type-2 fuzzy system for parameter adaptation in ACO

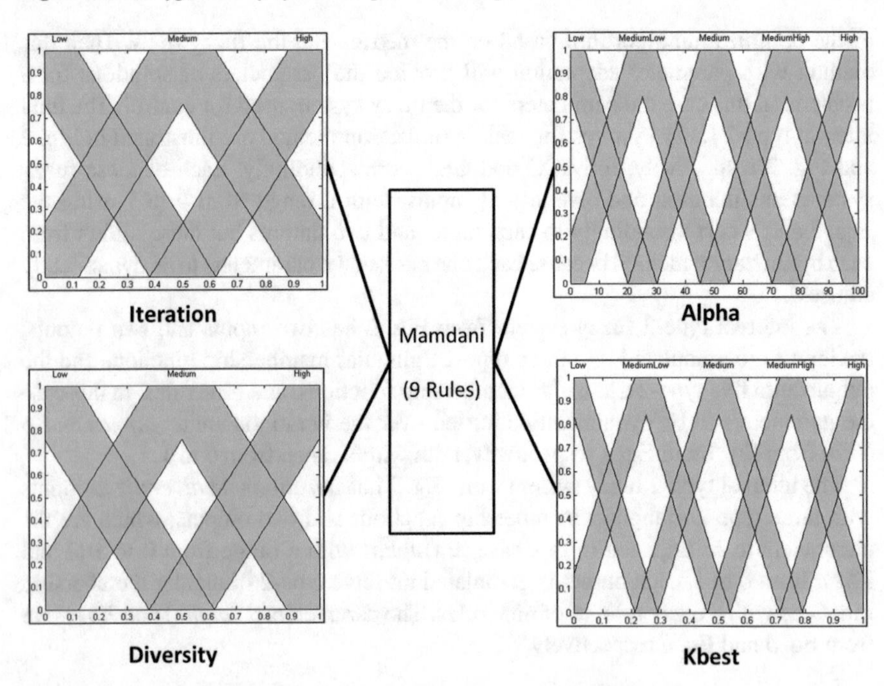

Fig. 3 Interval type-2 fuzzy system for parameter adaptation in GSA

4 Problems Statement

The comparison of ACO and GSA is through the optimization of a fuzzy controller from two different non-linear complex plants, where these two problems use a fuzzy system for control. The first problem is the optimization of the trajectory of an autonomous mobile robot and the objective is to minimize the error in the trajectory, the robot has two wheeled motors and one stabilization wheel, it can move in any direction. The desired trajectory is illustrated in Fig. 4, where first the robot must start from point (0, 0) and it needs to follow the reference using the fuzzy system from Fig. 5 as a controller. The reference illustrated in Fig. 4 helps in the design of a good controller because it uses only nonlinear trajectories, to assure that the robot can follow any trajectory.

The fuzzy system used for control is illustrated in Fig. 5, and uses the linear and angular errors to control the motorized wheels of the robot. In this problem the optimization methods will aim at finding better parameters for the membership functions, using the same fuzzy rule set. The second problem is the automatic temperature control in a shower, and the optimization method will optimize the fuzzy controller illustrated in Fig. 6, which will try to follow the flow and temperature reference values. The fuzzy system used as control is illustrated in Fig. 6 and has two input variables, temperature and flow; a fuzzy rule set of nine rules and two outputs cold and hot is presented. The fuzzy system uses the inputs and with the fuzzy rules to control the open-close mechanism of the cold and hot water.

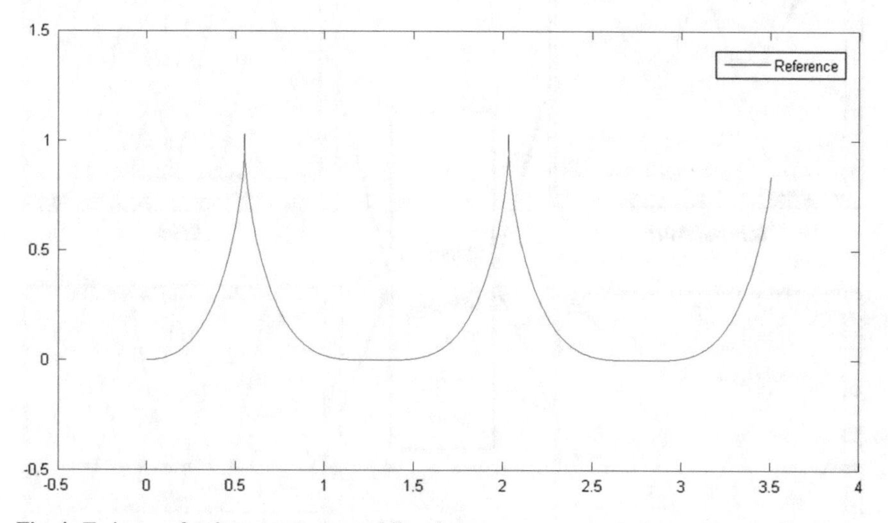

Fig. 4 Trajectory for the autonomous mobile robot

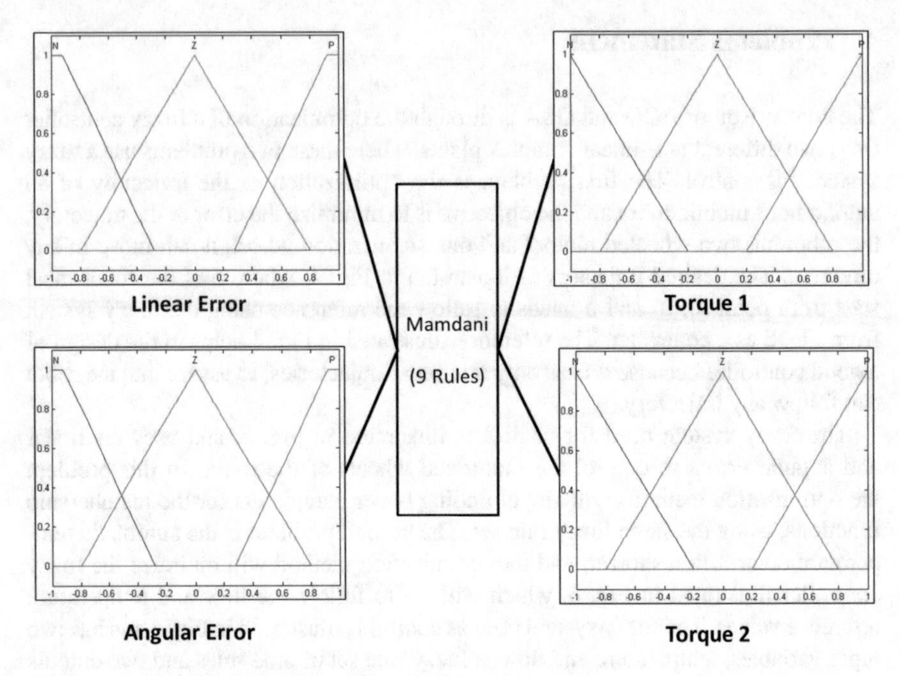

Fig. 5 Fuzzy controller for the autonomous mobile robot

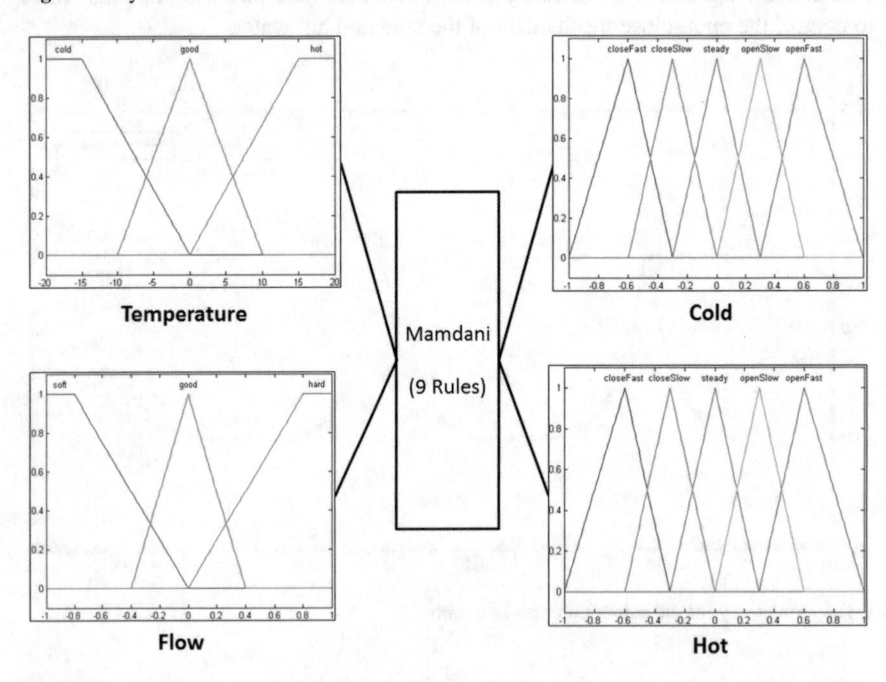

Fig. 6 Fuzzy controller for the automatic temperature control in a shower

5 Simulations, Experiments and Results

The optimization methods were applied to the optimization of the membership functions of the fuzzy system used as controllers for the two problems described in Sect. 4. In this case, using the parameters from Table 1, each method was applied to both problems. In the case of the problem of the trajectory of an autonomous mobile robot there are 40 points to be search for all the membership functions, and in the problem of the automatic temperature control in a shower there are 52 points. The methods to be compared are: the original ACO method, ACO with parameter adaptation, original GSA method and GSA with parameter adaptation.

The parameters from Table 1 are a challenge for the optimization methods, because there are only 50 iterations to find the best possible fuzzy controller for each problem. This is a good manner to show the advantages of the proposed methodology for parameter adaptation using an interval type-2 fuzzy system. Table 2 contains the results of applying all the optimization methods to the optimization of the fuzzy controller for an autonomous mobile robot, the average is from 30 experiments (with Mean Square Error (MSE)) and results in bold are best, the 30 experiments means that each method was applied to the fuzzy controller optimization for 30 times resulting in 30 different fuzzy controllers for each method.

Table 1 Parameters for each optimization method

Parameter	Original ACO	ACO with parameter adaptation	Original GSA	GSA with parameter adaptation
Population	30	30	30	30
Iterations	50	50	50	50
α (Alpha)	1	Dynamic	40	Dynamic
β (Beta)	2	2		
ρ (Rho)	0.1	Dynamic		
Kbest			Linear decreasing (100–2%)	Dynamic
G_0			100	100

Table 2 Results of the simulations with the robot problem

MSE	Original ACO	ACO with parameter adaptation	Original GSA	GSA with parameter adaptation
Average	0.4641	**0.0418**	36.4831	15.4646
Best	0.1285	**0.0048**	10.4751	3.2375
Worst	0.9128	**0.1276**	76.0243	30.8511
Standard deviation	0.2110	**0.0314**	15.8073	8.6371

Table 3 Results of the simulations with the shower problem

MSE	Original ACO	ACO with parameter adaptation	Original GSA	GSA with parameter adaptation
Average	0.6005	0.4894	3.8611	**0.1151**
Best	0.5407	0.3980	1.9227	**0.0106**
Worst	0.9036	0.5437	6.5659	**0.3960**
Standard deviation	0.0696	0.0378	1.0860	**0.0913**

Table 4 Parameters for the statistical Z-test

Parameter	Value
Level of significance	95%
Alpha (α)	5%
Alternative hypothesis (H_a)	$\mu_1 < \mu_2$ (claim)
Null hypothesis (H_0)	$\mu_1 \geq \mu_2$
Critical value	-1.645

From Table 2 the optimization method that obtains better results is ACO with parameter adaptation using the proposed methodology with an interval type-2 fuzzy system, also it can be seen that the results of GSA with parameter adaptation are better that the original GSA, but ACO is better. The results in Table 3 are from applying all the methods to optimize the fuzzy controller for the automatic temperature control in a shower, the average is from 30 experiments (with the Mean Square Error (MSE)) and also the results in bold are best, same as the first problem the 30 experiments means that each method was applied to the optimization of the fuzzy controller and obtaining 30 different fuzzy controller for each method.

From the results in Table 3 in this case the GSA with parameter adaptation using the proposed methodology using an interval type-2 fuzzy system obtains better results than the other methods. Also it can be seen that ACO with parameter adaptation can obtain better results than the original ACO method and the original GSA method.

6 Statistical Comparison

The Z-test is a tool to prove that the methods with parameter adaptation can obtain on average better results than its counterparts the original methods, also to know what method is better on certain problem by comparing its results with all of the other methods. The comparison between the methods is using the statistical test Z-test, using the parameters from Table 4 and the results of the comparisons are in Table 5 and Table 6 for the robot and shower problems, respectively.

Table 5 Results of the Z-test for comparison in the robot problem

μ_1	μ_2			
	Original ACO	ACO with parameter adaptation	Original GSA	GSA with parameter adaptation
Original ACO		10.8415	**−12.4795**	**−9.5098**
ACO with parameter adaptation	**−10.8415**		**−12.6269**	**−9.7803**
Original GSA	12.4795	12.6269		6.3911
GSA with parameter adaptation	9.5098	9.7803	**−6.3911**	

Table 6 Results of the Z-test for comparison in the shower problem

μ_1	μ_2			
	Original ACO	ACO with parameter adaptation	Original GSA	GSA with parameter adaptation
Original ACO		7.6813	**−16.4115**	23.1516
ACO with parameter adaptation	**−7.6813**		**−16.9950**	20.7332
Original GSA	16.4115	16.9950		18.8264
GSA with parameter adaptation	**−23.1516**	**−20.7332**	**−18.8264**	

The results in Table 5 are obtained using the parameters in Table 4 for the Z-test, where it claims that a method (μ_1) has on average better results (we are comparing errors, so minimum is better) than the other method (μ_2), in Tables 5 and 6 the first column correspond to the methods as μ_1 and the first row correspond to the methods as μ_2, also we are not comparing the same method with itself, results in bold means that there are enough evidence to reject the null hypothesis.

From the results in Table 5, which correspond to the optimization of a fuzzy controller for the trajectory of an autonomous mobile robot, there is enough evidence that ACO method with parameter adaptation can obtain on average better results than all of the other methods. There is enough evidence that the original ACO method can obtain on average better results than the original GSA and GSA with parameter adaptation. There is also enough evidence that GSA with parameter adaptation can obtain on average better results than the original GSA method.

From the results in Table 6, which correspond to the optimization of a fuzzy controller for the automatic temperature control in a shower, there is enough evidence that GSA with parameter adaptation can obtain on average better results than all of the other methods. There is enough evidence that ACO with parameter adaptation can obtain on average better results than the original ACO method and the original GSA method. There is also enough evidence that the original ACO method can obtain on average better results than the original GSA method.

7 Conclusions

The optimization of fuzzy controllers is a complex task, because it requires the search of values for several parameters in a set of infinite possibilities in the range of each input or output variables. Bio-inspired optimization methods help in the search because they are guided by some kind of intelligence, from swarm intelligence or from laws of physics and can make a better search of parameters. With the inclusion of a fuzzy system in this case based on interval type-2 fuzzy logic, the bio-inspired methods can search even in a better way, because is guided by the knowledge of an expert system that models a proper behavior in determined states of the search, in the beginning improves the global search or exploration of the search space and in final improves the local search or the exploitation of the best area found so far of the entire search space. From the results with the MSE there is clearly that ACO with parameter adaptation has the best results in the robot problem, and GSA with parameter adaptation has the best results in the shower problem, but with the statistical test it confirm these affirmations. The statistical comparison shows that the methods with parameter adaptation are better than their counterparts the original methods. Also ACO is a better method with the robot problem, but GSA is better in the shower problem (which is simpler). As future work, we envision optimizing type-2 fuzzy controllers, like in [25–28]. Also, consider other types of applications, like in [29–31].

References

1. Valdez, F., Melin, P., Castillo, O.: Evolutionary method combining particle swarm optimization and genetic algorithms using fuzzy logic for decision making. In: IEEE International Conference on Fuzzy Systems, pp. 2114–2119 (2009)
2. Wang, B., Liang, G., Chan Lin, W., Yunlong, D.: A new kind of fuzzy particle swarm optimization fuzzy_PSO algorithm. In: 1st International Symposium on Systems and Control in Aerospace and Astronautics, ISSCAA 2006, pp. 309–311 (2006)
3. Hongbo, L., Abraham, A.: A fuzzy adaptive turbulent particle swarm optimization. Int. J. Innov. Comput. Appl. **1**(1), 39–47 (2007)
4. Taher, N., Ehsan, A., Masoud, J.: A new hybrid evolutionary algorithm based on new fuzzy adaptive PSO and NM algorithms for distribution feeder reconfiguration. Energy Convers. Manag. (Elsevier) **54**, 7–16 (2012)
5. Melin, P., Olivas, F., Castillo, O., Valdez, F., Soria, J., Garcia, J.: Optimal design of fuzzy classification systems using PSO with dynamic parameter adaptation through fuzzy logic. Expert Syst. Appl. (Elsevier) **40**(8), 3196–3206 (2013)

6. Neyoy, H., Castillo, O., Soria, J.: Dynamic Fuzzy Logic Parameter Tuning for ACO and Its Application in TSP Problems. Studies in Computational Intelligence, vol. 451, pp. 259–271. Springer (2012)
7. Olivas, F., Valdez, F., Castillo, O., Melin, P.: Dynamic parameter adaptation in particle swarm optimization using interval type-2 fuzzy logic. Soft. Comput. **20**(3), 1057–1070 (2016)
8. Olivas, F., Valdez, F., Castillo, O., Gonzalez, C., Martinez, G., Melin, P.: Ant colony optimization with dynamic parameter adaptation based on interval type-2 fuzzy logic systems. Appl. Soft Comput. **53**, 74–87 (2017)
9. Olivas, F., Valdez, F., Castillo, O., Melin, P.: Interval type-2 fuzzy logic for dynamic parameter adaptation in a modified gravitational search algorithm. Inf. Sci. **476**, 159–175 (2019)
10. Shi, Y., Eberhart, R.: Fuzzy adaptive particle swarm optimization. In: Proceeding of IEEE International Conference on Evolutionary Computation, Piscataway, NJ, pp. 101–106. IEEE Service Center, Seoul, Korea (2001)
11. Sombra, A., Valdez, F., Melin, P., Castillo, O.: A new gravitational search algorithm using fuzzy logic to parameter adaptation. In: 2013 IEEE Congress on Evolutionary Computation (CEC), pp. 1068–1074. IEEE, June 2013
12. Amador-Angulo, L., Castillo, O.: Statistical analysis of type-1 and interval type-2 fuzzy logic in dynamic parameter adaptation of the BCO. In: 2015 Conference of the International Fuzzy Systems Association and the European Society for Fuzzy Logic and Technology (IFSA-EUSFLAT-15). Atlantis Press, June 2015
13. Guerrero, M., Castillo, O., Garcia, M.: Fuzzy dynamic parameters adaptation in the Cuckoo Search Algorithm using fuzzy logic. In: 2015 IEEE Congress on Evolutionary Computation (CEC), pp. 441–448. IEEE, May 2015
14. Ochoa, P., Castillo, O., Soria, J.: Differential evolution with dynamic adaptation of parameters for the optimization of fuzzy controllers. In: Recent Advances on Hybrid Approaches for Designing Intelligent Systems, pp. 275–288. Springer International Publishing (2014)
15. Peraza, C., Valdez, F., Castillo, O.: An improved harmony search algorithm using fuzzy logic for the optimization of mathematical functions. In: Design of Intelligent Systems Based on Fuzzy Logic, Neural Networks and Nature-Inspired Optimization, pp. 605–615. Springer International Publishing (2015)
16. Perez, J., Valdez, F., Castillo, O., Melin, P., Gonzalez, C., Martinez, G.: Interval type-2 fuzzy logic for dynamic parameter adaptation in the bat algorithm. Soft Comput. 1–19 (2016)
17. Solano-Aragon, C., Castillo, O.: Optimization of benchmark mathematical functions using the firefly algorithm with dynamic parameters. In: Fuzzy Logic Augmentation of Nature-Inspired Optimization Metaheuristics, pp. 81–89. Springer International Publishing (2015)
18. Zadeh, L.: Fuzzy sets. Inf. Control **8** (1965)
19. Zadeh, L.: Fuzzy logic. IEEE Comput. **83–92** (1965)
20. Zadeh, L.: The concept of a linguistic variable and its application to approximate reasoning—I. Inf. Sci. **8**, 199–249 (1975)
21. Kuntsevich, V.M.: Control Under Uncertainty: Guaranteed Results in Control and Identification Problems. Naukova Dumka, Kyiv (2006) (in Russian)
22. Kuntsevich, V.M., Gubarev, V.F., Kondratenko, Y.P., Lebedev, D.V., Lysenko, V.P. (eds.): Control Systems: Theory and Applications. Series in Automation, Control and Robotics. River Publishers (2018)
23. Dorigo, M.: Optimization, Learning and Natural Algorithms. Ph.D. thesis, Dipartimento di Elettronica, Politechico di Milano, Italy (1992)
24. Rashedi, E., Nezamabadi-pour, H., Saryazdi, S.: GSA: a gravitational search algorithm. Inf. Sci. (Elsevier) **179**(13), 2232–2248 (2009) (Iran)
25. Leal Ramírez, C., Castillo, O., Melin, P., Rodríguez Díaz, A.: Simulation of the bird age-structured population growth based on an interval type-2 fuzzy cellular structure. Inf. Sci. **181**(3), 519–535 (2011)
26. Cázarez-Castro, N.R., Aguilar, L.T., Castillo, O.: Designing type-1 and type-2 fuzzy logic controllers via fuzzy Lyapunov synthesis for nonsmooth mechanical systems. Eng. Appl. AI **25**(5), 971–979 (2012)

27. Castillo, O., Melin, P.: Intelligent systems with interval type-2 fuzzy logic. Int. J. Innov. Comput. Inf. Control **4**(4), 771–783 (2008)
28. Mendez, G.M., Castillo, O.: Interval type-2 TSK fuzzy logic systems using hybrid learning algorithm. In: The 14th IEEE International Conference on Fuzzy Systems, 2005, FUZZ'05, pp. 230–235
29. Melin, P., Castillo, O.: Intelligent control of complex electrochemical systems with a neuro-fuzzy-genetic approach. IEEE Trans. Ind. Electron. **48**(5), 951–955
30. Melin, P., Sánchez, D., Castillo, O.: Genetic optimization of modular neural networks with fuzzy response integration for human recognition. Inf. Sci. **197**, 1–19 (2012)
31. Melin, P., Sánchez, D.: Multi-objective optimization for modular granular neural networks applied to pattern recognition. Inf. Sci. **460–461**, 594–610 (2018)

A Status Quo Biased Multistage Decision Model for Regional Agricultural Socioeconomic Planning Under Fuzzy Information

Janusz Kacprzyk, Yuriy P. Kondratenko, Jos'e M. Merigó, Jorge Hernandez Hormazabal, Gia Sirbiladze and Ana Maria Gil-Lafuente

Abstract We proposed a novel fuzzy multistage control model of sustainable regional agricultural development which better reflects specific features of human stakeholders. First, we use fuzzy logic for the modeling of imprecision in human judgments, intentions, preferences, evaluations, etc. Second, we propose to reflect in the model the so called status quo bias of the humans which basically stands for a common propensity of the humans to stay with known and already employed procedures and courses of action, avoiding larger changes. We develop therefore a

J. Kacprzyk (✉)
Systems Research Institute, Polish Academy of Sciences,
ul. Newelska 6, 01-447 Warsaw, Poland
e-mail: kacprzyk@ibspan.waw.pl

WIT – Warsaw School of Information Technology, ul. Newelska 6,
01-447 Warsaw, Poland

Y. P. Kondratenko
Department of Intelligent Information Systems, Petro Mohyla Black Sea
National University, 10, 68th Desantnykiv Str., Mykolaiv 54003, Ukraine
e-mail: yuriy.kondratenko@chmnu.edu.ua

J. M. Merigó
Department of Management Control and Information Systems, University of Chile,
Av. Diagonal Paraguay 257, 8330015 Santiago, Chile
e-mail: jmerigo@fen.uchile.cl

J. H. Hormazabal
Management School, University of Liverpool, Chatham Street,
Liverpool L69 7ZH, UK
e-mail: J.E.Hernandez@liverpool.ac.uk

G. Sirbiladze
Department of Computer Sciences, Ivane Javakhishvili Tbilisi State University,
University St. 13, 0186 Tbilisi, Georgia
e-mail: gia.sirbiladze@tsu.ge

A. M. Gil-Lafuente
Department of Business Administration, University of Barcelona,
Av. Diagonal 690, 08034 Barcelona, Spain
e-mail: amgil@ub.edu

Y. P. Kondratenko et al. (eds.), *Advanced Control Techniques in Complex Engineering Systems: Theory and Applications*, Studies in Systems, Decision and Control 203, https://doi.org/10.1007/978-3-030-21927-7_10

human centric model. We also indicate that the inclusion of the status quo bias can be viewed as a way to mitigate risk which is crucial, notably in the case of agriculture. We present some simple example of how a best (optimal) investment policy can be obtained under different development scenarios, and indicate what change the inclusion of the status quo bias brings.

1 Introduction

The purpose of this conceptual paper is many fold. First, we wish to revisit an open loop control model for the planning of sustainable agricultural regional development model proposed by Kacprzyk and Straszak [34] developed while the authors were working for the International Institute for Applied Systems Analysis (IIASA) in Laxenburg, Austria (www.iiasa.ac.at), and then used over the years for the modeling of many agricultural regions all over the world. The model is based on a multistage control (decision making) model (cf. [22, 29]). The model is based on life quality indicators attained as results of some investments (expenditures, outlays). Second, we wish to propose an attempt at using this model for the purpose of The European Union Project "RUC-APS: Enhancing and implementing Knowledge based ICT solutions within high Risk and Uncertain Conditions for Agriculture Production Systems" (www.ruc-aps.eu), funded by the EU under H2020-MSCA-RISE-2015, notably in the sense of handling the broadly perceived risk in the agricultural value chain. Third, and presumably the most important, in the paper a novel attempt is presented to take into account some human specific characteristic features, notably the so called *status quo bias* the essence of which is that the humans usually and on the average prefer traditional, well established procedures, courses of action and solution—cf. Beedell and Rehman [3], Feola and Binder [9], Hermann, Musshoff and Agethen [13], Rossi Borges, Oude Lansink, Marques Ribeiro and Lutke [52], Senger, Rossi Borges and Desimon Machadao [54], to just cite a few. Needless to say that farmers and other stakeholders in agricultural systems may often be conservative and exhibit the status quo bias. In this status quo based direction we mention a very interesting, and relevant for our purposes, direction of research which related the "intensity" of status quo bias in particular stakeholders (e.g. farmers or their groups) with their risk friendliness, so that a relation to the very topic of the RUC-APS Project can be established. Finally, since the aggregation of partial scores is crucial for the evaluation of goodness of agricultural regional development trajectories, we will advocate the use of some more sophisticated aggregation operators, basically from a rich family of the OWA (ordered weighted averaging) operators presented in a comprehensive review and analysis in Kacprzyk, Yager and Merigó [38].

Our point of interest is a decision making problem, meant in the traditional Bellman and Zadeh's [4] sense, in which there are some imprecisely specified constraints, the so called fuzzy constraints, on the set of possible options (variants, choices, alternatives, …) and some imprecisely specified goals, the so called fuzzy goals, on the resulting consequences, i.e. outcome implied by the selection of a particular option. A relation (mapping) from the set of options to the set of consequences is known, maybe also in an imprecise (fuzzy) manner. The goodness (quality) of applying an option, i.e making a decision, is evaluated by an aggregation of the degrees to which the fuzzy constraints and fuzzy goals are satisfied, and this is called a fuzzy decision. In our case, we assume a dynamic setting in which the fuzzy constraints and fuzzy goals are assumed at subsequent planning stages over some planning horizon, so that we have a multistage decision making (control) problem under fuzziness which can be conveniently formulated and solved in terms of dynamic programming, more specifically fuzzy dynamic programming originally introduced in the source paper by Bellman and Zadeh's [4], and then considerably extended in Kacprzyk's books [19, 22]. The fuzzy dynamic programming will be presented as a powerful tool for the solution of multistage decision making and control problems under imprecise (fuzzy) information, i.e. under fuzzy goals and constraints. Of course, dynamic programming is plagued by the so called curse of dimensionality so that the solution of very large problems may be computationally difficult but this does not concern our model of sustainable agricultural regional development.

Many fuzzy dynamic programing models have found applications in diverse areas but for our purposes the use of fuzzy dynamic programming for solving sustainable agricultural regional development planning problems are relevant. These works have been initiated by Kacprzyk and Straszak [31–34] in the end of the 1970s and beginning of the 1980s at the International Institute for Applied Systems Analysis (IIASA) in Laxenburg, Austria (www.iiasa.at), and then the model has been employed in numerous agricultural regional planning projects in various countries, exemplified by the Upper Noteć Region in Poland, Tisza Region in Hungary, Kinki Region in Japan, and many other ones. The works have resulted in many publications as mentioned above and one should mention in our context Kacprzyk, Francelin and Gomide [37] in which some models of both objective and subjective aspects of evaluations have been proposed. These models proposed and application results have been mentioned as one of the most successful examples of fuzzy systems modeling in a Special Volume on the Fiftieth Anniversary of the British Operational Research Society published in 1987 by Pergamon Press—cf. Thomas [58].

Recently, the above models have been extended along many lines, notably by using elements of Wang's [61, 62] *cognitive informatics* to better represent the role of human cognition for the modeling of sustainable agricultural development (cf. [28]). Moreover, as possible extensions the inclusion of elements of a deeper analysis of uncertainly and imprecision (cf. [42]), extended analyses of information granulation (cf. [51]), fuzzy rule based modeling of systems and decision processes (cf. [41, 47]). Some elements of the use of fuzzy models of economic analyses, are exemplified by Anselin-Avila and Gil-Lafuente [2]. The use more sophisticated aggregation operators well suited for economic problems, exemplified by Merigó and Gil-Lafuente [45].

A brief account of various approaches to a more general problem of the modeling can be found in, for instance, in Jones et al. [16] in which a history of agricultural systems modeling is presented in a short yet comprehensive way. On the other hand, Domptail and Nuppeneau [6] consider the role of uncertainties (risks) and expectations and how to model them in the context of dynamic modeling; this is relevant for our discussion.

In the first part of this paper, to set the stage, we will briefly present a brief survey of fuzzy sets theory, to fix the notation to be used, followed by fuzzy dynamic programming and then a short description of the multistage control model for sustainable agricultural regional development planning considered in terms of expenditures, subsidies, life qualities, etc.

The next issue considered in the paper is a possible extension which may be viewed as following an interesting and often advocated direction in the mathematic modeling of various human centric systems, i.e. systems in which the human being with his/her judgments, intentions, preferences, evaluations, etc. plays an important role. In our context we consider the so called *cognitive biases* the essence of which, in our context, is that humans deviate in their behavior form what can be expected from a rational behavior that can be expected from traditional models of rationality based on the utility maximization. The cognitive biases have been for years an area of intensive research by psychologists, cognitive scientists, decision scientists, economists, etc. This field, in our decision making setting, has presumably been initiated by Kahneman, a winner of the Nobel Prize in economics, and his late collaborator Tversky (cf. [59]).

In our context, usually, a decision is called rational if it is in some sense *optimal*, and individuals or organizations are often called *rational* if they tend to act *optimally* in pursuit of their goals, thus rationality simply refers to the success of goal attainment, and often has a self-interest or selfish type. The optimality is basically in the sense of utility maximization. It is easy to see that the humans not always behave following these patterns. One cannot say that the existence of cognitive biases does not mean that the humans make mistakes or are irrational, but this only means that the humans make judgments or decisions that are systematically different from what the traditional economic models (i.e. of the homo economicus type) say, and the cognitive biases do not necessarily lead to bad or sub-optimal outcomes (cf. [12, 50, 59], to just cite a few).

One of the most widely encountered and important, rich in consequences, cognitive biases is the famous *status quo bias* (cf. [53]) which is basically a preference for the current state of affairs, and a perception of any deviation from the status quo as a loss. A similar, though different from the cognitive, psychological and behavioral points of view is the so called *psychological inertia* boiling down to a lack of intervention in a current course of affairs.

One can easily see that the agricultural systems in virtually all countries are highly human centric in the sense that their crucial elements are human beings whose judgments, intentions, preferences, decisions, behavior, etc. determine the structure and operation of the system. Who play the role of various stakeholders. Moreover, the agricultural systems are traditional in the sense that the production process has usually been known for decades, maybe centuries, so that a natural consequence is that it proceeds rather traditionally. A natural consequence is that the stakeholders usually operate in a traditional way so that their behavior can for sure be said to follow to a large extent the status quo bias. There are very many confirmation for this in the economic and agricultural literature. There is a very rich literature on this topic, and many works are concerned with the use if the so called *theory of planned behavior* which is basically a socio-psychological theory that links beliefs and behavior of humans, and states that attitude toward the behavior, subjective norms, and perceived behavioral control determine together behavioral intentions and behaviors of humans. The papers mentioned below as inspirations for our work, usually use at least some ideas and concepts form the theory of planned behavior. Presumably, the paper by Beedell and Rehman [3] is the most inspiring and, briefly speaking, the authors present results of research on attitudes and motivations of farmers and shown examples of their conservative attitude. Hermann, Musshoff and Agethen [13] show results of experiments on small farms, in the context of rural development, and an analysis of which and how underlying psychological factors affect farmers' intention to diversify their agricultural production. Borges, Lansink, Ribeiro and Lutke [52] consider the intention of farmers to adopt improved natural grassland taking into account a so called behavioral beliefs: increased cattle weight gains, increased number of animals per hectare, availability of pasture the whole year, increased pasture resistance, decreased soil erosion, and decreased feeding costs. Senger, Borges and Machado [54] and Rossi Borges, Oude Lansink, Marques Ribeiro and Lutke [52] consider a similar problem of intentions of small farmers to diversify their agricultural production. All these works have shown that the use of the theory of planned behavior is an effective tool for the analysis of farmers' intentions and behavior.

As one can see from many examples, the farmers often exhibit a conservative approach which can be simply attributed to the very human characteristics, i.e. a propensity to follow the status quo bias. As we have already mentioned, the status quo bias may be justified, notably in the presence of insufficient information which prohibits the use of more sophisticated decision making tools and techniques. The status quo bias, as shown by Samuelson and Zeckhauser [53], is consistent with *loss aversion*, a very human specific feature, and *regret avoidance*, and as stated by Kahneman, Slovic and Tversky [39] people feel greater regret for bad outcomes that result from their new actions than for bad consequences of no actions undertaken.

In this context of the status quo bias and risk, which is relevant for the RUC-APS H2020 Project, the work of Cordaro and Desdoigt [5] is important for our considerations. The authors thoroughly investigate small cocoa farmers in Ivory Coast in Africa who suffer from economic difficulties in the sense of a low yield per hectare, stagnation of cocoa production, etc. Surely, they are small, do not have enough resources, and are conservative so that rarely use newer technologies, and prefer status quo solutions over proactive behavior. As opposed to the mainstream of economic literature which would suggest in this case as possible reasons some external factors exemplified by a lack of capital, the authors analyze internal factors and show interesting results of questioning farmers about reasons for not trying to change technology. The results indicate that such an attitude may change as a result of learning and attempts to increase the social capital. The study mentions also the role of these undertaking to deal with risk, which is an inherent characteristic feature in agriculture, also in cocoa production in the case of the paper considered. An attempt to reduce, or mitigate, risk implies the status quo based behavior. This is very relevant for the RUC-APS project and many other modeling projects in human centric systems, not only in agriculture.

2 Brief Introduction to Fuzzy Sets Theory

The fuzzy set, introduced by Zadeh in 1965 [67], may be viewed as a class of objects with unsharp boundaries, i.e. in which the transition from the belongingness to non-belongingness is gradual rather than abrupt, i.e. elements of a fuzzy set may belong to it to *partial degrees*, from the full belongingness to the full non-belongingness through all intermediate values. This may formally be represented by a *membership function* defined as

$$\mu_A : X \longrightarrow [0,1] \tag{1}$$

such that $\mu_A(x) \in [0,1]$ is the degree to which an element $x \in X$ belongs to A: from $\mu_A(x) = 0$ for the full non-belongingness to $\mu_A(x) = 1$ for the full belongingness, through all intermediate $(0 < \mu_A(x) < 1)$ values.

In practice the membership function is usually assumed to be piecewise linear and it will also be the case here. Moreover, the form of the membership function is *subjective* as opposed to an *objective* one of a characteristic function.

A *fuzzy set* A in a universe of discourse $X = \{x\}$, written A in X, can be conveniently defined as a set of pairs

$$A = \{(\mu_A(x), x)\} \tag{2}$$

where $\mu_A : X \longrightarrow [0,1]$ is the *membership function* of A and $\mu_A(x) \in [0,1]$ is the *grade of membership* (or a *membership grade*) of an element $x \in X$ in a fuzzy set A; this is clearly equivalent to (2).

In practice it is often assumed X is finite as, e.g., $X = \{x_1, \ldots, x_n\}$, and then A in X is written as

$$\begin{aligned} &= \{(\mu_A(x), x)\} = \{\mu_A(x)/x\} = \\ &= \mu_A(x_1)/x_1 + \cdots + \mu_A(x_n)/x_n = \sum_{i=1}^{n} \mu_A(x_i)/x_i \end{aligned} \tag{3}$$

where "+" and "$\sum$" are meant in the set-theoretic sense, and by convention, the pairs "$\mu_A(x)/x$" with $\mu_A(x) = 0$ are omitted.

As to the basic definitions and properties related to fuzzy sets, we can list the following ones:

- A fuzzy set A is *empty*, written $A = \emptyset$, if and only if $\mu_A(x) = 0, \forall x \in X$;

- Two fuzzy sets A and B in X are *equal*, written $A = B$, if and only if $\mu_a(x) = \mu_b(x), \forall x \in A$;
- A fuzzy set A in X is *contained in*, or is a *subset of* a fuzzy set B in X, written $A \subseteq B$, if and only if $\mu_A(x) \leq \mu_B(x), \forall x \in X$;
- A fuzzy set A in X is *normal* if and only if $\max_{x \in X} \mu_A(x) = 1$.

There are also some important nonfuzzy sets associated with a fuzzy set, notably the α*-cut*, or the α*-level set*, of A in X, written A_α, defined as the following (nonfuzzy) set

$$A_\alpha = \{x \in X : \mu_A(x) \geq \alpha\}, \qquad \text{for each } \alpha \in (0, 1] \tag{4}$$

which play an extremely relevant role in both formal analyses and applications as they make it possible to uniquely replace a fuzzy set by a sequence of nonfuzzy sets (cf. [7, 22]).

An important concept is the *cardinality* of a fuzzy set which, in the simplest case, is the *nonfuzzy cardinality* of $A = \mu_A(x_1)/x_1 + \cdots + \mu_A(x_n)/x_n$, the so-called *sigma-count*, denoted $\sum\text{Count}(A)$, defined as (cf. [70, 71]) $\sum\text{Count}(A) = \sum_{i=1}^{n} \mu_A(x_i)$,.

The distance between two fuzzy sets, A and B, both defined in $X = \{x_1, \ldots, x_n\}$, is very relevant for our purposes and the two basic (normalized) distances are:

- the *normalized linear* (Hamming) *distance* between A and B in X defined as $l(A, B) = \frac{1}{n} \sum_{i=1}^{n} \mid \mu_A(x_i) - \mu_B(x_i) \mid$
- the *normalized quadratic* (Euclidean) *distance* between A and B in X defined as $q(A, B) = \sqrt{\frac{1}{n} \sum_{i=1}^{n} [\mu_A(x_i) - \mu_B(x_i)]^2}$

Among the basic operations on fuzzy sets, one can quote the following ones:

- The *complement* of a fuzzy set A in X, written $\neg A$, is defined as

$$\mu_{\neg A}(x) = 1 - \mu_A(x), \qquad \text{for each } x \in X \tag{5}$$

  and the complement corresponds to the negation "not."
- The *intersection* of two fuzzy sets A and B in X, written $A \cap B$, is defined as

$$\mu_{A \cap B}(x) = \mu_A(x) \wedge \mu_B(x), \qquad \text{for each } x \in X \tag{6}$$

  where "$\wedge$" is the minimum operation, i.e. $a \wedge b = \min(a, b)$; the intersection of two fuzzy sets corresponds to the connective "and."
- The *union* of two fuzzy sets A and B in X, written $A + B$, is defined as

$$\mu_{A+B}(x) = \mu_A(x) \vee \mu_B(x), \qquad \text{for each } x \in X \tag{7}$$

  where "$\vee$" is the maximum operation, i.e. $a \vee b = \max(a, b)$; the union of two fuzzy sets corresponds to the connective "or".

The above definitions can be generalized to the t-norms and s-norms, for the intersection and union, respectively defined as follows.

A *t-norm* is defined as:

$$t : [0, 1] \times [0, 1] \longrightarrow [0, 1] \quad (8)$$

such that, for each $a, b, c \in [0, 1]$:

1. it has 1 as the unit element, i.e. $t(a, 1) = a$,
2. it is monotone, i.e. $a \leq b \Longrightarrow t(a, c) \leq t(b, c)$,
3. it is commutative, i.e. $t(a, b) = t(b, a)$, and
4. it is associative, i.e. $t[a, t(b, c)] = t[t(a, b), c]$.

Some relevant examples of t-norms are:

- the minimum (which is the most widely used)

$$t(a, b) = a \wedge b = \min(a, b) \quad (9)$$

- the algebraic product

$$t(a, b) = a \cdot b \quad (10)$$

- the Łukasiewicz t-norm

$$t(a, b) = \max(0, a + b - 1) \quad (11)$$

An *s-norm* (or a *t-conorm*) is defined as

$$s : [0, 1,] \times [0, 1] \longrightarrow [0, 1] \quad (12)$$

such that, for each $a, b, c \in [0, 1]$:

1. it has 0 as the unit element, i.e. $s(a, 0) = a$,
2. it is monotone, i.e. $a \leq b \Longrightarrow s(a, c) \leq s(b, c)$,
3. it is commutative, i.e. $s(a, b) = s(b, a)$, and
4. it is associative, i.e. $s[a, s(b, c)] = s[s(a, b), c]$.

Some relevant examples of s-norms are:

- the maximum (which is the most widely used, also here)

$$s(a, b) = a \vee b = \max(a, b) \quad (13)$$

- the probabilistic product

$$s(a, b) = a + b - ab \quad (14)$$

- the Łukasiewicz s-norm

$$s(a, b) = \min(a + b, 1) \quad (15)$$

Notice that a t-norms is *dual* to an s-norms in that $s(a, b) = 1 - t(1 - a, 1 - b)$. Among other important concepts related to fuzzy sets, one can quote:

- a *fuzzy relation* R between two (nonfuzzy) sets $X = \{x\}$ and $Y = \{y\}$ defined as a fuzzy set in the Cartesian product $X \times Y$, i.e.

$$R = \{(\mu_R(x, y), (x, y))\} = \\ = \{\mu_R(x, y)/(x, y)\}, \qquad \text{for each } (x, y) \in X \times Y \qquad (16)$$

where $\mu_R(x, y) : X \times Y \longrightarrow [0, 1]$ is the membership function of the fuzzy relation R, and $\mu_R(x, y) \in [0, 1]$ gives the degree to which the elements $x \in X$ and $y \in Y$ are in relation R between each other;
- the *max-min composition* of two fuzzy relations R in $X \times Y$ and S in $Y \times Z$, written $R \circ_{\max-\min} S$, defined as a fuzzy relation in $X \times Z$ such that

$$\mu_{R \circ_{\max-\min} S}(x, y) = \\ = \max_{y \in Y}[\mu_R(x, y) \wedge \mu_S(y, z)], \qquad \text{for each } x \in X, z \in Z \qquad (17)$$

- the *Cartesian product* of two fuzzy sets A in X and B in Y, written $A \times B$, defined as a fuzzy set in $X \times Y$ such that

$$\mu_{A \times B}(x, y) = [\mu_A(x) \wedge \mu_B(y)], \qquad \text{for each } x \in X, y \in Y \qquad (18)$$

These are basic definitions and operations to be used in the paper, and for more details we refer the reader, for instance, to Kacprzyk [22]. Moreover, we have assumed a purely fuzzy perspective though a possibilistic setting can also be used—cf. Dubois and Prade [8].

3 Brief Introduction to Multistage Control Under Fuzziness and Fuzzy Dynamic Programming

Our point of departure is the famous Bellman and Zadeh's [4] model of decision making under fuzziness in which if $X = \{x\}$ is some set of possible *options* (alternatives, variants, choices, decisions, …), then the *fuzzy goal* is defined as a fuzzy set G in X, characterized by its membership function $\mu_G : X \longrightarrow [0, 1]$ such that $\mu_G(x) \in [0, 1]$ specifies the grade of membership of a particular option $x \in X$ in the fuzzy goal G, and the *fuzzy constraint* is similarly defined as a fuzzy set C in the set of options X, characterized by $\mu_C : X \longrightarrow [0, 1]$ such that $\mu_C(x) \in [0, 1]$ specifies the grade of membership of a particular option $x \in X$ in the fuzzy constraint C.

The problem is to: "Attain G and satisfy C" which leads to the *fuzzy decision*

$$\mu_D(x) = \mu_G(x) \wedge \mu_C(x), \qquad \text{for each } x \in X \qquad (19)$$

where "$\wedge$" stands for the minimum that may be replaced, for instance, by a t-norm.

The *maximizing decision*, which is the non-fuzzy solution sought, is defined as an $x^* \in X$ such that

$$\mu_D(x^*) = \max_{x \in X} \mu_D(x) \tag{20}$$

The human cognition related aspect (cf. [63]) is that, first, the strict optimization in (20) may be viewed too rigid and unnecessary and some sort of a satisfactory, good enough solution could well do; notice that this idea parallels the famous Simon's satisficing decision making paradigm.

The problem formulation is clearly to

$$\text{"Attain } [G_o \text{ and } G_s] \ \underline{\text{and}} \text{ satisfy } [C_o \text{ and } C_s]\text{"} \tag{21}$$

which leads to the fuzzy decision

$$\mu_D(x) = [\mu_{G_o}(x) \wedge \mu_{G_s}(x)] \wedge [\mu_{C_o}(x) \wedge \mu_{C_s}(x)], \qquad \text{for each } x \in X \tag{22}$$

and the *maximizing*, or *optimal* decision is defined as in (20).

This framework can be extended to handle multiple fuzzy constraints and fuzzy goals, and also fuzzy constraints and fuzzy goals defined in different spaces, cf. Kacprzyk's [22] book. Namely, if we have: $n_o > 1$ objective fuzzy goals—$G_o^1, \ldots, G_o^{n_o}$ defined in Y, $n_s > 1$ subjective fuzzy goals—$G_s^1, \ldots, G_s^{n_s}$ defined in Y, $m_o > 1$ objective fuzzy constraints—$C_o^1, \ldots, C_o^{m_o}$ defined in X, $m_s > 1$ subjective fuzzy constraints—$C_s^1, \ldots, C_s^{m_s}$ defined in X, and a function $f : X \longrightarrow Y$, $y = f(x)$, then

$$\begin{aligned} &\mu_D(x) = \\ &= (\mu_{G_o^1}[f(x)] \wedge \cdots \wedge \mu_{G_o^{n_o}}[f(x)]) \wedge (\mu_{G_s^1}[f(x)] \wedge \cdots \wedge \mu_{G_s^{n_s}}[f(x)]) \wedge \\ &\quad \wedge [\mu_{C_o^1}(x) \wedge \cdots \wedge \mu_{C_o^{m_o}}(x)] \wedge [\mu_{C_s^1}(x) \wedge \cdots \wedge \mu_{C_s^{m_s}}(x)] \wedge \\ &\quad \wedge [\mu_{C_s^1}(x) \wedge \cdots \wedge \mu_{C_s^{m_s}}(x)], \qquad \text{for each } x \in X \end{aligned} \tag{23}$$

and the *maximizing decision* is defined as (20), i.e. $\mu_D(x^*) = \max_{x \in X} \mu_D(x)$.

In our context of *multistage control* the decision (control) space is $U = \{u\} = \{c_1, \ldots, c_m\}$, the state (equated with the output) space is $X = \{x\} = \{s_1, \ldots, s_n\}$, and both are assumed here finite. We start from an initial state $x_0 \in X$, apply a decision (control) $u_0 \in U$, which is subjected to a fuzzy constraint $\mu_{C^0}(u_0)$, and attain a state $x_1 \in X$ via a known state transition equation of the system under control S; a fuzzy goal $\mu_{G^1}(x_1)$ is imposed on x_1. Next, we apply u_1, subjected to $\mu_{C^1}(u_1)$, and attain x_2, subjected to $\mu_{G^2}(x_2)$, etc.

The (deterministic) system under control is described by a *state transition equation*

$$x_{t+1} = f(x_t, u_t), \qquad t = 0, 1, \ldots \tag{24}$$

where $x_t, x_{t+1} \in X = \{s_1, \ldots, s_n\}$ are the states at t and $t+1$, respectively, and $u_t \in U = \{c_1, \ldots, c_m\}$ is the decision (control) at t.

At $t, t = 0, 1, \ldots, u_t \in U$ is subjected to a *fuzzy constraint* $\mu_{C^t}(u_t)$, and on $x_{t+1} \in X$ a *fuzzy goal* is imposed, $\mu_{G^{t+1}}(x_{t+1})$. The fixed and specified in advance *initial state* is $x_0 \in X$, and the *termination time* (planning horizon), $N \in \{1, 2, \ldots\}$, is finite, and fixed and specified in advance.

The *performance* of the particular decision making (control) stage $t, t = 0, 1, \ldots, N-1$, is evaluated by

$$v_t = \mu_{C^t}(u_t) \wedge \mu_{G^{t+1}}(x_{t+1}) = \mu_{C^t}(u_t) \wedge \mu_{G^{t+1}}[f(x_t, u_t)] \tag{25}$$

while the *performance* of the whole multistage decision making (control) process is given by the fuzzy decision

$$\begin{aligned} \mu_D(u_0, \ldots, u_{N-1} \mid x_0) &= v_0 \wedge v_1 \wedge \cdots \wedge v_{N-1} = \\ &= [\mu_{C^0}(u_0) \wedge \mu_{G^1}(x_1)] \wedge \cdots \wedge [\mu_{C^{N-1}}(u_{N-1}) \wedge \mu_{G^N}(x_N)] \end{aligned} \tag{26}$$

The problem is to find an optimal sequence of controls $u_0^*, \ldots, u_{N-1}^*$ such that

$$\mu_D(u_0^*, \ldots, u_{N-1}^* \mid x_0) = \max_{u_0, \ldots, u_{N-1} \in U} \mu_D(u_0, \ldots, u_{N-1} \mid x_0) \tag{27}$$

Kacprzyk's [22] book provides a wide and comprehensive coverage of various aspects and extensions to this basic formulation.

The trajectory of the multistage decision making (control) process from $t = 0$ to a current stage $t = k$ is

$$H_k = (x_0, u_0, C^0, x_1, G^1, \ldots, u_{k-1}, C^{k-1}, x_k, G^k) \tag{28}$$

that is, it involves all what has occurred in terms of decisions applied, states attained, and evaluations of how well the fuzzy constraints have been satisfied and fuzzy goals attained.

However, for simplicity, it is often sufficient to take into account the *reduced trajectory*

$$h_k = (x_{k-2}, u_{k-2}, C^{k-2}, x_{k-1}, G^{k-1}, u_{k-1}, C^{k-1}, x_k, G^k) \tag{29}$$

which only takes into account the current, $t = k$, and previous stage, $t = k - 1$. Notice a long tradition of such an attitude, e.g. in the Markov decision processes.

A further simplification is that with a trajectory, or a reduced trajectory, an evaluation function is associated, $E : S(H_k) \longrightarrow [0, 1]$ or $e : S(h_k) \longrightarrow [0, 1]$, where $S(H_k)$ and $S(h_k)$ are the sets of trajectories and reduced trajectories, respectively, such that $E(H_k) \in [0, 1]$ and $e(h_k) \in [0, 1]$ denote the satisfaction of the past development, from 1 for full satisfaction to 0 for full dissatisfaction, through all intermediate values.

It should be noted that since the two main stakeholders in the agricultural regional development process are the authorities and inhabitants, and all judgments, preferences, evaluations, etc. it often makes sense to introduce the objective and subjective fuzzy constraints and goals which are related to the satisfaction of some values of, for instance, life quality indicators set by higher level authorities, and a perception of that satisfaction and their dynamics by the inhabitant. This important problem will not be discussed in this paper, and details can be found in the source paper by Kacprzyk, Francelin and Gomide [37] or in Kacprzyk's book [23]. For simplicity, since the purpose of this paper is to present a conceptually new formulation of the multistage agricultural regional development problem, we will use the concept of a fuzzy constraint and fuzzy goal without distinguishing the objective and subjective ones.

Problem (27) can be solved using the following two basic traditional techniques: dynamic programming (cf. [4, 19, 22]), and branch-and-bound [17], and also using the two new ones: a neural network (cf. [10, 11]), and a genetic algorithm (cf. [23, 24, 29]).

We will only consider here the use of dynamic programming, and refer the reader for an extensive coverage on this and other solution techniques to Kacprzyk's [22] book.

First, we rewrite (27) as to find $u_0^*, \ldots, u_{N-1}^*$ such that

$$\begin{aligned}
&\mu_D(u_0^*, \ldots, u_{N-1} \mid x_0) = \\
&= \max_{u_0,\ldots,u_{N-1}} [\mu_{C^0}(u_0) \wedge \mu_{G^1}(x_1) \wedge \cdots \\
&\quad \cdots \wedge \mu_{C^{N-1}}(u_{N-1}) \wedge \mu_{G^N}(f(x_{N-1}, u_{N-1}))]
\end{aligned} \tag{30}$$

and then, since

$$\mu_{C^{N-1}}(u_{N-1}) \wedge \mu_{G^N}(f(x_{N-1}, u_{N-1}))$$

depends only on u_{N-1}, then the maximization with respect to $u_0, \ldots, u_{N-1}$ in (30) can be split into:

- the maximization with respect to $u_0, \ldots, u_{N-2}$, and
- the maximization with respect to u_{N-1},

written as

$$\begin{aligned}
&\mu_D(u_0^*, \ldots, u_{N-1}^* \mid x_0) = \\
&= \max_{u_0,\ldots,u_{N-2}} \{\mu_{C^0}(u_0) \wedge \mu_{G^1}(x_1) \wedge \cdots \\
&\quad \cdots \wedge \mu_{C^{N-2}}(u_{N-2}) \wedge \mu_{G^{N-1}}(x_{N-1}) \wedge \\
&\quad \wedge \max_{u_{N-1}} [\mu_{C^{N-1}}(u_{N-1}) \wedge \mu_{G^N}(f(x_{N-1}, u_{N-1}))]\}
\end{aligned} \tag{31}$$

which may be continued for u_{N-2}, u_{N-3}, etc.

This backward iteration leads to the following set of fuzzy dynamic programming recurrence equations:

$$\begin{cases} \mu_{\overline{G}^{N-i}}(x_{N-i}) = \\ \qquad = \max_{u_{N-i}}[\mu_{C^{N-i}}(u_{N-i}) \wedge \mu_{G^{N-i}}(x_{N-i}) \wedge \mu_{\overline{G}^{N-i+1}}(x_{N-i+1})] \\ x_{N-i+1} = f(x_{N-i}, u_{N-i}); \qquad i = 0, 1, \ldots, N \end{cases} \tag{32}$$

where $\mu_{\overline{G}^{N-i}}(x_{N-i})$ is viewed as a fuzzy goal at control stage $t = N - i$ induced by the fuzzy goal at $t = N - i + 1, i = 0, 1, \ldots, N$; $\mu_{\overline{G}^N}(x_N) = \mu_{G^N}(x_N)$.

The $u_0, \ldots, u_{N-1}$ sought is given by the successive maximizing values of u_{N-i}, $i = 1, \ldots, N$ in (32) which are obtained as functions of x_{N-i}, i.e. as an *optimal policy*, $a_{N-i} : X \longrightarrow U$, such that $u_{N-i} = a_{N-i}(x_{N-i})$.

4 Sustainable Socioeconomic Agricultural Regional Development Planning Under Fuzziness Using a Multistage Control Model

We are concerned here with a fuzzy multistage control model for agricultural regional development planning proposed by Kacprzyk and Straszak [31–34], and used for various real regional development modeling projects at the International Institute for Applied Systems Analysis (IIASA) in Laxenburg, Austria (www.iiasa.at), and then employed for various regions (cf. [20, 26, 28, 29, 37], etc.).

Basically, these works concern a (rural) region plagued by severe difficulties mainly related to a poor *life quality* perceived by various actors involved like the inhabitants and authorities. Hence, life quality (or, in fact, a perception thereof) should be improved by some (mostly external) funds (investments) the amount of which and their temporal distribution should be found.

First, the essence of socioeconomic regional development may be depicted as in Fig. 1.

The region is here represented by a *socioeconomic dynamic system under control* the state of which at the development (planning) stage $t - 1$, X_{t-1}, is characterized

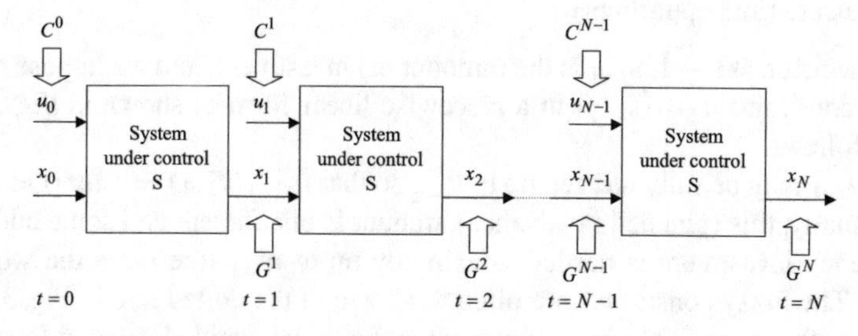

Fig. 1 Essence of socioeconomic regional development

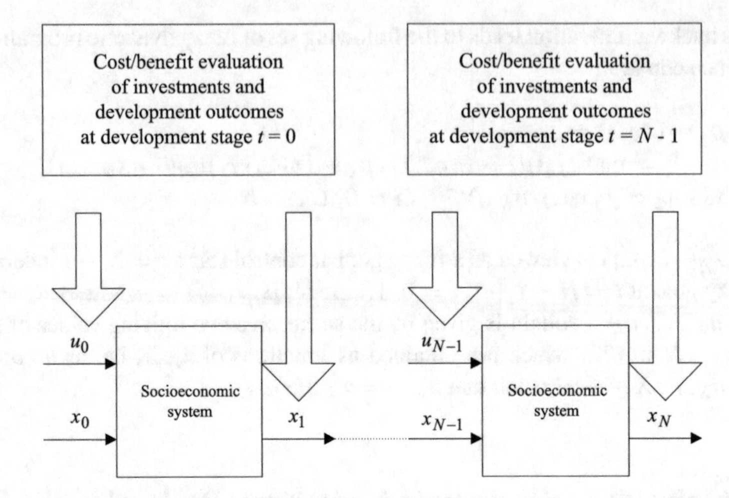

Fig. 2 The socioeconomic system under control

by a set of relevant *socioeconomic life quality indicators*. Then, the decisions or controls (i.e. investments), at $t-1$, u_{t-1}, change X_{t-1} to X_t; $t = 1, \ldots, N$; N is a finite, fixed and specified planning horizon.

The evaluation of a planning stage t, $t = 1, \ldots, N$, is via a cost–benefit type analysis, by accounting for both the "goodness" of the u_{t-1} applied (i.e. costs), and the "goodness" of the X_t attained (i.e. benefits); the former concerns how well some constraints are satisfied, and the latter how well some goals are attained.

The socioeconomic system is represented as in Fig. 2. Its state (output) X_t is equated with a *life quality index* that consists of the following seven *life quality indicators* (i.e. $X_t = [x_t^1, \ldots, x_t^7]$):

- x_t^1—economic quality (e.g., wages, salaries, income, ...),
- x_t^2—environmental quality,
- x_t^3—housing quality,
- x_t^4—health service quality,
- x_t^5—infrastructure quality,
- x_t^6—work opportunity,
- x_t^7—leisure time opportunity,

The decision at $t-1$, u_{t-1} is the (amount of) investment, and we impose on u_{t-1} a fuzzy constraint $\mu_{C^{t-1}}(u_{t-1})$ in a piecewise linear form as shown in Fig. 3 to be read as follows.

The u_{t-1} may be fully utilized up to u_{t-1}^p so that $\mu_{C^{t-1}}(u_{t-1}) = 1$ for $0 < u_{t-1} < u_{t-1}^p$. Usually, this (planned in advance) amount is insufficient and some additional contingency investment is needed, maximally up to u_{t-1}^c (the more the worse, of course). The fuzzy constraints are often as shown in the dotted line in Fig. 3 in that too low a use of available investments should also be avoided, since a lower than planned use of funds may also be not welcome.

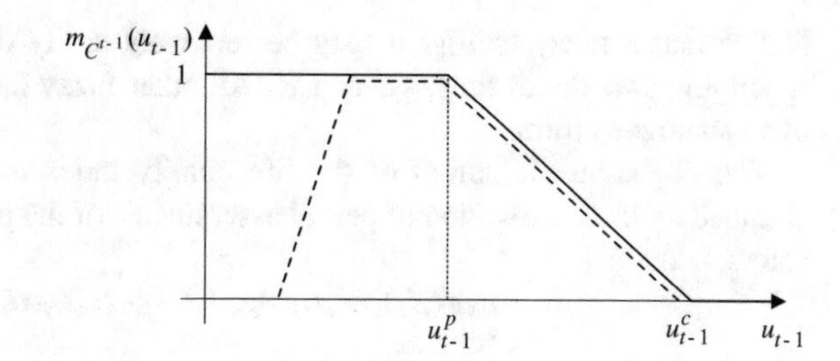

Fig. 3 Fuzzy constraints on investment u_{t-1}

The $t-1$, u_{t-1} is partitioned into $u^1_{t-1}, \ldots, u^7_{t-1}$, devoted to an improvement of the respective life quality indicators; the partitioning problem is difficult and will not be considered here, for simplicity.

The temporal evolution of the particular life quality indicators is governed by the state transition equation

$$x^i_t = f^i_{t-1}(x^i_{t-1}, u^i_{t-1}), \qquad i = 1, \ldots, 7; t = 1, \ldots, N \tag{33}$$

which may be derived by, e.g., using experts' opinions, past experience, knowledge or even data driven modeling, etc.

The evaluation of development takes into account how well some predetermined goals are fulfilled, i.e. *effectiveness*, then related to the investment spent, i.e. *efficiency*—cf. Kacprzyk's [22] book. The effectiveness of regional development involves two aspects: the *effectiveness of a particular development stage*, and the *effectiveness of the whole development*.

The effectiveness of a particular development stage is basically the determination of how well the fuzzy constraints are fulfilled, and fuzzy goals are attained, and they concern desired values of the life quality indicators. The goal attainment is clearly not clear-cut so that a fuzzy goal is employed.

For each life quality indicator at $t = 1, \ldots, N$, x^i_t, we define a *fuzzy subgoal* $G^{t,i}$ characterized by $\mu_{G^{t,i}}(x^i_t)$ as shown in Fig. 4 to be read as follows: $G^{t,i}$ is fully satisfied for $x^i_t \geq \overline{x}^u_t$, where $\overline{x}^i$ is some *aspiration level* for the indicator x^i_t; therefore, $\mu_{G^{t,i}}(x^i_t) = 1$, for $x^i_t \geq \overline{x}^i_t$. Less preferable are $\underline{x}^i_t < x^i_t < \overline{x}^i_t$ for which $0 < \mu_{G^{t,i}}(x^i_t) < 1$, and $x^i_t \leq \underline{x}^i_t$ are assumed to be impossible, hence $\mu_{G^{t,i}}(x^i_t) = 0$.

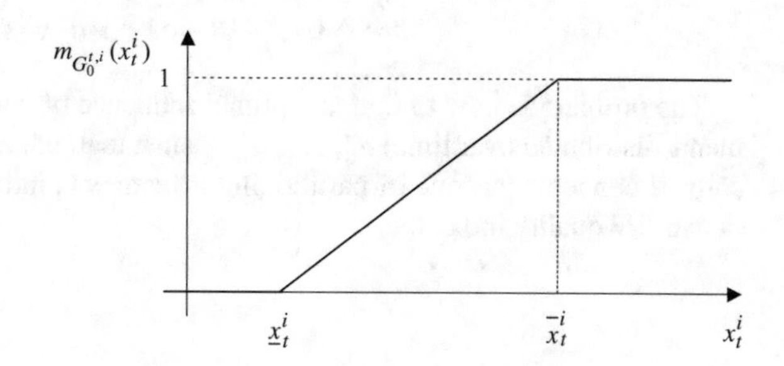

Fig. 4 Fuzzy subgoal

Notice that a fuzzy (sub)goal may be relatively easily determined by experts by specifying two values only, $\underline{x}t_t^i$ and $\overline{x}_t^i$. All other fuzzy (sub)goals will be assumed in an analogous form.

The objective evaluation of the life quality index at t, $X_t = [x_t^1, \ldots, x_t^7]$, is obtained by the aggregation of partial assessments of the particular life quality indicators, i.e.

$$\mu_{G^t}(X_t) = \mu_{G^{t,1}}(x_t^1) \wedge \cdots \wedge \mu_{G^{t,7}}(x_t^7) \tag{34}$$

and "$\wedge$" may be replaced here and later on by another suitable operation as, e.g., a t-norm (cf. [22, 23]) but this will not be considered here. Basically, the use of "$\wedge$" (minimum) reflects a pessimistic, safety-first attitude, and a lack of substitutability (i.e. that a low value of one life quality indicator cannot be compensated by a higher value of another), which is often adequate, in particular in the type or regions dealt with in our works, that is, predominantly rural regions plagued by serious difficulties as the out-migration of younger population, aging of the society, and a general economic decay.

The effectiveness of a planning stage t is meant again as some cost–benefit relation of what has been attained (the life quality indices) to what has been "paid for" (the respective investments).

Formally, the (fuzzy) effectiveness of stage t, $t = 1, \ldots, N$, is expressed as

$$\mu_{E^t}(u_{t-1}, X_t, s_t) = \mu_{C^{t-1}}(u_{t-1}) \wedge \mu_{G_o^t}(X_t) \wedge \mu_{G_s^t}(s_t) \tag{35}$$

and the aggregation using the minimum reflects a safety-first attitude which is quite justified for our troubled regions considered.

Then, the effectiveness measures of the particular $t = 1, \ldots, N$, $\mu_{E^t}(u_{t-1}, X_t, s_t)$ given by (35), are aggregated to yield the *fuzzy effectiveness measure for the whole development*

$$\mu_E(H_N) = \mu_{E^1}(u_0, X_1, s_1) \wedge \cdots \wedge \mu_{E^N}(u_{N-1}, X_N, s_n) \tag{36}$$

The fuzzy decision is

$$\begin{aligned} &\mu_D(u_0, \ldots, u_{N-1} \mid X_0, B_N) = \\ &= [\mu_{C^0}(u_0) \wedge \mu_{G^1}(X_1)] \wedge \cdots \\ &\cdots \wedge [\mu_{C^{N-1}}(u_{N-1}) \wedge \mu_{G_o^N}(X_N)] \end{aligned} \tag{37}$$

The problem is now to find an optimal sequence of controls (amounts of investments distributed over time) $u_0^*, \ldots, u_{N-1}^*$ such that, under a given policy B_N (basically, it concerns the rule of partitioning investments into parts devoted to the particular life quality indicators):

$$\mu_D(u_0^*, \ldots, u_{N-1}^* \mid X_0, B_N) =$$
$$= \max_{u_0, \ldots, u_{u_{N-1}}} \{[\mu_{C^0}(u_0) \wedge \mu_{G^1}(X_1)] \wedge \cdots$$
$$\cdots \wedge [\mu_{C^{N-1}}(u_{N-1}] \wedge \mu_{G^N}(X_N)]\} \quad (38)$$

For illustration we will show a simple example that in its initial form was shown first in Kacprzyk's [22] book but will be changed with respect to numbers to account for different economic conditions in the present time. It should be emphasized that the results of using the model to solve a real regional development problem are voluminous and contain, for instance, many partial results, an extended visualization and analyzes. This cannot be presented in this short article and, above all, could also be unclear to a novice user. We have therefore chosen, for clarity and comprehensiveness, to present a very simple example that would just show the essence, maybe on simplified, not fully realistic data. **Example**: The region, which is—as in virtually all our works in this area—is predominantly agricultural, has a population of ca. 120,000 inhabitants, and its arable land is ca. 450,000 acres, and is plagued by serious difficulties, maybe even an economic decay, due to a low perception of life quality, out-migration of younger population to towns, aging of the population, etc.

For simplicity, the region's development will be considered over the next 3 development stages (years, for simplicity) which is clearly, for this type of problems and a very low reaction of the system to any changes clearly too short and can only serve as illustration.

The life quality index, that is again very simplified, consists of the four life quality indicators:

- x_t^{I}—average subsidies in US$ per acre (per year),
- x_t^{II}—sanitation expenditures (water and sewage) in US$ per capita (per year),
- x_t^{III}—health care expenditures in US$ per capita (per year), and
- x_t^{IV}—expenditures for paved roads (new roads and maintenance of the existing ones) in US$ (per year).

Suppose now that the investments are partitioned into parts devoted to the improvement of the above life quality indicators due to the fixed partitioning rule $A_{t-1}(u_{t-1}, i)$: 5% for subsidies, 25% for sanitation, 45% for health care, and 25% for infrastructure. It is obvious that in practice such a type of a fixed partitioning rule can occur, which is often implied by intricacies of the legal systems, tradition, etc., but is usually somehow corrected in the course of analysis, and notably the implementation of the model for a particular region. However, this is a separable problem which, as already mentioned is out of scope of our analysis motivated by the length of the paper.

Let the initial, at $t = 0$, values of the life quality indicators be:

$$x_0^{\mathrm{I}} = 0.5 \quad x_0^{\mathrm{II}} = 15 \quad x_0^{\mathrm{III}} = 27 \quad x_0^{\mathrm{IV}} = 1{,}700{,}000$$

Needless to say that the particular values of the life quality indicators do not correspond, for obvious reasons to real valued most of which are not meant to be disclosed.

Though in our fuzzy dynamic programming model finite sets of controls and states are assumed, in the practical solution of such large scale socio-economic problems a scenario analysis is usually employed. For clarity, we will only take into account here the following two *scenarios* (policies):

- Policy 1: $u_0 = \$16,000,000$ $u_1 = \$16,000,000$ $u_2 = \$16,000,000$
- Policy 2: $u_0 = \$15,000,000$ $u_1 = \$16,000,000$ $u_2 = \$17,000,000$

Under the particular scenarios (Policy 1 and Policy 2), the values of the life quality indicators attained are:

Policy 1: Year(t)	u_t	x_t^{I}	x_t^{II}	x_t^{III}	x_t^{IV}
0	\$16,000,000				
1	\$16,000,000	0.88	16.7	30	\$4,000,000
2	\$16,000,000	0.88	16.7	30	\$4,000,000
3		0.88	16.7	30	\$4,000,000

Policy 2: Year(t)	u_t	x_t^{I}	x_t^{II}	x_t^{III}	x_t^{IV}
0	\$15,000,000				
1	\$16,000,000	0.83	15.6	28.1	\$3,500,000
2	\$17,000,000	0.88	16.7	30	\$8,000,000
3		0.94	17.7	31.9	\$2,250,000

For the evaluation of the above two development trajectories, for simplicity and readability we will only take into account the *effectiveness* of development, and the objective evaluation only. The consecutive fuzzy constraints and objective fuzzy subgoals are assumed piecewise linear, i.e. their definition requires two values only (cf. Figs. 3, and 4): the aspiration level (i.e. the fully acceptable value) and the lowest (or highest) possible (still acceptable) value) which are:

t

$$
\begin{array}{llll}
0 & C^0 : u_0^p = \$15,000,000 & & \\
 & u_0^c = \$17,000,000 & & \\
1 & C^1 : u_1^p = \$16,500,000 & & \\
 & u_1^c = \$18,000,000 & G_o^{1,\mathrm{I}} : \underline{x}_1^{\mathrm{I}} = 0.6 & \overline{x}_1^{\mathrm{I}} = 0.85 \\
 & & G_o^{1,\mathrm{II}} : \underline{x}_1^{\mathrm{II}} = 14 & \overline{x}_1^{\mathrm{II}} = 16 \\
 & & G_o^{1,\mathrm{III}} : \underline{x}_1^{\mathrm{III}} = 27 & \overline{x}_1^{\mathrm{III}} = 29 \\
 & & G_o^{1,\mathrm{IV}} : \underline{x}_1^{\mathrm{IV}} = \$3,600,000 & \overline{x}_1^{\mathrm{IV}} = \$3,800,000
\end{array}
$$

$$
\begin{array}{llll}
2 & C^2: u_2^p = \$16{,}000{,}000 & & \\
 & u_1^c = \$20{,}000{,}000 & G_o^{2,\mathrm{I}}: \underline{x}_2^{\mathrm{I}} = 0.7 & \overline{x}_1^{\mathrm{I}} = 0.9 \\
 & & G_o^{2,\mathrm{II}}: \underline{x}_2^{\mathrm{II}} = 15 & \overline{x}_1^{\mathrm{II}} = 17 \\
 & & G_o^{2,\mathrm{III}}: \underline{x}_2^{\mathrm{III}} = 28 & \overline{x}_1^{\mathrm{III}} = 30 \\
 & & G_o^{2,\mathrm{IV}}: \underline{x}_2^{\mathrm{IV}} = \$3{,}800{,}000 & \overline{x}_2^{\mathrm{IV}} = \$4{,}000{,}000 \\
3 & & G_o^{3,\mathrm{I}}: \underline{x}_3^{\mathrm{I}} = 0.75 & \overline{x}_3^{\mathrm{I}} = 1 \\
 & & G_o^{3,\mathrm{II}}: \underline{x}_3^{\mathrm{II}} = 16 & \overline{x}_1^{\mathrm{II}} = 18.5 \\
 & & G_o^{3,\mathrm{III}}: \underline{x}_3^{\mathrm{III}} = 29 & \overline{x}_1^{\mathrm{III}} = 31 \\
 & & G_o^{3,\mathrm{IV}}: \underline{x}_3^{\mathrm{IV}} = \$3{,}800{,}000 & \overline{x}_3^{\mathrm{IV}} = \$4{,}200{,}000
\end{array}
$$

Using the "$\wedge$" (minimum) to reflect a safety-first attitude, which is clearly preferable in the situation considered (a rural region plagued by aging of the society, out-migration to neighboring urban areas, economic decay, etc.), the evaluation of the two investment policies is:

- Policy 1

$$
\begin{aligned}
&\mu_D(\$16{,}000{,}000; \$16{,}000{,}000; \$16{,}000{,}000 \mid .) = \\
&= \mu_{C^0}(\$16{,}000{,}000) \wedge (\mu_{G_o^{1,\mathrm{I}}}(0.88) \wedge \\
&\quad \wedge \mu_{G_o^{1,\mathrm{II}}}(16.7) \wedge \mu_{G_o^{1,\mathrm{III}}}(30) \wedge \mu_{G_o^{1,\mathrm{IV}}}(\$4{,}000{,}000)) \wedge \\
&\quad \wedge \mu_{C^1}(\$16{,}000{,}000) \wedge (\mu_{G_o^{2,\mathrm{I}}}(0.88) \wedge \\
&\quad \wedge \mu_{G_o^{2,\mathrm{II}}}(16.7) \wedge \mu_{G_o^{2,\mathrm{III}}}(30) \wedge \mu_{G_o^{2,\mathrm{IV}}}(\$4{,}000{,}000)) \wedge \\
&\quad \wedge \mu_{C^2}(\$16{,}000{,}000) \wedge (\mu_{G_o^{3,\mathrm{I}}}(0.88) \wedge \\
&\quad \wedge \mu_{G_o^{3,\mathrm{II}}}(16.7) \wedge \mu_{G_o^{3,\mathrm{III}}}(30) \wedge \mu_{G_o^{3,\mathrm{IV}}}(\$4{,}000{,}000)) = \\
&= 0.5 \wedge (1 \wedge 1 \wedge 1 \wedge 1) \wedge 0.8 \wedge \\
&\quad \wedge (0.9 \wedge 0.85 \wedge 1 \wedge 1) \wedge 1 \wedge (0.52 \wedge 0.28 \wedge 0.5 \wedge 0.33) = \\
&= 0.5 \wedge 0.8 \wedge 0.28 = 0.28
\end{aligned}
$$

- Policy 2

$$
\begin{aligned}
&\mu_D(\$15{,}000{,}000; \$16{,}000{,}000; \$15{,}500{,}000 \mid .) = \\
&= \mu_{C^0}(\$15{,}000{,}000) \wedge (\mu_{G_o^{1,\mathrm{I}}}(0.83) \wedge \\
&\quad \wedge \mu_{G_o^{1,\mathrm{II}}}(15.6) \wedge \mu_{G_o^{1,\mathrm{III}}}(28.1) \wedge \mu_{G_o^{1,\mathrm{IV}}}(\$3{,}750{,}000)) \wedge \\
&\quad \wedge \mu_{C^1}(\$16{,}000{,}000) \wedge (\mu_{G_o^{2,\mathrm{I}}}(0.88) \wedge \\
&\quad \wedge \mu_{G_o^{2,\mathrm{II}}}(16.7) \wedge \mu_{G_o^{2,\mathrm{III}}}(30) \wedge \mu_{G_o^{2,\mathrm{IV}}}(\$4{,}000{,}000)) \wedge \\
&\quad \wedge \mu_{C^2}(\$17{,}000{,}000) \wedge (\mu_{G_o^{3,\mathrm{I}}}(0.94) \wedge \\
&\quad \wedge \mu_{G_o^{3,\mathrm{II}}}(17.7) \wedge \mu_{G_o^{3,\mathrm{III}}}(31.9) \wedge \mu_{G_o^{3,\mathrm{IV}}}(\$4{,}250{,}000)) = \\
&= 1 \wedge (0.92 \wedge 0.8 \wedge 0.55 \wedge 0.75) \wedge 0.8 \wedge \\
&\quad \wedge (0.9 \wedge 0.85 \wedge 1 \wedge 1) \wedge 0.75 \wedge (0.76 \wedge 0.68 \wedge 1 \wedge 1) = \\
&= 0.55 \wedge 0.8 \wedge 0.68 = 0.55
\end{aligned}
$$

The second policy is therefore better.

As it can be seen, the inclusion in the basic fuzzy dynamic programming based multistage model for sustainable regional development planning, has yielded a new insight and quality mainly by making it possible to reflect both objective and subjective judgments, interests of various parties (actors) involved, etc. In general, this can be viewed as helping reflect many psychological and cognitive aspects that play a considerable role in the case of such a highly human centric system considered.

5 Including the Status Quo Bias into the Multistage Sustainable Agricultural Regional Development Model

As we have indicated in Sect. 2, the so called cognitive biases are common in human decision making and they imply that the human beings can very often prefer both solutions and procedures which do not differ too much from those which have been used for a long time. This phenomenon happens in particular in those areas of human activities which are more traditional and in which the technological progress occurs but is less rapid than in, for instance, high tech industries, and also when there is some particular attachment to the economic activities exemplified by the ownership of the arable land. Therefore, agriculture is clearly such a specific area in which the human cognitive biases are more pronounced, in particular the status quo bias.

We will now briefly show the ideas of introducing the status quo bias into our multistage model of agricultural regional development. Basically, this is essentially an optimization type model and optimization may produce (sub)optimal results which differ too much from the status quo, i.e. the present ones. Therefore, we should introduce into our model some correction terms which would prohibit from the attainment of such solutions differing too much from the status quo. Of course, we will still maintain the very essence of optimization, i.e. seek a best (maybe sometimes just good enough) solution.

We will now reformulate the model presented in Sect. 4. The evaluation of a planning stage $t, t = 1, \ldots, N$, is via a cost–benefit type analysis, by accounting for both the "goodness" of the u_{t-1} applied (i.e. costs), and the "goodness" of the X_t attained (i.e. benefits); the former concerns how well some constraints are satisfied, and the latter how well some goals are attained. However, trying to account for the status quo bias, we involve in the evaluation of the above goodness some penalty for too high a deviation from the status quo.

First, in the socioeconomic system presented in Fig. 2 the state is generally denoted by X_t and equated with the life quality indicators: $x_t^1, x_t^2, x_t^3, x_t^4, x_t^5, x_t^6$, and x_t^7, but we also take into account now their status quo values, i.e.: X_t^{sq}, and $x_t^{sq,1}$, $x_t^{sq,2}$, $x_t^{sq,3}$, $x_t^{sq,4}$, $x_t^{sq,5}$, $x_t^{sq,6}$6, and $x_t^{sq,7}$7,

The decision at $t-1$, u_{t-1} is again the (amount of) investment, and we impose on u_{t-1} a fuzzy constraint $\mu_{C^{t-1}}(u_{t-1})$ in a piecewise linear form as shown in Fig. 3. However, the evaluation of how well the fuzzy constraint is fulfilled is now extended to take into account the status quo and deviation thereof. A simple approach is here as follows: first, in the case of investment for which a natural ordering is "the lower the better", we introduce the status quo value of the (amount of) investment at $t-1$, $\underline{u}_{t-1}^{sq}$, for instance defined as:

$$u_{t-1}^{sq} = u_{t-1} + \alpha(| u_{t-1} - \underline{u}_{t-1}^{sq} |) \tag{39}$$

which should be understood as follows: to the "real" (i.e. status quo related) value of u_{t-1} the goodness of which should be evaluated as shown in Fig. 3, is added a term proportional to the deviation from the status quo, i.e. $\underline{u}_{t-1}^{sq}$, and the parameter $\alpha \in [0, 1]$ makes it possible to control the influence of the status quo. For instance, as mentioned in Sect. 2 a lower value of α can be assumed when the farmers are more informed, less traditional, have a higher social capital, etc.

And similarly, for the life quality indicators, we have

$$x_t^{sq} = x_t + \beta(| x_t - \underline{x}_{t-1}^{sq} |) \tag{40}$$

with an analogous meaning as for (39) for the control (investment).

At $t-1$, u_{t-1}^{sq} is then partitioned into $u_{t-1}^{sq,1}, \ldots, u_{t-1}^{sq,7}$, devoted to the improvement of the respective life quality indicators. The temporal evolution of the particular life quality indicators is governed by the status quo related state transition equation

$$x_t^{sq,i} = f_{t-1}^{sq,i}(x_{t-1}^{sq,i}, u_{t-1}^{sq,i}), \qquad i = 1, \ldots, 7; t = 1, \ldots, N \tag{41}$$

which may be derived by, e.g., using experts' opinions, past experience, knowledge or even data driven modeling, etc.

The evaluation of development trajectory takes into account how well some predetermined goals are fulfilled, i.e. *effectiveness*, then related to the investment spent, i.e. *efficiency*—cf. Kacprzyk's [22] book. Of course, now they are related to the status quo values. The effectiveness of regional development involves two aspects: the *effectiveness of a particular development stage*, and the *effectiveness of the whole development*.

The effectiveness of a particular development stage is again the determination of how well the fuzzy constraints are fulfilled, and fuzzy goals are attained, and they concern desired values of the life quality indicators, with a reference to the status quo.

For each life quality indicator at $t = 1, \ldots, N$, $x_t^{sq,i}$, we define a *fuzzy subgoal* $G_{sq}^{t,i}$ characterized by $\mu_{G_{sq}^{t,i}}(x_t^{sq,i})$ as in Fig. 4, and with a similar meaning, that is: $G_{sq}^{t,i}$ is fully satisfied for $x_t^{sq,i} \geq \overline{x}_t^{sq,u}$, where $\overline{x}^{sq,i}$ is some *aspiration level* for the life quality indicator $x_t^{sq,i}$; therefore, $\mu_{G_{sq}^{t,i}}(x_t^{sq,i}) = 1$, for $x_t^{sq,i} \geq \overline{x}_t^{sq,i}$. Less preferable are $\underline{x}_t^{sq,i} < x_t^{sq,i} < \overline{x}_t^{sq,i}$ for which $0 < \mu_{Gsq^{t,i}}(x_t^{sq,i}) < 1$, and $x_t^{sq,i} \leq \underline{x}_t^{sq,i}$ are assumed to be impossible, hence $\mu_{G_{sq}^{t,i}}(x_t^{sq,i}) = 0$. Notice that a fuzzy (sub)goal may be relatively easily determined by experts by specifying two values only, $\underline{xt}_t^{sq,i}$ and

$\overline{x}_t^{sq,i}$, as we assume for simplicity a piecewise representation of the membership functions.

The objective evaluation of the life quality index at t, $X_t sq = [x_t^{sq,1}, \ldots, x_t^{sq,7}]$, is obtained by the aggregation of partial assessments of the particular life quality indicators, i.e.

$$\mu_{G_{sq}^t}(X_t^{sq}) = \mu_{G_{sq}^{t,1}}(x_t^{sq,1}) \wedge \cdots \wedge \mu_{G_{sq}^{t,7}}(x_t^{sq,7}) \quad (42)$$

and "$\wedge$" may be replaced here and later on by another suitable operation as, e.g., a t-norm (cf. [10, 11]). Basically, the use of "$\wedge$" (minimum) reflects a pessimistic, safety-first attitude, and a lack of substitutability (i.e. that a low value of one life quality indicator cannot be compensated by a higher value of another), which is often adequate, in particular for regions with serious socio-economic difficulties dealt with in our works.

The effectiveness of a planning stage t is meant again as some cost–benefit relation of what has been attained (the life quality indices) to what has been "paid for" (the respective investments).

Formally, the (fuzzy) effectiveness of stage t, $t = 1, \ldots, N$, is expressed as

$$\mu_{E_{sq}^t}(u_{t-1}^{sq}, X_t^{sq}) = \mu_{C_{eq}^{t-1}}(u_{t-1}^{sq}) \wedge \mu_{G_{sq}^t}(X_t^{sq}) \quad (43)$$

and the aggregation using the minimum reflects a safety-first attitude and a lack of substitutability between the life quality indicators which is quite justified for our troubled regions considered.

Then, the effectiveness measures of the particular $t = 1, \ldots, N$, $\mu_{E_{sq}^t}(u_{t-1}^{sq}, X_t^{sq})$ given by (43), are aggregated to yield the *fuzzy effectiveness measure for the whole development trajectory*

$$\mu_{E_{sq}}(H_N) = \mu_{E_{sq}^1}(u_0^{sq}, X_1^{sq}) \wedge \cdots \wedge \mu_{E_{sq}^N}(u_{N-1^{eq}}, X_N^{sq}) \quad (44)$$

The fuzzy decision is then

$$\begin{aligned} &\mu_{D_{sq}}(u_0^{sq}, \ldots, u_{N-1}^{sq} \mid X_0^{sq}, B_N^{sq}) = \\ &= [\mu_{C_{sq}^0}(u_0^{sq}) \wedge \mu_{G_{sq}^1}(X_1^{sq})] \wedge \cdots \\ &\cdots \wedge [\mu_{C_{sq}^{N-1}}(u_{N-1}^{sq}) \wedge \mu_{G_{sq}^N}(X_N^{sq})] \end{aligned} \quad (45)$$

The problem is now to find an optimal sequence of controls (amounts of investments distributed over time) $u_0^{sq,*}, \ldots, u_{N-1}^{sq,*}$ such that, under a given policy B_N^{sq}:

$$\begin{aligned} &\mu_{D_{sq}}(u_0^{sq,*}, \ldots, u_{N-1}^{sq,*} \mid X_0^{sq}, B_N^{sq}) = \\ &= \max_{u_0^{sq}, \ldots, u_{u_{N-1}}^{sq}} \{[\mu_{C_{sq}^0}(u_0^{sq}) \wedge \mu_{G_{sq}^1}(X_1^{sq})] \wedge \cdots \\ &\cdots \wedge [\mu_{C_{sq}^{N-1}}(u_{N-1}^{sq}] \wedge \mu_{G_{sq}^N}(X_N^{sq})]\} \end{aligned} \quad (46)$$

Since the inclusion of the status quo bias into the model considered makes it more complicated, with more parameters to be identified, we will just mention that in the case of Example presented in Sect. 4, for lack of space, we cannot present all calculations. Briefly speaking, now, when the status quo bias is included into the model, among the two policies (scenarios) considered, i.e.

- Policy 1: $u_0 = \$16{,}000{,}000$ $u_1 = \$16{,}000{,}000$ $u_2 = \$16{,}000{,}000$
- Policy 2: $u_0 = \$15{,}000{,}000$ $u_1 = \$16{,}000{,}000$ $u_2 = \$17{,}000{,}000$

the first policy is better because the second one involves too large deviations in the consecutive investments (expenditure), which results in too large a changeability of the respective values of life quality indicators which implies a lower value of the fuzzy decision, that is the fuzzy effectiveness measure for the whole development trajectory.

6 Concluding Remarks

We proposed a novel fuzzy multistage control model of sustainable regional agricultural development in which a well known human cognitive bias, the so called status quo bias, is reflected. We advocated the use of such a human centric and realistic model for the modeling of agricultural systems and value chains. The model employs fuzzy logic to reflect imprecision in the specification of human judgments, intentions, evaluations, etc. which is common, in particular in our case of agricultural modeling. We mentioned that the inclusion of the status quo bias can be viewed as a way to mitigate risk. This problem will be a subject of further works. Moreover, we will discuss in next works the crucial problem of how to choose proper aggregation operators, in particular we will consider operators from a large class of the ordered weighted averaging (OWA) operators, notably we will discuss the use of non-standard OWA operators which are comprehensibly presented in Kacprzyk, Yager and Merig'o [38]. The above extensions and analyses will be illustrated on examples of real studies of agricultural regional modeling.

Acknowledgements The contribution of the Project 691249, RUC-APS: Enhancing and implementing Knowledge based ICT solutions within high Risk and Uncertain Conditions for Agriculture Production Systems (www.rucaps.eu), funded by the European Union under their funding scheme H2020-MSCARISE-2015 is acknowledged by Janusz Kacprzyk and Jorge Hernandez Hormazabal.

References

1. Alfaro-García, V.G., Merigó, J.M., Gil-Lafuente, A., M, Kacprzyk J.: Logarithmic aggregation operators and distance measures. Int. J. Intell. Syst. **33**(7), 1488–1506 (2018)
2. Anselin-Avila, E., Gil-Lafuente, A.M.: Fuzzy logic in the strategic analysis: impact of the external factors over business. Int. J. Bus. Innov. Res. **3**(5), 515–534 (2009)

3. Beedell, J.I., Rehman, T.: Using social psychology models to understand farmers's conservation behaviour. J. Rural Stud. **16**(1), 117–127 (2000)
4. Bellman, R.E., Zadeh, L.A.: Decision making in a fuzzy environment. Manag. Sci. **17**, 141–164 (1970)
5. Desdoigts, A., Cordaro, F.: Learning versus status quo bias and the role of social capital in technology adoption: the case of cocoa farmers in Cx̂ote d'Ivoire. Working paper 20160005, Université Paris 1 Panthéon Sorbonne, UMR Développement et Sociétés (2016)
6. Domptail, S., Nuppenau, E.-A.: The role of uncertainty and expectations in modeling (range) land use strategies: an application of dynamic o ptimization modeling with recursion. Ecol. Econ. **69**(12), 2475–2485 (2010)
7. Dubois, D., Prade, H.: Fuzzy Sets and Systems: Theory and Applications. Academic Press, New York (1980)
8. Dubois, D., Prade, H.: Possibility Theory: An Approach to Computerized Processing of Uncertainty. Plenum Press, New York (1988)
9. Feola, G., Binder, C.P.: Towards an improved understanding of farmers' behaviour: the integrative agent-centred (IAC) framework. Ecol. Econ. **69**(12), 2323–2333 (2010)
10. Francelin, R.A., Kacprzyk, J., Gomide, F.A.C.: Neural network based algorithm for dynamic system optimization. Asian J. Control **3**(2), 131–142 (2001)
11. Francelin, R.A., Gomide, F.A.C., Kacprzyk, J.: A biologically inspired neural network for dynamic programming. Int. J. Neural Syst. **11**(6), 561–572 (2001)
12. Haselton, M.G., Nettle, D., Andrews, P.W.: The evolution of cognitive bias. In: Buss, D.M. (ed.) Handbook of Evolutionary Psychology, pp. 724–746. Wiley, Hoboken (2005)
13. Hermann, D., Musshoff, O., Agethen, K.: Investment behavior and status quo bias of conventional and organic hog farmers: an experimental approach. Renew. Agric. Food Syst. **31**(4), 318–329 (2016)
14. Hilbert, M.: Toward a synthesis of cognitive biases: how noisy information processing can bias human decision making. Psychol. Bull. **138**(2), 211–237 (2012)
15. Hotaling, J.M., Busemeyer, J.R.: DFT-D: a cognitive-dynamical model of dynamic decision making. Synthese **189**(1), 67–80 (2012)
16. Jones, J.W., et al.: Brief history of agricultural systems modeling. Agric. Syst. **155**, 240–254 (2017)
17. Kacprzyk, J.: A branch-and-bound algorithm for the multistage control of a nonfuzzy system in a fuzzy environment. Control Cybern. **7**, 51–64 (1978)
18. Kacprzyk, J.: A branch-and-bound algorithm for the multistage control of a fuzzy system in a fuzzy environment. Kybernetes **8**, 139–147 (1979)
19. Kacprzyk, J.: Multistage Decision Making under Fuzziness. Verlag TÜV Rheinland, Cologne (1983)
20. Kacprzyk, J.: Design of socio-economic regional development policies via a fuzzy decision making model. In: Straszak, A. (ed.) Large Scale Systems Theory and Applications, Proceedings of Third IFAC/IFORS Symposium (Warsaw, Poland, 1983), pp. 228–232. Pergamon Press, Oxford (1984)
21. Kacprzyk, J.: Multistage control under fuzziness using genetic algorithms. Control Cybern. **25**, 1181–1215 (1996)
22. Kacprzyk, J.: Multistage Fuzzy Control. Wiley, Chichester (1997)
23. Kacprzyk, J.: A genetic algorithm for the multistage control of a fuzzy system in a fuzzy environment. Mathw. Soft Comput. **IV**, 219–232 (1997)
24. Kacprzyk, J.: Multistage control of a stochastic system in a fuzzy environment using a genetic algorithm. Int. J. Intell. Syst. **13**, 1011–1023 (1998)
25. Kacprzyk, J.: Including socio-economic aspects in a fuzzy multistage decision making model of regional development planning. In: Reznik, L., Dimitrov, V., Kacprzyk, J. (eds.) Fuzzy Systems Design, pp. 86–102. Physica-Verlag (Springer), Heidelberg (1998)
26. Kacprzyk, J.: Towards perception-based fuzzy modelling: an extended multistage fuzzy control model and its use in sustainable regional development planning. In: Sinčak, P., Vašçak, J., Hirota, K. (eds.) Machine Intelligence? Quo Vadis? pp. 301–337. World Scientific, Singapore (2004)

27. Kacprzyk, J.: Fuzzy dynamic programming: interpolative reasoning for an efficient derivation of optimal control policies. Control Cybern. **42**(1), 63–84 (2013)
28. Kacprzyk, J.: Cognitive informatics: a proper framework for the use of fuzzy dynamic programming for the modeling of regional development? In: Tamir, D.E., Rishe, N.D., Kandel, A. (eds.) Fifty Years of Fuzzy Logic and its Applications, pp. 183–200. Springer, Heidelberg (2015)
29. Kacprzyk, J.: Multistage fuzzy control of a stochastic system using a bacterial genetic algorithm. In: Grzegorzewski, P., Gagolewski, M., Hryniewicz, O., Gil, M.Á. (eds.) Strengthening Links Between Data Analysis and Soft Computing, pp. 273–283. Springer, Heidelberg (2015)
30. Kacprzyk, J., Esogbue, A.O.: Fuzzy dynamic programming: main developments and applications. Fuzzy Sets Syst. **81**, 31–46 (1996)
31. Kacprzyk, J., Straszak, A.: Application of fuzzy decision making models for determining optimal policies in 'stable' integrated regional development. In: Wang, P.P., Chang, S.K. (eds.) Fuzzy Sets Theory and Applications to Policy Analysis and Information Systems, pp. 321–328. Plenum, New York (1980)
32. Kacprzyk, J., Straszak, A.: A fuzzy approach to the stability of integrated regional development. In: Lasker, G.E. (ed.) Applied Systems and Cybernetics, vol. 6, pp. 2997–3004. Pergamon Press, New York (1982)
33. Kacprzyk, J., Straszak, A.: Determination of 'stable' regional development trajectories via a fuzzy decision making model. In: Yager, R.R. (ed.) Recent Developments in Fuzzy Sets (2019) and Possibility Theory, pp. 531–541. Pergamon Press, New York (1982)
34. Kacprzyk, J., Straszak, A.: Determination of stable trajectories for integrated regional development using fuzzy decision models. IEEE Trans. Syst. Man Cybern. **SMC-14**, 310–313 (1984)
35. Kacprzyk, J., Sugianto, L.F.: Multistage fuzzy control involving objective and subjective aspects. In: Proceedings of the 2nd International Conference on Knowledge-Based Intelligent Electronic Systems KES-98, Adelaide, Australia, pp. 564–573 (1998)
36. Kacprzyk, J., Owsiński, J.W., Straszak, A.: Agricultural policy making for integrated regional development in a mixed economy. In: Titli, A., Singh, M.G. (eds.) Proceedings of Second IFAC Large Scale Systems Theory and Applications Symposium (Toulouse, France, 1979), pp. 9–21. Pergamon Press, Oxford (1980)
37. Kacprzyk, J., Francelin, R.A., Gomide, F.A.C.: Involving objective and subjective aspects in multistage decision making and control under fuzziness: dynamic programming and neural networks. Int. J. Intell. Syst. **14**, 79–104 (1999)
38. Kacprzyk, J., Yager, R.R., Merigó, J.M.: Towards human-centric aggregation via ordered weighted aggregation operators and linguistic data summaries: a new perspective on Zadeh's inspirations. IEEE Comput. Intell. Mag. **14**(1), 16–30 (2019)
39. Kahneman, D., Slovic, P., Tversky, A.: Judgment Under Uncertainty: Heuristics and Biases. Cambridge University Press, New York (1982)
40. Kahnemen, D., Knetsch, J.L., Thaler, R.H.: Anomalies: the endowment effect, loss aversion and status quo bias. J. Econ. Perspect. **5**(1), 193–2006 (1991)
41. Kondratenko, Y.P., Sidenko, I.V.: Design and reconfiguration of intelligent knowledge-based system for fuzzy multi-criteria decision making in transport logistics. J. Comput. Optim. Econ. Financ. **6**(3), 229–242 (2014)
42. Lorkowski, J., Kreinovich, V.: Likert-type fuzzy uncertainty from a traditional decision making viewpoint: how symmetry helps explain human decision making (including seemingly irrational behavior) (survey). Appl. Comput. Math. **3**(3), 275–298 (2014)
43. MerigóJ, M., Casanovas, M.: The fuzzy generalized OWA operator and its application in strategic decision making. Cybern. Syst. **41**(5), 359–370 (2010)
44. Merigó, J.M., Casanovas, M.: Decision making with distance measures and induced aggregation operators. Comput. Ind. Eng. **60**, 66–76 (2011)
45. Merigó, J.M., Gil-Lafuente, A.M.: Decision making techniques in business and economics based on the OWA operator. SORT Stat. Oper. Res. Trans. **36**, 81–101 (2012)
46. Merigó, J.M., Yager, R.R.: Generalized moving averages, distance measures and OWA operators. Int. J. Uncertain. Fuzziness Knowl. Based Syst. **21**, 533–559 (2013)

47. Merigó, J.M., Gil-Lafuente, A.M., Gil-Aluja, J.: Decision making with the induced generalized adequacy coefficient. Appl. Comput. Math. **10**(3), 321–339 (2011)
48. Merigó, J.M., Lobato-Carral, C., Carrilero-Castillo, A.: Decision making in the European Union under risk and uncertainty. Eur. J. Int. Manag. **6**(5), 590–609 (2012)
49. Merigó, J.M., Palacios-Marqués, D., Soto-Acosta, P.: Distance measures, weighted averages, OWA operators and Bonferroni means. Appl. Soft Comput. **50**, 356–366 (2017)
50. Oechssler, J., Roider, A., Schmitz, P.W.: Cognitive abilities and behavioral biases. J. Econ. Behav. Organ. **72**(1), 147–152 (2009)
51. Pedrycz, W.: Granular computing: concepts and algorithmic developments. Appl. Comput. Math. **10**(1), 175–194 (2011)
52. Rossi Borges, J.A., Oude Lansink, A.G.J.M., Marques, Ribeiro C., Lutke, V.: Understanding farmers' intention to adopt improved natural grassland using the theory of planned behavior. Livest. Sci. **169**, 163–174 (2014)
53. Samuelson, W., Zeckhauser, R.: Status quo bias in decision making. J. Risk Uncertain. **1**, 7–59 (1988)
54. Senger, I., Rossi Borges, J.A., Dessimon Machado, J.A.: Using the theory of planned behavior to understand the intention of small farmers in diversifying their agricultural production. J. Rural Stud. **49**, 32–40 (2017)
55. Simon, H.A.: A behavioral model of rational choice. Q. J. Econ. **69**(1), 99–118 (1955)
56. Sirbiladze, G., Khutsishvili, I., Ghvaberidze, I.: Multistage decision-making fuzzy methodology for optimal investments based on experts? evaluations. Eur. J. Oper. Res. **232**(1), 169–177 (2014)
57. Sirbiladze, G., Khutsishvili, I., Badagadze, O., Kapanadze, M.: More precise decision-making methodology in the temporalized body of evidence. Application in the information technology management. Int. J. Inf. Technol. Decis. Mak. **15**(6), 1469–1502 (2016)
58. Thomas, L.C. (ed.): Golden Developments in Operational Research. Pergamon Press, New York (1987)
59. Tversky, A., Kahneman, D.: Judgement under uncertainty: heuristics and biases. Science **185**(4157), 1124–1131 (1974)
60. Torra, V.: The weighted OWA operator. Int. J. Intell. Syst. **12**, 153–166 (1997)
61. Wang, Y.: On cognitive informatics, brain and mind. A Transdiscipl. J. Neurosci. Neurophilos. **4**(2), 151–167 (2003)
62. Wang, Y.: The theoretical framework of cognitive informatics. Int. J. Cogn. Inf. Nat. Intell. **1**(1), 1–27 (2007)
63. Wang, Y., Ruhe, G.: The cognitive process of decision making. Int. J. Cogn. Inf. Nat. Intell. **1**(2), 73–85 (2007)
64. Yager, R.R.: Modeling, querying and mining social relational networks using fuzzy set techniques (Survey). Appl. Comput. Math. **13**(1), 3–17 (2014)
65. Yager, R.R., Kacprzyk, J. (eds.): The Ordered Weighted Averaging Operators: Theory, Methodology and Applications. Kluwer, Boston (1996)
66. Yager, R.R., Kacprzyk, J., Beliakov, G.: Recent Developments in the Ordered Weighted Averaging Operators - Theory and Practice. Springer, New York (2011)
67. Zadeh, L.A.: Fuzzy sets. Inf. Control **8**, 338–353 (1965)
68. Zadeh, L.A.: Probability measures of fuzzy events. J. Math. Anal. Appl. **23**, 421–427 (1968)
69. Zadeh, L.A.: Outline of a new approach to the analysis of complex systems and decision processes. IEEE Trans. Syst. Man Cybern. **SMC-2**, 28–44 (1973)
70. Zadeh, L.A.: Fuzzy sets as a basis for a theory of possibility. Fuzzy Sets Syst. **1**, 3–28 (1978)
71. Zadeh, L.A.: A computational approach to fuzzy quantifiers in natural languages. Comput. Math. Appl. **9**, 149–184 (1983)
72. Zadeh, L.A., Kacprzyk, J. (eds.): Fuzzy Logic for the Management of Uncertainty. Wiley, New York (1992)
73. Zadeh, L.A., Kacprzyk, J. (eds.): Computing with Words in Information/Intelligent Systems. Part 1: Foundations, Part 2: Applications. Physica–Verlag (Springer), Heidelberg (1999)
74. Zhang, S.X., Cueto, J.: The study of bias in entrepreneurship. Enterpreneursh. Theory Pract. **44**(3), 419–454 (2015)

Part III
Advanced Control Techniques for Industrial and Collaborative Automation

Part III
Advanced Control Techniques for Industrial and Collaborative Automation

Holonic Hybrid Supervised Control of Semi-continuous Radiopharmaceutical Production Processes

Theodor Borangiu, Silviu Răileanu, Ecaterina Virginia Oltean and Andrei Silişteanu

Abstract The paper applies the holonic paradigm to the supervised hybrid control of semi-continuous processes, which are exemplified by the production of radio-pharmaceutical substances. The supervisor of the control system fulfils two main functionalities: (i) optimization of global process planning that includes all client orders most recently received (the values of process parameters and operations timing are initially computed to maximize the number of accepted orders—optimal state trajectory); (ii) reconfiguring the parameters of the optimal state trajectory whenever unexpected events occur (in the production sub processes or in the environment parameters), providing thus robustness at disturbances. The implementation of the holonic supervised hybrid control system uses the multi-agent framework in semi-heterarchical topology. Two scenarios validating the optimization of planning and experimental results are reported.

Keywords Semi-continuous process control · Hybrid supervised control · Holonic paradigm · CP optimization · Radiopharmaceuticals production

T. Borangiu (✉) · S. Răileanu · E. V. Oltean · A. Silişteanu
Department of Automation and Applied Informatics, University Politehnica of Bucharest, Bucharest, Romania
e-mail: theodor.borangiu@cimr.pub.ro

S. Răileanu
e-mail: silviu.raileanu@cimr.pub.ro

E. V. Oltean
e-mail: ecaterina.oltean@aii.pub.ro

A. Silişteanu
e-mail: silisteanua@gmail.com

Y. P. Kondratenko et al. (eds.), *Advanced Control Techniques in Complex Engineering Systems: Theory and Applications*, Studies in Systems, Decision and Control 203, https://doi.org/10.1007/978-3-030-21927-7_11

1 Introduction. The Radiopharmaceuticals Production Process

Manufacturing systems are related conventionally to three basic classes of processes of discrete, semi-continuous or continuous nature [1, 2]. In the manufacturing domain, discrete-event processes execute on systems where separable, distinct products are executed on specific resources in sequences of discrete loosely coupled operations (machining, finishing or assembling). The control of *discrete processes* includes product planning, scheduling and resource allocation for the optimization of global cost functions (makespan, mean tardiness) at batch level. Current research is oriented towards semi-heterarchical control in which optimal plans and schedules, calculated off-line at centralized level by a hierarchical System Scheduler (SS) are taken as recommendations, and applied as long as the initial operating conditions are preserved. Whenever a technical (resource breakdown), environmental (parameter alert) or business (rush order) perturbation occurs, pre-computed schedules are abandoned and new schedules are issued in real time by collaborative agents representing the work-in-progress [3–6]. Such an approach is facilitated by the holonic concept, which extends physical entities (products, resources and orders) with software agents that represent their information counterparts. A Holonic Manufacturing System (HMS) is composed by a centralized layer (SS) and a decentralized MES layer with distributed intelligence (dMES) implemented in multi-agent framework [7, 8].

The class of *continuous processes* is related to continuous flows of material transformations and services grouped in generic classes (processing, conditioning, control) that use structures, utilities and environment facilities interconnected through piping streams. These processes are described by continuous-time differential equations. The optimization of such multivariable systems generally assumes model-based control, local observers and switching strategy compensators.

In the domain of *semi-continuous processes*, the continuous conventional process control is extended with discrete planning of ordered products [9]. The radiopharmaceutical production process belongs to the class of semi-continuous manufacturing processes characterized by continuous flows of materials, utilities and services similar to continuous processes, but which are not run in pure time-invariant, steady-state mode [10]. For global control of such processes at batch level the interaction of discrete control algorithms and continuous processes (dispensing and quality control) is treated as **hybrid control** (HC).

The production of radiopharmaceuticals involves handling large quantities of radioactive materials (radioisotopes) and chemical processing. While still on a relatively small scale in comparison to the production of conventional pharmaceuticals, it involves a number of aspects that can be quite demanding for small-scale manufacturers: the operation and maintenance of processing facilities complying with the codes of good manufacturing practices (GMP), effective quality assurance and quality control systems and transport means for radioactive products to hospitals. Radiopharmaceuticals produced using a cyclotron and dedicated radiochemistry equipment and labs are used for positron emission tomography (PET) and single photon emission computed tomography (SPECT), [11, 12].

The production of radiopharmaceuticals needs: (1) obtaining the radionuclides on which the pharmaceutical is based, i.e. the radioactive isotopes of elements with atomic numbers less than that of bismuth; (2) chemical processing, dispensing and packaging product doses (vials) with the final radiopharmaceutical product having a desired activity. Neutron-deficient radionuclides (having fewer neutrons in the nucleus than those required for stability) are radioactive isotopes of elements produced using a proton accelerator, such as a medical cyclotron [13].

The main objective of the cyclotron-based radiopharmaceutical production line (including: production facilities, process control and environment conditioning systems) is to manufacture valid nuclear medicine products by grouping as many orders received in the last 24 h as possible in daily standard production time slots (e.g. 2.5 h), safely for the personnel and surrounding environment. Such a production line is specialized in producing batches of products in small volumes, according to the orders received from hospitals and PET centres. While having a specific chemical structure, radioactivity and usage, each product follows the same manufacturing path: radioisotopes are produced in a particle accelerator (cyclotron)—sub process sp1, then transferred into technology isolators for chemical synthesis—sub process sp2 followed by (eventually diluting) and portioning (vial dispensing) the bulk product—sub process sp3, and quality control of the final product samples by conformity tests on multiple parameters—sp4; in the last stage, valid products are packed and transported to clients (hospitals, PET centres).in shielded containers—sub process sp5. Sub processes sp1 and sp2 are continuous, while sp3 and sp4 are controlled by discrete algorithms (e.g., valve opening/closing at constant time intervals to portion the bulk product; valve opening/closing at constant time intervals to assure identical dilutions of products in vials; robotized vial manipulating and packing, a.o.). The manufacturing sub-processes for a cyclotron-based production line of radiopharmaceuticals are presented in Fig. 1.

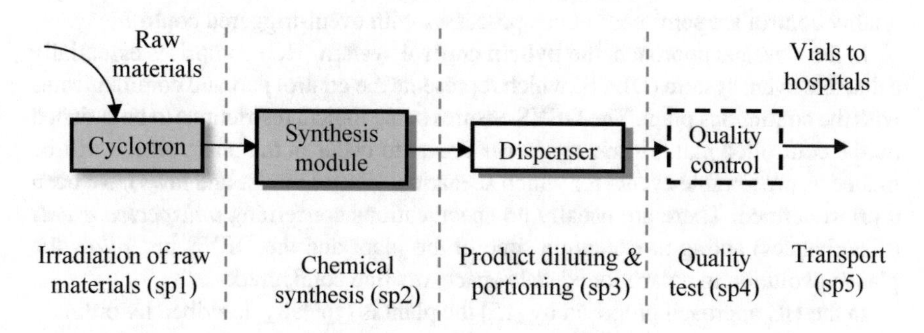

Fig. 1 Continuous (sp1, sp2) and semi-continuous (sp3, sp4) sub processes in the production of radiopharmaceuticals; dilution is optional in sp3 and sp4 is performed on sample vials

The production resources together with the quality testing equipment are disposed in *flow shop* processing mode: products flow in one single direction [14]. Raw materials enter the cyclotron and are irradiated, the active substance being transferred pipe to the synthesis module where chemical reactions occur; the resulting bulk radiopharmaceutical product is then portioned and eventually diluted in a robotized dispenser box; samples of vials are sent to the quality control lab for multi-parameter tests; finally, the vials are packed and labelled for transport to the client. In this plant, the product in different stages passes through capillary tubes from one resource to the next and receives an operation/service. The processing mode is non-preemptive, and the fixed set of precedence constraints among operations is a priori defined.

The rest of the article is organized as follows: Sect. 2 introduces the hybrid control for continuous production processes and extends it to supervisory control of the multi-stage radiopharmaceuticals production process in order to add new functionalities: optimal off-line planning of the production orders received from clients, and reconfiguring process parameters for robustness at disturbances. Section 3 presents a multi-agent implementing solution of the holonic hybrid supervised control. Section 4 describes how the supervisor (staff) holon acts as System Scheduler to optimize the planning of radiopharmaceuticals production at batch level. Finally, Sect. 5 reports two types of experiments that have been performed to validate the optimization of production planning in Constraint Programming approach and the results obtained.

2 The Supervised Hybrid Control Model of Radiopharmaceutical Production Processes

The irradiation and chemical synthesis processes of the radiopharmaceuticals production plant are described by nonlinear differential equations that involve continuous valued variables depending on continuous time $t \in \mathbf{R}^+$ and will be controlled in discrete time $t \in \mathbf{Z}^+$ in the plant, while bulk product portioning, dilution and sample quality control are semi-continuous processes with event-triggered control.

In the classical approach, the **hybrid control system** (HC) comprises essentially a discrete event system (DEVS) which represents the control part and communicates with the continuous plant. The DEVS ensures some logical restrictions to be satisfied by the controlled plant; it merely *forces events* to occur in the plant, which will be treated as predictable events for which scenarios, control modes and laws have been a priori defined. There are usually no specifications concerning *unexpected events* (disturbances) acting at execution time in the plant and the DEVS just *pilots* the plant's evolution in a partitioned state space, off-line configured.

In the HC approach proposed by [15] the plant is typically described by ordinary differential equations, the DEVS controller is modelled as a Moore machine and the interface contains the converters between the two systems [16]. Considering a given subspace of the continuous state space (which reflects control objectives and state restrictions) the system plant-interface is formalized as a DEVS-plant; then the DEVS controller is built, using the discrete event system theory.

In closed-loop dynamics, the DEVS controller outputs *control symbols* $\boldsymbol{r}$ that force predicted events in the DEVS-plant to occur and receives in response *plant events* z representing time sampled data from sensors that monitor the continuous evolution of the plant in the partitioned state space. The *continuous processes* (the *plant*) producing radiopharmaceuticals is modelled by the nonlinear differential equations:

$$\boldsymbol{x}^{(l)}(t) = \boldsymbol{f}(\boldsymbol{x}(t), \boldsymbol{u}(t))$$

where time $t \in \mathbf{R}$, the state vector $\boldsymbol{x}(t) \in X \subseteq \mathbf{R}^n$ is in continuous state space X, the control vector $\boldsymbol{u}(t) \in U \subseteq \mathbf{R}^m$ is in a set of allowable control values $U = \{u_1, \ldots, u_M\}$. U is mapped bijectively to an alphabet of control symbols $R = \{r_1, \ldots, r_M\}$.

The *plant interface* as intermediate layer consists from resources used to effectively control the processes (irradiation, chemical synthesis, dilution and portioning) that transform raw materials according to predefined or adapted recipes in order to obtain radiopharmaceutical products with imposed technical characteristics (purity, radioactivity). On its basic *sensing and control operating layer* the plant interface is a signal convertor between the plant and the controller by means of the information (logical) counterpart of the execution system and event generator. The *execution system* converts a string of control symbols $\omega_r \in \mathbf{R}^*$ to a vector of piecewise constant control signals $\boldsymbol{u}(t)$ for the plant. The *event generator* converts the state trajectory $x(\cdot)$ of the plant, evolving from the a priori known initial state in the desired partition of the state space (with the set of objectives and constraints), into a string of plant symbols z.

The model of the DEVS-plant is the automaton $G_p = \{P, R, f_p, Z, g_p\}$ where P is the set of discrete states, R is the input alphabet of control symbols, Z is the output alphabet of the plant symbols, $f_p : P \times R \to 2^P$ is the state transition function and $g_p : P \times P \to Z$ is the output function. The automaton G_p is the discrete state model of all possible evolutions of the plant, i.e., under all possible controls $\boldsymbol{u}(t)$ from all possible initial conditions $\boldsymbol{x}(t_0)$ for the plant's state equations, arbitrary located in the chosen state space partition. Although the continuous model $\boldsymbol{x}^{(l)}(t) = \boldsymbol{f}(\boldsymbol{x}(t), \boldsymbol{u}(t))$ is deterministic relative to $\boldsymbol{u}$, the DEVS-plant model is generally not deterministic to the control symbols because, depending on the initial conditions located in a given initial cell, different trajectories may transit, under the same control value, to different destination cells adjacent to the initial one [17].

The DEVS *plant controller* is modelled as Moore machine $G_c = \{S, Z, f_c, s_0, R, g_c\}$, with: S the finite set of discrete states, $s_0 \in S$ the initial state, Z the input alphabet, R the output alphabet, $f_c : S \times Z \to R$ the state transition function and $g_c : S \to R$ the output function. The *plant controller* is a discrete-event dynamic system, i.e., a discrete-state machine implemented by one industrial PC or, in the present case by several specialized computers each one controlling a plant sub process sp1, sp2, … (e.g. the cyclotron's irradiation controller). The plant controller receives input events represented by symbols (i) process data measured by sensors; (ii) descriptions of the current process state; (iii) events occurring in the process or environment. By processing these events, the plant controller produces output events that are represented by symbols and express piecewise constant commands sent to the execution subsystems (the actuators).

This HC approach is extended for the semi-continuous flow-shop radiopharmaceuticals production processes to **supervisory hybrid control** (SHC) to add new functionalities and operating modes:

1. *Optimal off-line planning of global production processes* for a time window of one day (2 working shifts of 8 h). The objective is to allow executing as much as possible orders received from clients (hospitals) which must be delivered during the current day. The optimization algorithm relies on the fact that radiopharmaceutical product with specified characteristics (e.g. level of activity) can be obtained in a specified quantity by customizing a master recipe according to an a priori formulated goal: minimizing the execution time of one production sequence or the waste of material.
2. *Reconfiguring on line the parameters* of the sub process currently performed and eventually of the process (es) that follow the current one, in response to unexpected events that may occur. The objective is to provide a reactive behaviour to unexpected events, without altering the generic flow-shop execution mode and characteristics of the ordered product.
3. *Constraining the process execution* by the evolution of the environment parameters: temperature, pressure, relative humidity, number of airborne particles and radioactivity in the production spaces, clean rooms and adjacent close spaces. The objective is to maintain permanently the proper environment conditions specified by the GMP for radiopharmaceutical production processes and personnel safety.

The SHC of the radiopharmaceutical production processes ensures these three new functionalities through optimization algorithms and high-level decision mechanisms. In the extended control mode, some events that were not predicted but are observed by the supervisor during process execution and classified as unexpected events will be forced to occur [18]. The supervised control concept proposed separates control from supervision according to the hierarchical scheme in Fig. 2.

The interface includes a mediation and integration middleware layer linking low-level I/O data protocols (such as: Modbus TCP, Profibus, OPC UA, binary OPC-UA of the PLCs) with high-level Web service protocols (such as DPWS). This middleware layer offers dynamic combination of web services and composite data, and handles value appearance and event notification; it is based on unified interfaces for the interoperability with protocols in the "I/O" and "Web service" classes.

As in the HC approach, the process is controlled by a logical (plant) controller which observes the events from the process and forces some events in the process to occur. The controller perceives the continuous process as a DEVS which produces events from Σ_{PR} (generated by the product-making processes and the working environment), and is itself a DEVS that evolves by producing *predictable events* from Σ_{PC}. The components of the global process coupled to the distributed plant controller

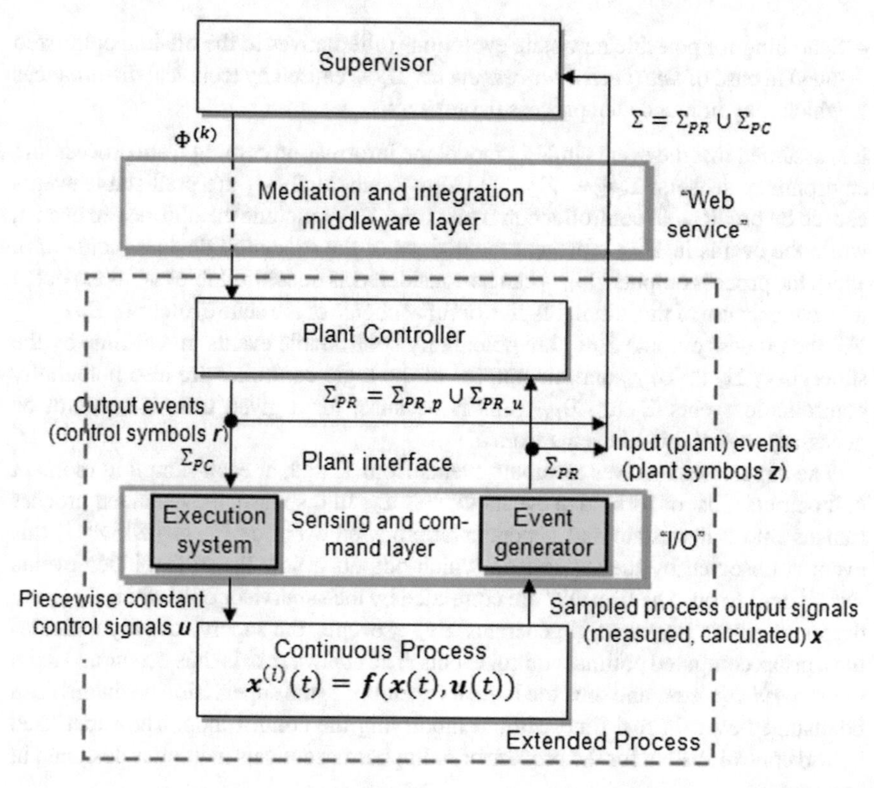

Fig. 2 The supervised hybrid control scheme of the radiopharmaceutical production process

(in each production stage a local process is performed under the real-time control of a dedicated resource), i.e., the feedback control loop, constitute the *extended process* which can be seen as a DEVS that generates events from $\Sigma = \Sigma_{PR} \cup \Sigma_{PC}$. The *supervision task* can be designed to restrict the evolution of the extended process by:

- Constraining the state trajectory of the extended process to one subspace so that the *optimization of a global cost function* is assured. The solution space for this optimization problem leads to a reduced number of possible state evolutions of the extended process; there are more than one such state trajectories, because a master recipe may provide the same product characteristics (radioactivity level, chemical properties, quantity, delivery time) with different process parameters (e.g. the same radioactivity of the bulk product can be obtained with different values of beam energy, beam current and exposure time; the required radioactivity of the portioned product can be reached from higher radioactivity levels of the bulk product and dilution percentages).
- Additionally restricting the evolution of the extended process by *environment manufacturing conditions* (values of environment parameters).

- Searching for possible new state evolutions (alternatives to the off-line optimized ones) in case of *unexpected process events* Σ_{PRu} caused by technical disturbances which may influence the process parameters.

It is assumed that the events in Σ_{PR} model the information coming from process and environment sensors: $\Sigma_{PR} = \Sigma_{PR\_p} \cup \Sigma_{PR\_u}$, where Σ_{PR_p} are predictable events caused by predictable controller commands and Σ_{PRu} include the unforeseen events, while the events in Σ_{PC} represent the actions of the execution devices (actuators). Thus, the process outputs Σ_{PR_p} can be considered as uncontrollable events expected to occur because of the initial, off-line optimal choice of the state evolution, $\Sigma_{PR\_p} \subseteq \Sigma_u$, the process outputs Σ_{PRu} are potentially controllable events in real-time by the supervisor, $\Sigma_c \subset \Sigma_{PRu}$ and the outputs of the logic controller are also potentially controllable events $\Sigma_c \subset \Sigma_{PC}$ (this is because, for a given task, it may not be necessary to prohibit all events from Σ_{PC}).

The supervisor receives as inputs events from Σ and, at each sampling moment k, it outputs a list of forbidden events $\Phi^{(k)} \subset \Sigma_c$. In response, the extended process transits into a new state and generates an allowed event $\sigma^{(k+1)} \in \left(\Sigma \backslash \Phi^{(k)}\right)$; this event is observed by the supervisor, which outputs a new list of forbidden events $\Phi^{(k+1)}$, and so on. The lists $\Phi^{(k)}$ are computed by the supervisor only when Σ_{PRu} are detected; otherwise, when Σ generates Σ_{PR_p} events, the supervisor only validates the a priori computed optimal control events. The *control task* is thus decoupled from the *supervision task*, and thus the logical restrictions (the supervision decisions) can be changed even in real time without modifying the control loop. The supervised hybrid control system for the process of radiopharmaceuticals making is designed in two steps:

1. Applying the *holonic theory* by defining reality-reflecting structural elements (resources, orders and products) with a high level of abstraction, and by distributing intelligence between the information counterparts of these holons to solve in multi-agent framework the global production planning and control problem optimally, with precision and in reactively to process disturbances [19].
2. Designing the *centralized supervisor* as a hierarchical System Scheduler for global process optimization and the production data base for resource and environment monitoring, and product traceability.

3 Multi-agent Implementing of the Holonic Hybrid Supervised Control (H^2SC)

Several approaches of applying the holonic control paradigm to continuous systems have been initiated in the last years. McFarlane [20] demonstrates that static and dynamic processes occur in manufacturing control. Holonic control solutions relate to global decisions of a System Scheduler being updated by the cooperation of distributed intelligent agents representing the local interests of their physical counterparts (resources, products, environment [21]). Borangiu et al. [22] report in their work a semi-heterarchical holonic control mechanism switching between optimal, centralized and reactive, heterarchical decision modes. Considering the hybrid behaviour, only the agents monitoring the work in process and the resources take part in the decision making process, while the product and its specifications represent merely constraints.

For the semi-continuous production process analysed there exists a master product recipe the specifications of which can be met with different values of: raw material characteristics, process parameters and timing. For example, different levels of bulk product irradiation obtained in the cyclotron with different time or beam current values impose different degrees of dilution in the product dispensing sub process. There are similarities between the SHC for discrete and semi-continuous processes:

- The supervised control optimizes order planning at complete time horizon (e.g., at product batch or workshift level) in case of production processes; for semi-continuous processes global optimization is only initially (off-line) done, while for discrete production processes real time updates of the initial planning is possible using a High Performance Computing problem solver. For this purpose, the supervisor acts as System Scheduler and uses a global problem solver running some type of optimization algorithm (e.g., based on production rules, heuristics or constraint programming).
- The plant controller distributes the intelligence for the composing sub processes; this approach is based on the agentification of above mentioned industrial artefacts and on the autonomy and collaboration of these agents.
- While supervision provides logical decisions selecting the best state trajectories of the controlled process, the plant controller applies these decisions through logical entities which are responsible with coordination and follow-up of the "work in process", at the local sub process For both discrete and continuous systems, the main goals of the control are similar: process stability and precision, timeliness, robustness at disturbances and reduced energy consumption.
- Service orientation and vertical integration of control and computing processes at plant level use a middleware layer interfacing low level ("I/O") and high level ("Web service") communication protocols [23, 24].

One main goal of the present research is to prove the applicability of the holonic control paradigm to hybrid supervised control of semi-continuous processes, taking as case study the H^2SC of radiopharmaceuticals flow-shop production process. For this purpose, the holonic reference architecture for discrete manufacturing control PROSA was considered. PROSA uses three classes of basic holons that communicate between them and collaborate in holarchies to reach a global goal at production process level: resource holon (RH), product holon (PH) and order holon (OH). The holon represents the association between such a physical artefact and its information counterpart (the agent); the holarchy created by the relationships between these holon classes is implemented in a multi-agent framework. The holonic hybrid supervised control model applied to the radiopharmaceutical production process is shown in Fig. 3.

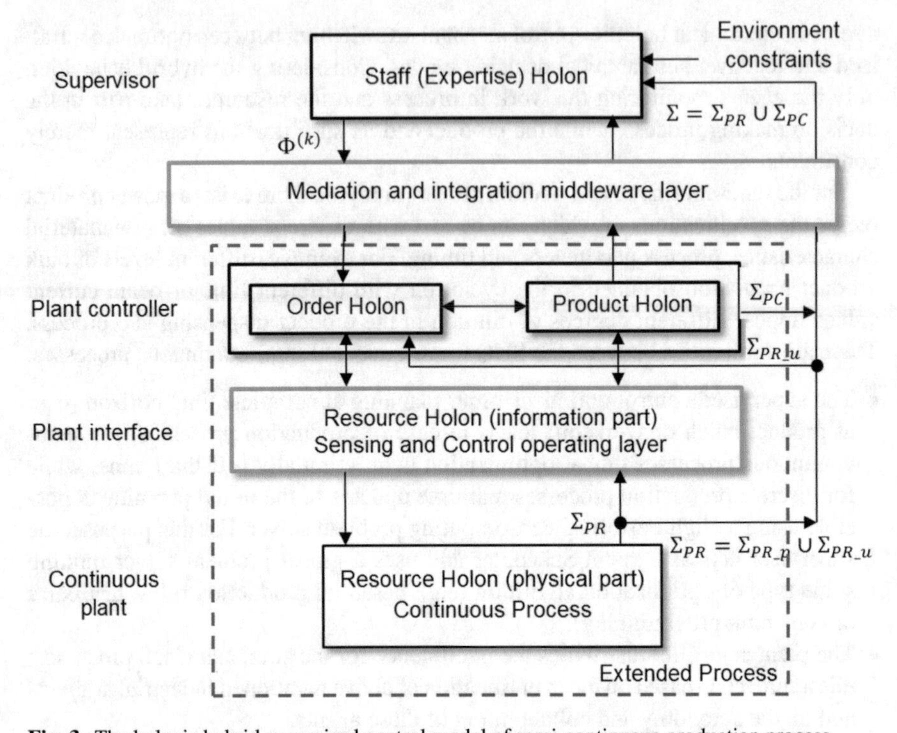

Fig. 3 The holonic hybrid supervised control model of semi-continuous production process

The components of the HSC in the scheme shown in Fig. 2 are reformulated in the holonic paradigm as follows:

- The plant controller is represented by the order holon and product holon, both having a logical (DEVS) nature.
- The plant interface is represented by the information part of the resource holons, being placed on the sensing and control operating layer.
- The plant is composed by the physical part of the resource holons (sensors, actuators, controllers, equipment) and by the continuous processes they perform.
- The supervisor is materialized by the staff (expertise) holon (SH) which provides logical decisions of: (i) off-line selecting optimal state trajectories for process control (predictive behaviour); (ii) real time update of the initially computed state trajectories when detecting unforeseen events (reactive behaviour).

For the semi-continuous production processes like the radiopharmaceuticals one, resources cannot be reassigned at process execution time; similar to the discrete case, they are non-preemptive, and the piping transfer network is initially configured and cannot be changed on line.

Figure 4 represents the building block diagram of the holonic hybrid supervised control of radiopharmaceutical processes, indicating the mapping of data (in the top

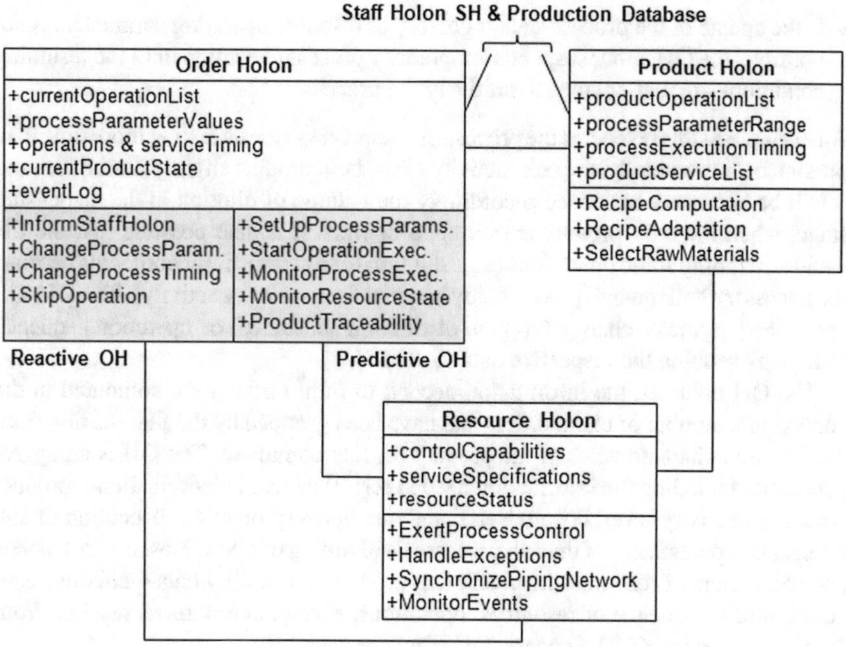

Staff Holon SH & Production Database

Order Holon
+currentOperationList
+processParameterValues
+operations & serviceTiming
+currentProductState
+eventLog
+InformStaffHolon
+ChangeProcessParam.
+ChangeProcessTiming
+SkipOperation
+SetUpProcessParams.
+StartOperationExec.
+MonitorProcessExec.
+MonitorResourceState
+ProductTraceability
Reactive OH
Predictive OH

Product Holon
+productOperationList
+processParameterRange
+processExecutionTiming
+productServiceList
+RecipeComputation
+RecipeAdaptation
+SelectRawMaterials

Resource Holon
+controlCapabilities
+serviceSpecifications
+resourceStatus
+ExertProcessControl
+HandleExceptions
+SynchronizePipingNetwork
+MonitorEvents

Fig. 4 Mapping of data and functions to the basic holons of the H^2SC system

boxes of the rectangles representing the holon types) and functions (in the bottom boxes) to the classes of basic holons.

The Order Holon is globally responsible for implementing the scheduling of product execution for a number of orders grouped and planned initially (off-line) by the Staff Holon. This means that OH sends to the RHs the values of: process, raw materials and control parameters to be applied, starts the execution of the continuous processes according to an optimally predefined timing, and monitors the occurrence of predicted events (maintaining an event log) and the status of the physical resources. These functions are performed by the *predictive part* of the OH.

Also, there is a dual behaviour and activity set of the OH, which is performed by its *reactive part*; this part receives from the RH information about unexpected events that may occur at process execution time. In such cases, the OH informs the Staff Holon about the necessity:

- To check whether an alternate state evolution of the faulty process (and of the subsequent processes) or of the resource that has generated the unexpected event can be computed to replace the existing one.
- If the update of the process state trajectory or resource operating parameters is possible, then the OH requests the SH to send the changes computed for these parameters, and applies the changes to the corresponding RHs.

- If the update of the process state trajectory or resource operating parameters is not possible, the OH either suspends temporarily process execution until the resuming conditions are met or stops definitively the process.

An example in this respect is the process of chemical synthesis lasting more time than predicted; consequently, the radioactivity of the bulk product slightly diminishes and it will be necessary to reduce accordingly the volume of dilution in the dispensing stage, when the bulk product is portioned in vials. It is also possible that the SH decides skipping some operations (e.g., dilution for a certain number of vials because the necessary bulk quantity has already reached the necessary activity). The OH will apply the parameter change (volume of dilution substance) or operation sequence change by sending the respective data to the RH.

The OH holds all the information needed to fulfil optimally a command in the context of a number of client orders that have been grouped by the SH, starting from the imposed characteristics of the product for that command. The OH is an aggregate entity including the information needed for: (i) demand identification—product type, radioactivity level, hospital #ID, and due delivery time; (ii) execution of sub processes—parameters of the sub processes and timing of operations, control modes and parameters of resources involved; (iii) product traceability data—effective execution times, sequence of resources, operations, execution reports as resulted from the physical process). The generic OH structure contains:

// Information computed in the planning process by SH and sent to resources:

- $\{\text{int}\}_i, i = 1\ldots\text{nr\_op\_order}$: *Resources*—the set of resources that must be visited in order to execute the operations stated above (Operations names);
- $\{\text{int}\}_i, i = 1\ldots\text{nr\_op\_order}$: *Parameters*—the set of parameters that configure the resources in order to receive the desired operations;
- $\{\text{int}\}_i, i = 1\ldots\text{nr\_op\_order}$: *Maximum execution time*—the time needed to execute the operation "Operations names *i*" on resource "Resources *i*".

// Information gathered during the sub processes executed and sent to the database:

- $\{\text{int}\}_i, i = 1\ldots\text{nr\_op\_order}$: *Timing*—the time spent to perform operation *i*;
- $\{\text{string}\}_i, i = 1\ldots\text{nr\_op\_order}$: *Parameters*—parameters used by resource *i* to execute operation *i*;
- int: *Quantity*—the quantity of the product that will be delivered to the hospital;
- int: *RFID*—the code which associates the physical product with the current in—formation.

The PH holds the process and product knowledge to assure the correct making of the product with sufficient quality. A Product Holon contains consistent and up-to-date information on the production cycle, requirements, recipes, process plans, bill of materials, quality assurance procedures, etc. As such, it contains the "product model" of the product type, not the "product state model" of one physical product instance being processed [25]. The PH acts as an information server to the other holons in the control system. The following information is held by the PH:

- string: *Product identification* (name, ID)—the name and ID of the current product;
- string: *Product description*;
- int: *Number of operations* (nr_op_prod)—the number of operations needed by the current product;
- {string}i, $i = 1\ldots$nr_op_prod: *Operations names*—the set of operations the current product needs;
- $\{\text{string}\}_i$, $i = 1\ldots$nr_op_prod: *Precedencies*—operations to precede operation i;
- {int}: *Activity level of the product*;
- The *relation between product activity and the irradiation activity* produced by the cyclotron;
- The *dependency between the activity level of the product and its dilution* with saline solution;
- {float}: The *halftime* of the irradiated raw material.

The RHs encapsulate both the informational/decisional part and the associated physical equipment used in the plant's sub processes: cyclotron, synthesis unit, dispenser and quality test equipment with their specialised controllers. Each resource is controlled independently, the role of the designed H^2SC system being to integrate in an efficient and robust manner these sub process automation islands by help of the SH via the OHs' control. The following attributes and methods were defined for an RH:

1. Attributes (information about system resources):

   - string: Resource identification (name, ID)—the name and identifier of the current resource;
   - string: *Resource description*;
   - int: *Number of operations* (nr_op_res)—the number of operations the current resource is capable of executing;
   - $\{\text{string}\}_i$, $i = 1\ldots$nr_op_res: *Operations names*—the set of operations the current resource is capable of executing;
   - $\{\text{int}\}_i$, $i = 1\ldots$nr_op_res: *Execution time* [seconds]—the set of execution times of the operations that can be done on the current resource;
   - string: *Current operation*—the operation the current resource is executing; this is one of the operations the resource is capable of executing;
   - int: *Maximum delay* [seconds]—the amount of time the product can be additionally held in the current sub process due to problems with resource or environment parameters at the next production sub process;
   - int: *Energy consumption* [Wh]—the energy consumed at the current stage;
   - $\{\text{string}\}_i$, $i = 1\ldots$nr_op_res: *Input*—the description of the current sub process input for operation i;
   - $\{\text{string}\}_i$, $i = 1\ldots$nr_op_res: *Output*—the description of the current sub process output for operation i;
   - date: *Online time*—the time and date starting from which the resource is online;
   - time: *Idle time*—the interval of time during which the resource was unused;
   - time: *Working time*—the interval of time during which the resource was used.

2. Methods to access the resource (types of configuration messages the current resource responds to):
   - void: *Configure operation* (target resource, parameter index, parameter value)—sets a specified parameter of the targeted resource to an imposed value;
   - void: *Start operation* (target resource, operation index)—triggers the start of the specified operation on the targeted resource;
   - int: *Request status* (target resource)—returns the current status of the target resource; the status can be offline, online and idle, online and working.

An instantiation for the attributes of the production resources (cyclotron, synthesis module and dispenser) modelled as RHs is given in Table 1:

Table 1 Instantiation of manufacturing resources modelled as RHs

Resource	Cyclotron	Synthesis modules	Dispenser
Description	Irradiate two liquid targets simultaneously	Automatically execute a prebuilt synthesis algorithm to produce one of two types of products: FDG and NaF; recover enriched water after synthesis; measure input and output radioactivity of raw material/ radiopharmaceutical compound (bulk solution)	Sterile dispense radiopharmaceutical compound into vials with final product using a robotic arm inside a technical isolator; fill, cap and crimp final product into vials and then into shielded containers to minimize the exposure of the operators; create dispensing recipe according to orders; read bar codes; measure initial and final product activity
Operation	Irradiation	Synthesis	Dispensing
Execution time	Irradiation time: 1–2 h	Approx. 23 min. (FDG) Approx. 7 min. (NaF)	5 min for the 1st vial, then 2 min for each next vial
Maximum delay	1 h FDG half-life is 1 h 9 min	30 min due to many dust particles in dispenser	15 min
Energy	100 KWh	300 Wh	400 Wh
Input	Enriched water	Irradiated enriched water from cyclotron	Radiopharmaceutical compound (bulk solution)
Output	Irradiated enriched water	FDG/NaF radiopharmaceutical compound	Final radiopharmaceutical product (delivered)

For this type of plant and semi-continuous processes PHs are related to product recipes managed by the H^2SC, and which are derived:

- From a *master recipe* indicating: (i) the sequence of operations/sub processes necessary to be performed in a predefined order; (ii) the desired values of the product characteristics (e.g., level of radioactivity); (iii) the portioning data (e.g., number of vials, volume of prepared product in each vial).
- Using well-established *dependence relationships* (diagrams, mathematical formula, lists of pre-computed values) between product characteristics and the corresponding process parameters which must be used by resources to reach the necessary values of product characteristics in each sub process.
- By *optimizing a cost function* (e.g. maximizing the number of client orders daily executed and delivered, reducing the energy consumed and/or product waste) with a constraint programming algorithm that computes the best combination of process parameters and timing for the due delivery date and product quality.

The Order Holon is decomposed in sub-process specifications, and distributed to resource holons at individual production stage levels (Fig. 5).

Figure 5 shows the iterative creation of OHs from client orders and master recipes received on daily basis, and grouped by the staff holon to be further optimized; as a result the resource operating modes, process parameters and timing data are computed and included in the OH's predictive part. The reconfiguring actions specified by the OH's reactive part are also represented.

The RHs are composed by hierarchically aggregating control and command devices and sub-processes that finally perform the four semi-continuous basic production processes performed in flow-shop mode. For example, the RH for bulk product dispensing results from the aggregation of: (1) an industrial robot; (2) the pipework for active product watering liquid mixing; (3) the programmable logic controller, pipes and valves for product diluting; (4) the robot manipulator handling vials and being in its turn the aggregation of: (4.1) the robot arm and (4.2) the end effector, etc. The RH aggregation hierarchy is not a fixed one; holons may belong, enter or leave more than one aggregations, e.g. a HVAC unit can be shared between several clean rooms where an unexpected process event requires air refreshment, temperature or humidity changes. Also, aggregated holons can dynamically change their content depending on current needs of the computed product recipe.

The H^2SC model is implemented in a multi-agent system, shown in Fig. 6.

By using the JADE framework, it is possible to replicate all agents in order to have a fault-tolerant platform. Concerning the communication layer all devices are interconnected at the physical level through a combination of wired (Ethernet) and wireless (WiFi) network which supports a TCP/IP communication protocol. On top of the communication protocol, the JADE framework offers a set of high-level interaction protocols which are based on message exchange according to the FIPA ACL standard (www.fipa.org). The execution of the production sub processes is authorized by an aggregated environment supervision (staff) agent (EA) that monitors permanently the values of the environment parameters in the production rooms (temperature, pressure, relative humidity and number of airborne particles) and the radioactivity

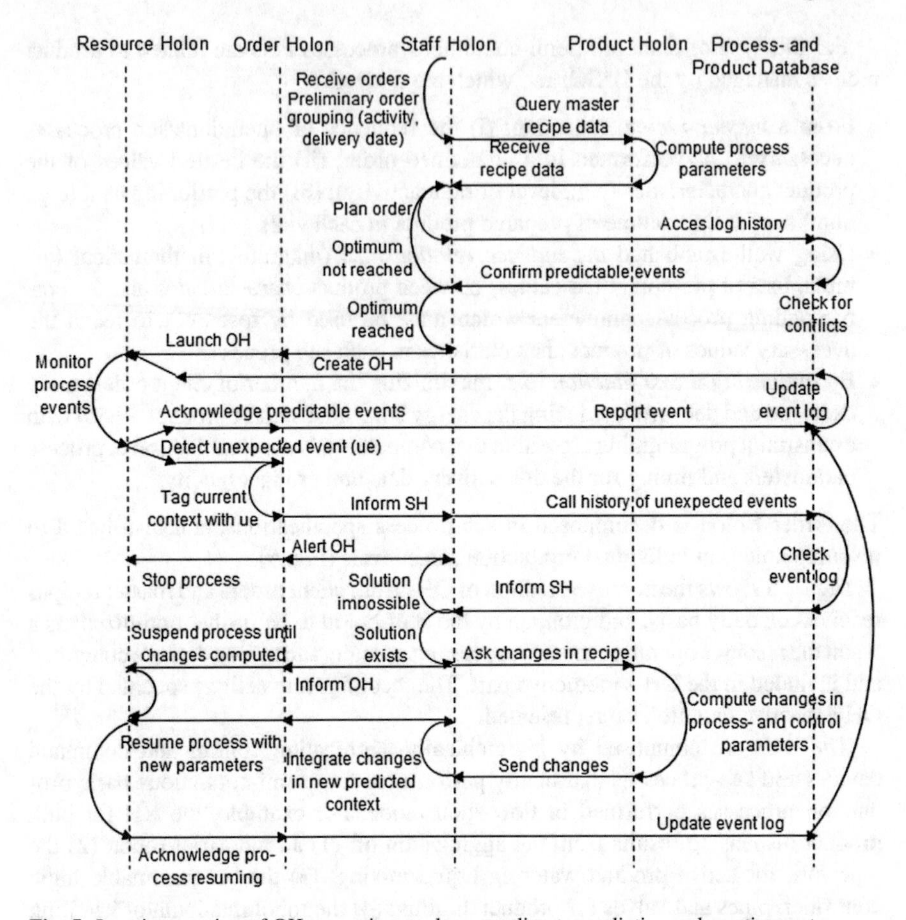

Fig. 5 Iterative creation of OH, operating modes according to predictive and reactive parts

in the plant's closed spaces; when the normal operating limits are exceeded, EA signals an unexpected event that may either stop production or alter timing and process parameters in reactive OH mode.

4 Optimizing Production Planning in Supervised Hybrid Control

One main function of the supervisor in the H^2SC model is to act as System Scheduler to optimize the planning of radiopharmaceuticals production processes. The objective is to group as many client orders received in the last 24 h for products of the same type as possible, irrespective of the products' level of requested radioactivity while

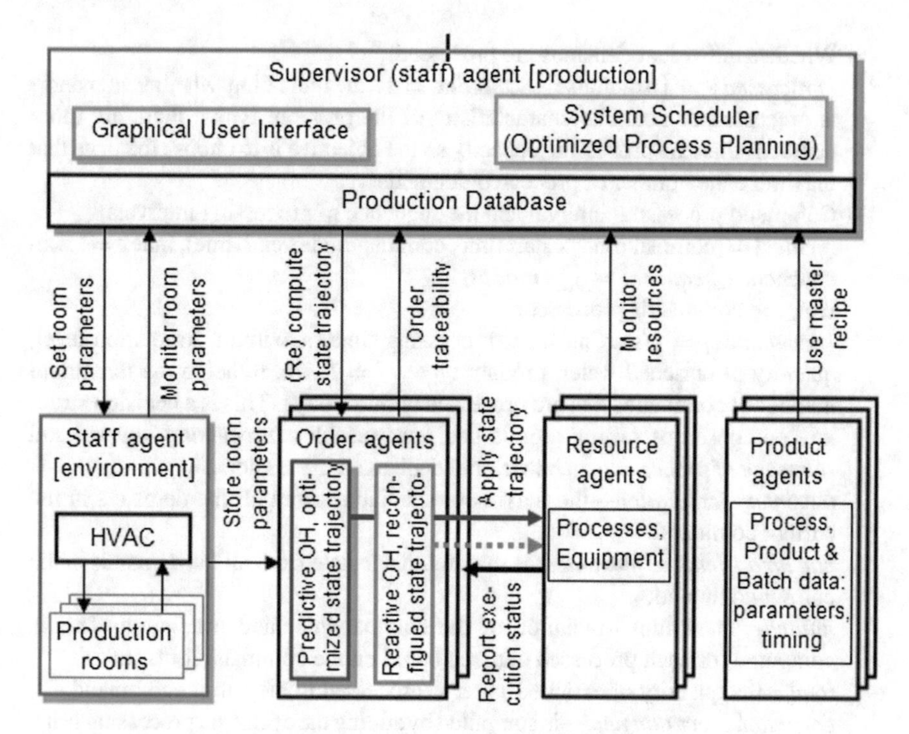

Fig. 6 Multi-agent system implementation of the holonic hybrid supervised control

respecting the due delivery dates and product characteristics, minimizing the product losses and limiting the energy consumption.

The following terms are used to formalize the production planning problem: *demand*: one vial, part or the client order, *command*: a set of demands that are produced together and contain the same type of product, and *batch*: the set of all demands that must arrive at the same hospital; a batch can contain different products. In this context the optimization problem is described by:

- An **input data set** representing the demands which are characterized by: requested product type, purity, radioactivity, and due delivery date:
  - {*demands*} = {(product type, requested irradiation, hospital, delivery time, requested volume)$_i$, i = 1…n, n being the number of vials requested for the current day}.
- The modelled **decision variables**:
  - How the individual demands are allocated to commands: {*where*} = {(command)$_{index}$, command = 1…m, *index* = 1…n, m being the maximum number of commands that are processed within a day, n being the total number of vials requested for the current day};

- Whether individual demands are processed:
  {*processed*} = {(true/false)$_{index}$, *index* = 1…n, indicating whether demand *i* is processed or not. The characteristic of the problem is that there are more demands than the processing capacity so the objective is to choose the ones that maximize the number of processed demands};
- Command processing interval and the sequence of processing intervals;
  [*modes*] = [commandindex start time, commandindex end time], index = 1…m
  $commands\_sequence = \bigcup_{i=1}^{m} modes_i$
- How are commands processed:
  {*commands*} = {(starting time, processing time, maximum irradiation level, quantity of enriched water, product type)$_j$, $j = 1 \ldots$m, m being the maximum number of commands that are processed within a day}. This is a decision structure composed of several atomic decision variables (*processing interval* and *sequence of processing intervals*) and decisional expressions:
  maximum irradiation—the maximum irradiation from all the demands in the current command;
  maximum demand due date–the maximum due date from all the demands in the current command;
  diluted—maximum irradiation of the command/demand irradiation. This is computed for each processed demand based on the command it fits in;
  load—the quantity of enriched water is processed in the current command;
  command execution time—is computed by adding the cyclotron processing time, the synthesis time, the dispense time and the imposed delay between commands.

While decision variables can be chosen freely in a given interval, decision expressions are computed based on decision variables using a fixed formula. The latter are derived from decision variables and are used to simplify the optimization model, especially to ease the way the constraints and objective function are written.

- The **constraints**:

- The weighted sum of all volumes of the demands within a command should be less than the volume of the irradiated target:

$$\sum_{i=1}^{n} V_i * \frac{desired\ activity_i}{\text{maximum activity}} \leq \text{target volume} - \text{loss}$$

where: i is the demand that will be produced within the current command, V_i is the requested volume of demand i, "desired activity$_i$" is the radioactivity of the product in demand i, "maximum activity" is the radioactivity at which is irradiated the raw material for the current command, "target volume" is the maximum raw material quantity that can be irradiated in a production cycle, and "loss" is the quantity of product that is lost when transporting from one stage to another.

- Within each command only demands with the same type of product should be processed;

- The dimension of the processing interval for each command should be equal to execution time of the command;
- The end time of the processing interval of each command should be equal to the maximum delivery date (including transport) of the demands contained;
- Do not overlap the processing intervals of the commands;
- Do not use the production facility over maintenance periods:

$$\text{production intervals} \cap \text{maintenance intervals} = \emptyset,$$

where "production intervals" are defined by the starting time of the command and its duration and "maintenance intervals" are predefined based on a fixed scheme which takes into account the resources' usage.

- Possible **objective functions**:
  - minimize production time at command level;
  - maximize the number of commands for daily client orders;
  - minimize the quantity of daily lost raw materials

By analysing the requirements stated above it can be seen that the optimization problem is first of all a matching problem (which demand is allocated to which command) that is subject to a set of constraints (delivery dates of the vials together with a given quantity and radioactivity level). Finally, since there is flexibility when irradiating the raw materials (less material can be irradiated at a higher level, and then diluted when dispensing), an objective function represented by production time can be added.

Thus, the planning problem deals with *combinatorial optimization* and detailed scheduling, both aspects being tacked in literature by Constraint Programming (CP) approaches [26]. A comprehensive list of CP solvers addressing various combinatorial problems can be found at (www.constrantsolving, [27]). Since the problem optimization option must be included as a functionality of the Supervisor Holon, the IBM ILOG OPL optimization engine was chosen because it can be easily integrated with separate applications using standard C ++, C# or JAVA interfaces. This is facilitated through the Concert technology [28].

The procedure described above was integrated in the radiopharmaceuticals production process both for *optimized offline planning* (minimize production duration while respecting imposed deadlines—**off-line run**) and for *online agile execution* (online adaptation to variations of environment parameters which can delay commands being executed within the same day—**online run**).

Off-line run

1. Demands are gathered for the next production day.
2. Maintenance restrictions are introduced as constraints into the CP model.
3. The ILOG model is called based on the set of demands for the next day:

(a) If a feasible solution is reached, the demands are accepted as received and the production plan is transmitted to the distributed operating control level in order to be implemented;
(b) If no feasible solution is reached (due to conflicting time constraints or tight deadlines) new deadlines are proposed based on the maximum load of the production system and on the rule "first came first serve" in order to fulfil as much as possible demands.

Online run

1. Apply process and environment parameter configuring according to the off line computed commands.
2. Measure environment parameters which affect production time (dust particles in dispenser chamber and radioactivity levels).
3. If parameters are out of range and the current command is delayed, re plan the next commands taking into account the new constraints. For any command, since the maximum allowed delay in production (30 min) is less than the maximum delay accepted for delivery (1 h) the worst case scenario is to use the off-line computed production plan and just delay it.

ILOG optimization sequence

The following optimization sequence will be run for a maximum amount of raw material max_irr_vol (capacity of max. irradiated target) processed for any production command:

1. Consider all demands valid for scheduling ($processed(d) = \text{true}, d = 1\ldots n$);
2. Order demands based on product type;
3. Choose the highest activity required (all other products will be diluted in order to obtain an inferior activity) for each product type (max_irr_level);
4. For all demands with the same product type (stp) compute the sum:

$$\text{sum\_prod} = \sum_{for\ all\ stp} requested\ volume \cdot \frac{desired\ activity}{max.activity}$$

   Clearly, sum_prod ≤ max_irr_target.
5. Based on the requested product quantity (sum_prod), on the maximum activity (max_irr_level) and on the number of vials, an estimated production time (makespan) for the demands is computed, considering Ti, $1 \le i \le 4$.
6. Test if the production intervals with the width computed at step 5 can be scheduled one after another, with a break between them of 1:30 h (resource checking periods between successive commands execution), without invalidating the delivery times for each hospital demand:

   - For all c in commands
   - For all d in demands
     - p = the command processed before (c)

– If p = null (c is the first command to be produced)
  YES: production starting time(c) = 6:00 (the plant begins to function at 6:00 in the morning);
  NO: otherwise production starting time(c) = production ending time(p) + 90 min.
– If (delivery_time(d)) < makespan(c) + production break + production starting time(c)
  NO: Eliminate demand d from the schedule (processed(d) = false) and Goto 2.

7. Compute the remaining raw material quantity (rem_raw_mat) that can be used with the associated deadline and maximum activity

   - For all c in commands
   - For all d in demands, *processed*(d) == true
     – rem_raw_mat(c) = max_irr_volume—sum_prod.

8. Test whether the demands eliminated can be produced using the remaining raw material quantity computed at step 7:

   - For all c in commands
   - For all d in demands, *processed*(d) == false
     – If the product of demand d is the same as the product manufactured in command c and there is enough remaining raw material in command c to produce demand d:
       Propose a new delivery time for demand d
       Set *processed*(d) == true
       Recompute sum_prod for command c with the new considered demand d
       Goto 7.

Process optimization takes place in semi-heterarchical control architecture, in which the configuration of process parameters initially established for global optimality is kept as long as there are no significant process disturbances; once such unexpected events occur, new order holons are computed from the master product recipe preserving the specification and the continuity of production. In the holonic paradigm used, the decisions of the supervisor are conditioned by external systems such as the one monitoring the environment parameters of the plant.

5 Experimental Results and Conclusions

The experiments carried out on a recently built radiopharmaceuticals plant in the IFIN-HH Radiopharmaceutical Research Centre (Bucharest) aimed at validating the three main characteristics of the global planning and control system proposed:

- Applying the holonic paradigm to the hybrid supervised control of the production process.
- Organizing in semi-heterarchical topology the supervised control architecture based on: (i) centralized System Scheduler—part of the supervisor organized as staff holon for optimization of multi-batch production planning and predicting the related state trajectory; (ii) delegate multi-agent system (D-MAS) composed by interconnected networks of order-, resource- and environment agents for decentralized monitoring of sub processes execution and parameter reconfiguring at the occurrence of unexpected events.
- Constraining the production process by decisions issued by a dual facility environment monitoring and control system and transferred to the H^2SC supervisor.

These validation objectives have been materialized in experiments that were designed to: (i) configure process parameters—how to assure the imposed product specification by deriving product holons from master product recipes; (ii), select the type of control—predictive (apply the control off-line computed for optimal state trajectory/reconfigure in real time the control for robustness at occurrence of unexpected events; (iii) confirm the global optimization method and algorithm.

The parameter values for the sub processes: cyclotron irradiation, chemical synthesis and dispensing are computed in the off-line production planning and parameter configuring phase, using theoretical or experimentally derived relationships (mathematical equations derived from physics laws, look-up tables) between the product's characteristics and the needed process parameters.

Experiments have been carried out to determine the dependencies of the product's radioactivity on the following process parameters: beam current in the cyclotron, bombardment time in the cyclotron, and volume of the dilution liquid in the dispensing sub process for all demands in the currently executed command having the same radioactivity.

Since the irradiation phase is based on complex particle reactions at higher levels of energies, the amount of activity obtained for each production batch cannot be calculated precisely in advance, and therefore it was estimated to a certain value by using the cyclotron irradiation curves. These dependency curves are particularly influenced by the machine type and energy used (19/24/30 meV), the target design (its capacity of handling high beam current) and the irradiation session parameters (the beam current used and bombardment time) [29].

In the case of [18F] fluoro-2-deoxy-D-glucose (FDG) production, the reaction chamber of the target was filled with ^{18}O enriched water (95%). The isotopic enrichment is necessary because of the reduced abundance of ^{18}O isotope within the natural Oxygen [30, 31]. The target holder was bombarded with a proton beam with W energy and beam current I_{target} allowing:

- the nuclear reaction $^{18}O(p,n)\ ^{18}F$ to take place;
- to calculate the newly obtained ^{18}F product; we considered two simultaneous processes: the creation of new radionuclides through the nuclear reaction and the decay of the existing radionuclides;
- to calculate the product activity after the bombardment time $t_{bombardment}$ as:

$$A_{product} = S * I_{target} * \left(1 - e^{-\ln 2 * \frac{t_{bombardment}}{T_{1/2}}}\right)$$

where S represents the saturation yield and is related to the target performance and $T_{1/2}$ represents the half-life of the ^{18}F isotope (~110 min.)

A series of irradiation session were performed using a 2.5 ml Niobium water target made by ACSI in order to define the irradiation curves for the ACSI TR-19 cyclotron. Experimental data results are presented in Table 2.

The experimental results obtained from the irradiation sessions performed with the TR-19 meV cyclotron of the IFIN-HH Radiopharmaceutical Research Centre have been used to configure the current and time constraints for the irradiation level satisfying all demands included in a production command optimally planned and scheduled by the ILOG CP-based production planning algorithm.

Experiments were performed concerning the optimization of radiopharmaceuticals production planning. The optimization model was structured in four parts:

1. Input data (fixed elements): include the number of demands, the maximum level of irradiation in the cyclotron's target, the maximum raw product quantity that can be irradiated and the demand data specifying the due delivery time and the necessary activity of each demand (in a vial).
2. Variables and decisional expressions (these elements are varied by the optimization engine in order to improve a criterion). The decisional expressions are relationships between decisional variables and are used to simplify the declaration of the objective function and of the constraints, and for the intuitive display of the results. There are three decisional variables for each command (the set of demands which will be planned for execution in the same production cycle):

   - The client orders (demands) which are grouped together to be processed in the respective command (a vector-type variable specifying whether the demand having the respective index will be processed or not in that command),

Table 2 Irradiation sessions for 2.5 ml irradiation target

Integrated beam current (μAh)	Activity concentration at EOB (GBq/ml)	Bombardment time (min)	Total Activity at EOB (normed to 2.5 ml)
25.2	24.26	30.24	60.65
37.5	33.54	45	83.85
50	41.94	60	104.85
62.5	45.83	75	114.575
75	56.87	90	142.175
99.5	78.82	119.4	197.05
119.1	82.06	142.92	205.15

- The delivery date and hour of each client order (scalar), and
- The activity level of the demand. For the second experimental optimization scenario the activity level is a constant (1500 MBq), as will be further explained.

There exist the following decisional expressions for each demand: (i) the difference between the due delivery time of each demand (imposed by the hospital) and the completion of the execution time of each demand (a vector whose dimension corresponds to the number of demands); (ii) the maximum activity at which the demand is irradiated (vector); (iii) the volume of non-diluted bulk product in the cyclotron target for the command that includes the considered demand (vector); (iv) the total bulk product resulting in the target for an executed command (scalar); (v) the number of demands grouped in a command (scalar); (vi) the total execution time for a command (scalar).

From these six decisional expressions, the total execution time T_{ex} is computed according to the following formula:

$$T_{ex} = t_{ir\_max} + t_{ch\_synt} + t_{vial\_fill} + t_{clean\_transp}$$

where:

- t_{ir_max}: the cyclotron irradiation time for the planned command (a linear function of the irradiation level);
- t_{ch_synt}: the time needed for the chemical synthesis of the product (a constant value of about 20 min);
- t_{vial_fill}: the time needed to dilute all the partial quantities of product in the bulk tank for dilution, corresponding to the sets of demands in the current command which must have the same radioactivity level (lower that the level of activity of the total bull product which has been irradiated in the cyclotron's target) plus the time needed to fill the vials (one vial per hospital demand);
- t_{clean_transp}: the time needed to clean the plants equipment—a constant value of about 90 min; plant cleaning and transport of products are simultaneously done;
- t_{transp}: the transportation time of all the vials corresponding to demands with products ordered and grouped in the same command to the specified hospital.

3. <u>Objective function</u>: maximizing the total number of accepted daily received demands from clients.
4. <u>Constraints:</u>

    (c1) For each working shift, the quantity of irradiated material must not exceed 3500 μL (the capacity of a vial is $\leq$ 500 μL and the product it contains has a lower activity than the maximum activity of the bulk product for that command,—hence the product must be eventually diluted for a vial

    (c2) There are 2 commands that must have time to execute daily:

$$360 + time_{processcommand\_1} + time_{processcommand\_2} < 1200[min]$$

(c3) The start time of processing the 1st command must be greater than the start time of the first workshift
(c4) The delivery hour of the first command must be before the start hour of the 2nd command
(c5) The 2nd command must be completed earlier than the end time of the daily work program (earlier than the end time of the 2nd workshift)
(c6) For the orders received from clients, the difference between the delivery time of ordered products (demands of products delivered in vials) and the completion time of a command must be positive and less or equal to 90 min
(c7) The irradiation level of the command must equal the maximum level of irradiation of all the demands included in that command (this constraint applies only for the variable level of irradiation applied in the scenario for Experiment 1).

Experimental Scenario 1: Irradiation of the cyclotron target *at the maximum activity requested* for the demands included in a planned command.

Scenario description: The CP optimization algorithm was executed for a set of 50 demands received with levels of activity of radiopharmaceutical products imposed in the range 600–1200 MBq, and due delivery times randomly grouped in two time intervals: 10:00–14:00 h (for execution in the 1st day's workshift) and 16:00–20:00 h (for execution in the 2nd workshift of the day), as presented in Table 3. The irradiation of the cyclotron's target is done at the maximum activity of 1200 MBq requested for the demands included in the two commands to be defined using ILOG OPL. This means that in this scenario, the maximum irradiation level is a decisional variable.

In this case, 21 client orders were accepted to be executed in the 2-shift work day as follows: 10 demands were grouped in a first command with execution in the 1st workshift, and 11 demands were grouped in a second command with execution in the 2nd workshift.

Figure 7 shows statistical data describing the iterative computation performed, and the time evolution of the solution for the optimization problem in Scenario 1.

Table 3 Description of demands for Scenarios 1 and 2

Max. possible activity in target	1600 MBq	Max. bulk product radioactivity that can be obtained in the target
Target capacity	3500 μl	Max. quantity of product that can be irradiated in the cyclotron in a command
No. of demands	50	Number of client demands received for one day
Requested radioactivity	590–1200	Radioactivity values for the 50 received demands
Due delivery time [min] from 0:00 h	600–1250	Delivery times in 2 shifts

Problems | Scripting log | Solutions | Conflicts | Relaxations | Engine log | Profiler | Statistics

Statistic	Value
CP	
Constraints	707
Variables	104
Memory usage	2181000
Number of solutions	7
Number of branches	385735
Number of fails	188669
Choice points	197667
Objective	21

(a) Statistical data of iterative computation for the optimization model in Scenario 1

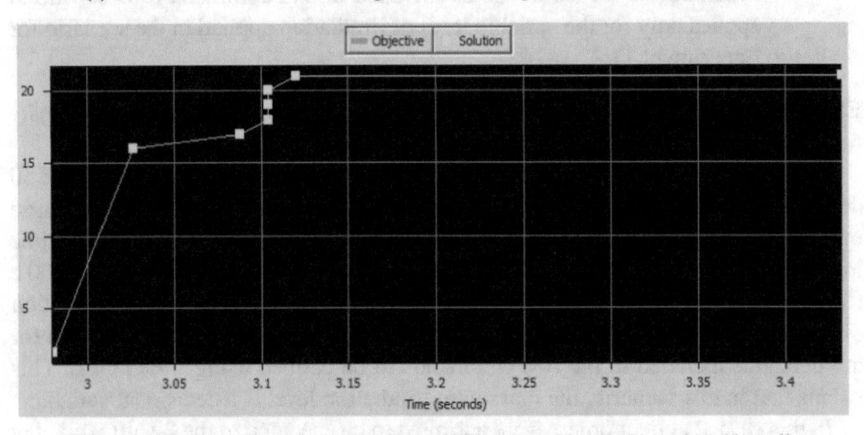

(b) Time evolution of the solution for the optimization problem in Scenario 1 (the yellow points represent solutions and the green curve shows the evolution of the solution)

Fig. 7 Characterisation of the iterative computation and evolution of the solution for the production planning optimization problem in experimental Scenario 1

Experimental Scenario 2: Irradiation of the cyclotron target at an activity level *higher than the maximum activity requested* for the demands included in a planned command.

Scenario description: The CP optimization algorithm was executed again for a set of 50 demands received with levels of activity of radiopharmaceutical products imposed in the same range of 590–1200 MBq and due delivery times randomly grouped in the same two time intervals: 10:00–14:00 h and 16:00–20:00 h (for execution in the two work shifts), as shown in Table 3. This time, the irradiation of the cyclotron's target is done at the activity level 1500 MBq, higher than the maximum activity of 1200 MBq requested for the demands included in the list of Table 3, still possible for this target. Thus, the irradiation level is a constant in this scenario.

In this second case, more client orders [24] than in the first scenario [11] were accepted to be executed in the 2-shift work day as follows: 12 demands were grouped in a first command with execution in the 1st workshift, and 14 demands were grouped in a second command with execution in the 2nd workshift.

After running the CP optimization model for the set of data specified in Table 3, it was noted that for Scenario 2 there remains a significant quantity of 500 μl irradiated liquid that is not used in command 1 (loss). In this situation, if the time interval for acceptance of the 1st command is increased from 90 min. to 150 min., the no. of satisfied demands increases from 26 to 28, as well as the quantity of used bulk product 3000–3433 μl.

Interpretation of experimental results and conclusions: the experiments performed in the two scenarios confirmed the correctness of the CP optimization method and validated thus the approach for supervised production planning and scheduling with objective function "maximization of the number of accepted client demands for execution in 2-shift daily work program".

Several conclusions could be derived concerning the adjustment of the optimization model parameters:

- The delivery time cannot be significantly increased, since this would cause an important reduction of the imposed product's radioactivity.
- With a moderate increase of up to 20% of the total command execution time, the degree of utilisation of the irradiated bulk product can be increased up to 98% for the actual capacity of the cyclotron's target. This minimizes the waste of irradiated material reducing in consequence the production cost.
- By comparing the results obtained in the basic versions of Scenarios 1 and 2, it results that irradiating the raw material in the cyclotron's target beyond the highest activity requested for the ordered demands allows accepting more client demands, but at higher production cost because of the increased energy consumption caused by the higher value of the programmed beam current in the cyclotron. Therefore, the irradiation level should be normally prescribed function of the maximal activity imposed by the client's demands as used in Scenario 1, and only in special situations of rush orders that have to be accepted without delay Scenario 2 should be used.

The output values obtained in the off-line optimized production planning are then applied during predictive process control by the order agents who send to the resource agents (implemented as process control tasks in the equipment controllers) the values of: process- and timing parameters to be used, acknowledge the occurrence of predicted events (maintaining an event log) and detect unexpected events in the current process- and environment context. The combinatorial search for optimal solutions based on ILOG OPL software proved to be efficient in consumption of time and computing resources.

The supervision task was defined and designed to restrict the evolution of the extended process, first by constraining the state trajectory of the extended process to one subspace so that the optimization of a global cost function is assured with additional operating constraints imposed by the environment and safety conditions, and then by searching for possible new state evolutions when unexpected process events occur.

A holonic hybrid supervised control model was developed and applied to the radio-pharmaceuticals production process. The principal advantage of the holonic approach for the supervised control of this class of semi-continuous processes consists in decoupling of the control layer from the process layer, which provides flexibility in reconfiguring the control at process execution time and robustness at disturbances. The supervisor was implemented by a staff (expertise) holon having a global view of all the plant's processes, which makes possible the optimization of the overall cost function for the greatest number of individual demands that can be produced from a master product recipe in a time interval limited by products' due delivery dates, and plant operating and maintenance constraints.

Using the holonic paradigm for the supervised production process planning and control brings the benefits that holonic organisations provide to living organisms: direct connectivity between physical entities of the plant (resources, products and orders—i.e. process timing and execution details) and their information counterparts (agents); robustness at perturbations; adaptability and flexibility at changes in product recipes and variations in customer orders (number of demands, quantities and type of products); and efficient utilisation of available resources. Implementing the holonic organization in the domain of semi-continuous production processes transposes the principles of organizing discrete manufacturing processes in the domain of hybrid supervised control, and applies the multi-agent system technology—MAS that is mainly adequate for modularity, decentralization, distributing intelligence and reusing active entities in the 'plug-in' technique.

Process optimization takes place in semi-heterarchical control architecture, in which the configuration of process parameters initially established for global optimality is kept as long as there are no significant process disturbances; once such unexpected events occur, new order holons are fed (if possible) with alternate information computed by the supervisor from the master product recipe to meet the imposed product specification. In the holonic paradigm used, the decisions of the supervisor can be determined by the external system monitoring the plant environment.

Further research will be directed towards using the digital twin concept and IoT gateway implementing solutions to improve process reality mirroring for event prediction and anomaly detection in real time.

References

1. Andreu, D., Pascal, J., Valette, R.: Interaction of discrete and continuous parts of a batch process control system, ADEDOPS Workshop, Imperial College, London (1995)
2. Ramadge, P.J., Wonham, W.M.: The control of discrete event systems. Proc. of the IEEE **77**(1), 81–89 (1989)
3. Aytug, H., Lawley, M.A., McKay, K., Mohan, S., Uzsoy, R.: Executing production schedules in the face of uncertainties: a review and some future directions. Eur. J. Oper. Res. **161**(1), 86–110 (2005)
4. Băbiceanu, R.F., Chen, F.: Development and applications of holonic manufacturing systems: a survey. J. Intell. Manuf. **17**(1), 111–131 (2006)

5. Borangiu, T., Silisteanu, A., Răileanu, S., Morariu, O.: Holonic facility environment monitoring and control for radiopharmaceutical agent-based production. Service Orientation in Holonic and Multi-Agent Manufacturing, Springer series Studies in Computational Intelligence, vol. 694, 269–286 (2017)
6. Mehta, S.V., Uzsoy, R.N.: Predictable scheduling of a job shop subject to breakdowns. IEEE Trans. Robot. Autom. **14**(3), 365–378 (1998)
7. Borangiu, T., Răileanu, S., Trentesaux, D., Berger, T., Iacob, I.: Distributed manufacturing control with extended CNP interaction of intelligent products. J. Intell. Manuf. **25**(5), 1065–1075 (2014)
8. Cardin, O., Trentesaux, D., Thomas, A., Castagna, P., Berger, T., El-Haouzi, H.B.: Coupling predictive scheduling and reactive control in manufacturing hybrid control architectures: state of the art and future challenges. J. Intell. Manuf. **28**(7), 1503–1517 (2017). https://doi.org/10.1007/s10845-015-1139-0
9. Indriago, C., Cardin, O., Rakoto, N., Castagna, P., Chacon, E.: H2CM: a holonic architecture for flexible hybrid control. Comput. Ind. **77**, 15–28 (2016)
10. Răileanu, S., Borangiu, T.: Centralized HMES with environment adaptation for production of radiopharmaceuticals. Service Orientation in Holonic and Multi-Agent Manufacturing, Studies in Computational Intelligence, vol. 640, chapter 1.1, pp. 3–18, Springer (2016). ISBN 978-0-443-07312-0
11. Ell, P., Gambhir, S.: Nuclear Medicine in Clinical Diagnosis and Treatment. Churchill Livingstone (2004)
12. Iverson, Ch., et al. (eds.): 15.9.2 Radiopharmaceuticals, AMA Manual of Style, 10th edn. Oxford, Oxfordshire: Oxford University Press (2007). ISBN 978-0-19-517633-9
13. Mas, J.C.: A Patient's Guide to Nuclear Medicine Procedures: English-Spanish. Society of Nuclear Medicine (2008). ISBN 978-0-9726478-9-2
14. Kusiak, A.: Intelligent Manufacturing Systems. Prentice Hall, Englewood Cliffs, NJ (1990). ISBN 0-13-468364-1
15. Antsaklis, P.J.: Hybrid control systems: an introductory discussion to the special issue. IEEE Trans. Autom. Control. **43**(4), 457–460 (1998)
16. Oltean, V.E.: Hybrid control systems—basic problems and trends. Revue Roumaine Sci. Techn.-Électrotechnique et Énergétique **46**(2), 225–238 (2001)
17. Kuntsevich, V., Gubarev, V., Kondratenko, Y., Lebedev, D., Lysenko, V.: Control Systems: Theory and Applications. River Publishers (2018). ISBN: 9788770220248
18. Charbonnier, F., Alla, H., David, R.: The supervised control of discrete event dynamic systems: a new approach. Proceedings of the 34th Conference on Decision and Control, New Orleans, LA, pp. 913–920 (1995)
19. Van Brussel, H., Wyns, J., Valckenaers, P., Bongaerts, L., P.: Reference Architecture for Holonic Manufacturing Systems: PROSA. Comput. Ind., Spec. Issue Intell. Manuf. Syst. **37**(3), 255–276 (1998)
20. McFarlane, D.: Holonic manufacturing systems in continuous processing: concepts and control requirements. In: Proceedings of ASI, vol. 95, 273–282 (1995)
21. Novas, J.M., Bahtiar, R, Van Belle, J., Valckenaers, P.: (2012). An approach for the integration of a scheduling system and a multiagent manufacturing execution system. Towards a collaborative framework. Proceedings of the 14th IFAC Symposium INCOM'12, Bucharest, pp. 728–733, IFAC PapersOnLine
22. Borangiu, Th., Răileanu, S., Berger, T., Trentesaux, D.: Switching mode control strategy in manufacturing execution systems. Int. J. Prod. Res. IJPR (Francis & Taylor, Oxfordshire, UK) **53**(7), 1950–1963 (2015). http://dx.doi.org/10.1080/00207543.2014.935825
23. Tsai, W.T. Service-Oriented System Engineering: A New Paradigm, Proceedings of the 2005 IEEE International Workshop on Service-Oriented System Engineering (SOSE'05), IEEE Computer Society, 0-7695-2438-9/05 (2005)
24. De Deugd, S., Carroll, R., Kelly, K.E., Millett, B., Ricker, J.: SODA: Service-Oriented Device Architecture. IEEE Pervasive Comput. **5**(3), 94–C3 (2006)

25. McFarlane, D., Giannikas, V., Wong, C.Y., Harrison, M.: Product intelligence in industrial control: theory and practice. Annu. Rev. Control. **37**, 69–88 (2013)
26. Răileanu, S., Anton, F., Iatan, A., Borangiu, Th., Anton, S., Morariu, O.: Resource scheduling based on energy consumption for sustainable manufacturing. J. Intell. Manuf. (Springer). https://doi.org/10.1007/s10845-015-1142-5 (2015) ISSN: 0956-5515
27. www.constraintsolving.com/solvers. Consulted in October 2018
28. ILOG: Read October 2017 (2017). https://www-01.ibm.com/software/info/ilog/
29. IAEA: Cyclotron produced radionuclides: guidance on facility design and production of [^{18}F] fluorodeoxyglucose (FDG). In: IAEA radioisotopes and radiopharmaceuticals series, no. 3. International Atomic Energy Agency, Vienna (2012). ISSN 2077–6462
30. Medema, J., Luurtsema G., Keizer, H., Tjilkema, S., Elsinga, P.H., Franssen, E.J.F., Paans, A.M.J., Vaalburg, W.: Fully automated and unattended [^{18}F] fluoride and [^{18}F] FDG production using PLC controlled systems. Proceedings of the 31st European Cyclotron Progress Meeting, Zürich (1997)
31. Hess, E., et al.: Excitation function of ^{18}O(p,n) ^{18}F nuclear reaction from threshold up to 30 MeV, Radiochim. Acta (Oldenbourg Wissenschaftsverlag, München) **89**, 357–362 (2001)

Hybrid Control Structure and Reconfiguration Capabilities in Bionic Assembly System

Branko Katalinic, Damir Haskovic, Ilya Kukushkin and Ilija Zec

Abstract This paper presents the research focused on the investigation of working scenarios and efficiency of next generation of modern assembly systems. These systems are known as Bionic Assembly System (BAS). It is based on biologically inspired principles of self-organisation, reduced centralized control, networking between units and natural parallel distribution of processes. BAS control system combines two principles: subordination from factory level to BAS control structure, and self-organization at the shop floor level. This concept is here called hybrid control structure. Informational interface between the subordinating and self-organizing subsystems is called the BAS Cloud. BAS is a human centric system which promotes the integration of workers in the working process. Human tasks on the shop floor are performed by the shop floor operators. Human tasks in the control system are performed by the system operator. He makes the final decisions. Main goal of BAS is to increase system efficiency and robustness. This can be achieved using reconfiguration. Various reconfiguration capabilities are described in this paper, where the investigation is limited to a normal working mode. The results show that production systems with high technical similarity with BAS can increase their efficiency using the proposed concepts. This represents a promising direction of development of future modern assembly systems.

Keywords Bionic assembly system · Efficiency · Hybrid control structure · Self-organization · Working scenarios · Reconfiguration

1 Introduction

The ever-increasing impact of science and technology introduces constant changes to our world [1]. A perfect example of this is globalization and its impact on the pro-

---

B. Katalinic · D. Haskovic (✉) · I. Kukushkin · I. Zec
EU, Vienna University of Technology, Institute for Production Engineering and Laser Technology, Getreidemarkt 9/311, BA 08, 1060 Vienna, Austria
e-mail: haskovic.damir@gmail.com

Y. P. Kondratenko et al. (eds.), *Advanced Control Techniques in Complex Engineering Systems: Theory and Applications*, Studies in Systems, Decision and Control 203, https://doi.org/10.1007/978-3-030-21927-7_12

duction industry. High competitiveness and a dynamic environment represent a challenge for today's companies [2]. Current product development is defined with shorter lifetime as well as increased variety and product complexity. To stay competitive, modern assembly systems need to respond to these challenges through adaptability, efficiency and robustness.

To realise such a system, there are multiple directions of development. One such direction is flexibility. It allows the system to quickly adapt to any changes during assembly. Another direction is self-organization. It allows the system to reduce the role of a centralised control system. It is based on networking and parallel distribution of tasks among the executing units on the shop floor. Another direction is intelligence. It allows the system to improve its performance over time. These development directions are a result of Industry 4.0 which facilitates the convergence of computers, networking, sensors, technology etc. [3].

However, at the same time, these modern assembly systems need to be human centric. That means they need to promote the integration of humans within the assembly process where the main decision maker is the system operator.

Therefore, in the frame of this work, the focus is the development of a next generation of modern, hybrid assembly system—the Bionic Assembly System (BAS). It is a part of a natural development within Industry 4.0 [4]. It is based on biologically inspired principles of self-organisation, reduced centralized control, networking between units and natural parallel distribution of processes. It is developed by the research group from the Institute for Production Engineering and Laser Technology at Vienna University of Technology.

Therefore, BAS needs to be founded on the following characteristics:

- **Modularity**—it allows the system to be expanded, reduced or recombined without the need to perform shutdowns.
- **Reconfigurability**—it allows the system to quickly accommodate to different conditions during the execution of working scenarios.
- **Decentralization**—by decentralizing the system, each component is a self-sufficient unit which interacts with other units and consequently reduces the overall complexity.
- **Flexibility**—the system is capable to process variable product volumes and types.
- **Robustness**—natural parallel distribution of tasks among units makes such a system robust against disturbances. If one element fails, the second one replaces it.
- **Ability to learn**—by storing all the data, the system can learn and or avoid a potential future disturbance.

The paper is organized in 5 sections. Section 1 chapter introduces the context of investigation and core references. Section 2 defines BAS layout and its elements. The Sect. 3 explains the BAS hybrid control structure. The Sect. 4 presents BAS reconfigurations capabilities. The Sect. 5 describes a normal BAS working scenario. The paper concludes with key points, limitations of investigation, and outlines future research direction.

2 BAS Layout and Elements

As shown in Fig. 1, the layout of BAS is divided into 2 subsystems: the core subsystem and the supplementary subsystem [5]. Dominating activities in the core subsystem are assembly, quality control repair and packaging. It contains stations, shop floor operators, pool of robots, pool of pallets. The supplementary subsystem surrounds the core subsystem. Its main function is the storage of parts, materials and components. The main elements of BAS subsystems are shown in Table 1.

The outputs from the supplementary subsystem are components and parts. The output from the core subsystem is a final, assembled product with a satisfactory level of quality. The functions of the core subsystem elements are:

- **Shop floor operators**—workers that perform tasks (which suitable for them) on the shop floor, next to mobile robots and assembly stations.
- **Mobile robots**—autonomous units for transporting assembly pallets from station to station until the final product is assembled with satisfactory quality. They can be turned off, idle, active or in the state of error/repair.
- **Assembly stations**—some stations are designed to complete multiple operations, and some are specifically designed for one type of assembly operation. Assembly stations can be manual (shop floor operator), semi-automatic and automatic.
- **Quality control station**—if the quality is positive, the mobile robot transports the product to further stations or in case of a finished product to the unloading station. In case that the quality is not satisfactory, the robot transports it to the repair station.
- **Repair station**—repair is conducted in case of an error during assembly or if a certain component was defective. If the product is not suitable for repair, it is recycled.
- **Loading/Unloading station**—the assembly procedure starts with the loading station. A first component is placed on an assembly pallet. If the assembled product satisfies all the quality control checks, it is transported to the unloading station as a finished product.
- **Packing station**—assembled product is unloaded, packed and sent for customer delivery.

3 BAS Hybrid Control Structure

As shown in Fig. 2, BAS control system combines two principles: First is subordination from factory level to BAS control structure. It is based on hierarchy, where only one leading element exists. The second is self-organization at the shop floor level. It is based on heterarchy where there is no leading element and the control is decentralised. This entire concept is here called hybrid control structure [6].

Hierarchy is commonly used in highly automated assembly systems, where each step is precisely controlled by a central computer. With heterarchy there is no apparent

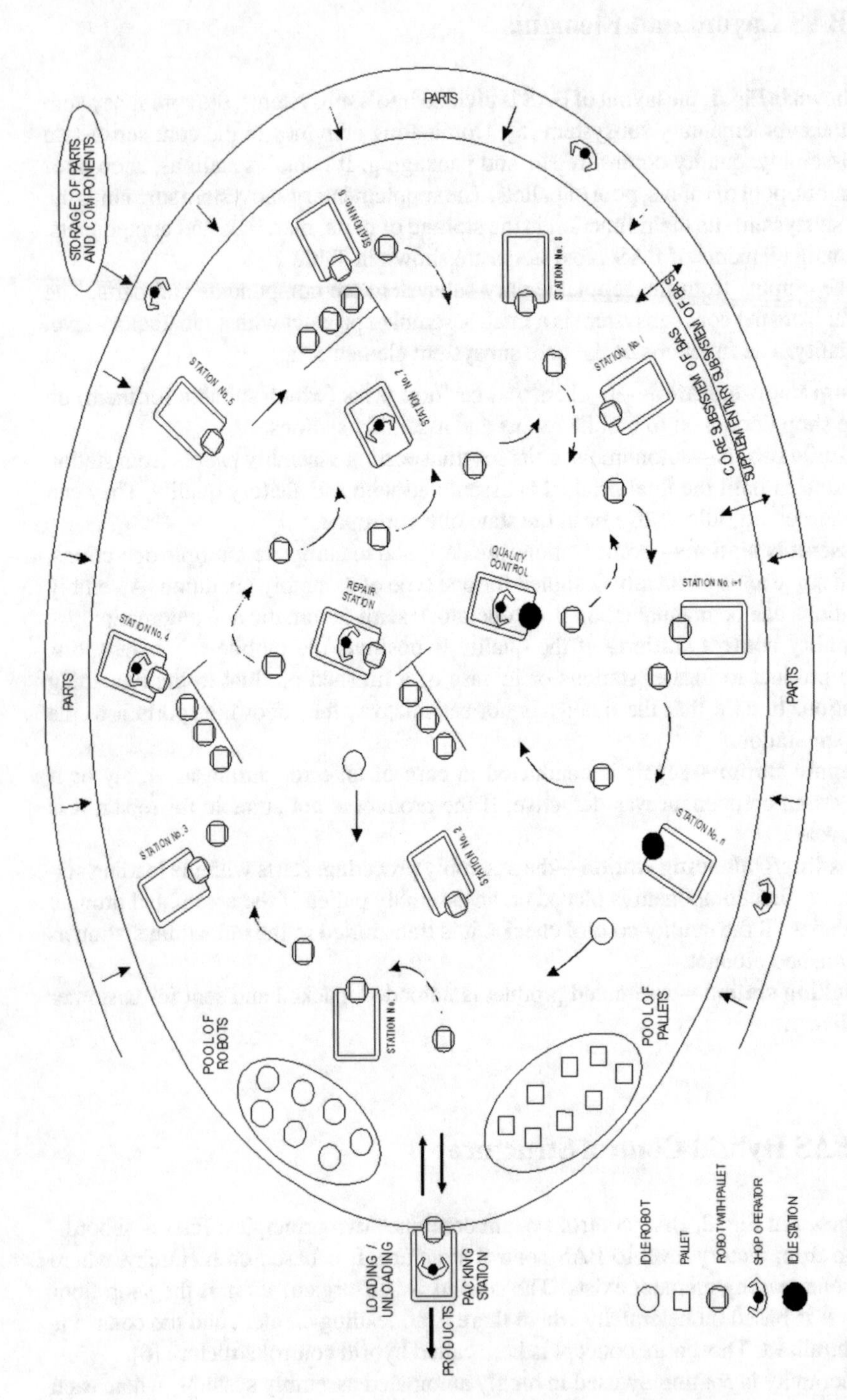

Fig. 1 Bionic assembly system layout

Table 1 Core and supplementary subsystem elements in BAS

Core subsystem	Supplementary subsystem/storage
Shop operator	(Lavatories, breakroom, operational rooms)
Mobile robots	(Service parts, batteries, replacements)
Assembly stations	(Operational fluids, tools, parts, service)
Quality control station	(Replacement tools, measurement devices)
Repair station	(Surplus parts, defective bin, recycling)
Loading/Unloading station	(Product components, assembled products)
Packing station	(Packing pallets, wrapping material)

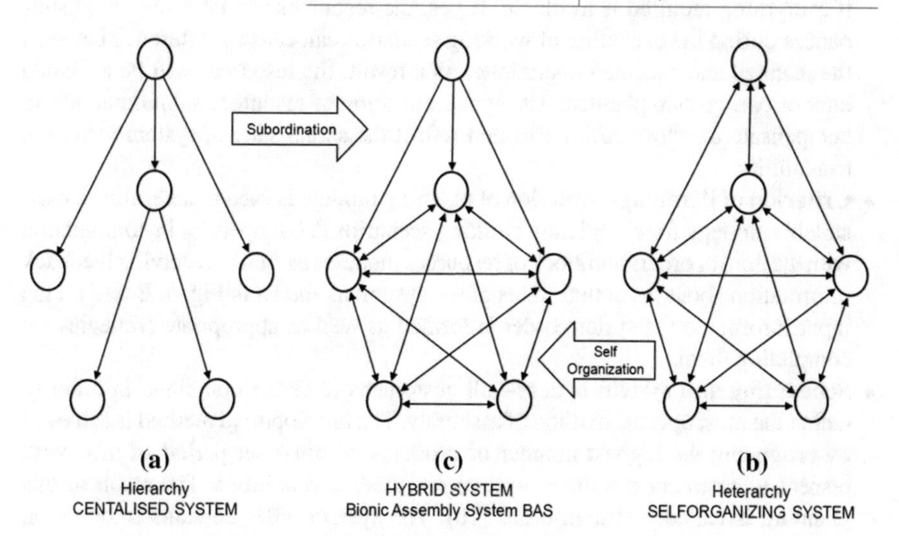

Fig. 2 BAS—hybrid system

source of commands but nevertheless, the common goals are completed (instinct, deployment of basic rules). This approach is harder to implement in assembly systems as it is very difficult to achieve global factory goals.

Hybrid system aims to combine the decentralised control simplicity and robustness of self-organization with the goal-oriented subordination. The main problem of introducing self-organization in the context of assembly systems is the conflict between non-compatible top–down concepts of orders at the factory level and self-organizing nature of the execution level.

The complete overview of the BAS hybrid control system is shown in Fig. 3. Its main elements are:

- **Factory level**—the highest level of planning for the entire assembly system. It is used to determine all BAS activities which should be completed. A long-term production strategy for the entire system is set at this level of planning. It defines what products in which quantity and by when need to be assembled. Planning goals are set according to the conditions of the system, desired results and methods of task completion [7, 8].
- **Pool of orders**—orders are coming from the factory level. One order is defined with the customer name, delivery deadline, type and number of products. All the orders are stored in pool of orders. A priority system is introduced as a method to expel the finished product from the system.
- **Stock of resources**—the primary function of this module is to track the status of all system resources which are necessary for the execution of working scenarios. During the formation of the assembly order, the stock of resources is checked if everything required is available. If yes, the resources can be reserved. Disturbances during the execution of working scenarios can cause a difference between the planned and executed operations. As a result, the resources will be available later or earlier than planned. The synchronisation of resources submodule has to compensate for these differences and reflect the actual state of system resources availability.
- **Criterion of Planning**—criterion of planning module is used to determine a most suitable strategy for completing working scenarios [9]. It operates in combination with the pool of orders and stock of resources modules as well as receiving feedback information about the actual status of the system as shown in Fig. 4. Based on the input information, a system order is formed as well as appropriate strategies for completing them.
- **Scheduling and system orders**—all customer orders are combined in order to define the most optimal method of assembly. The most optimal method is achieved by producing the highest number of products within a set period of time with respect to customer deadlines, system resources and abilities. The result of this planning is called a system order [10]. The system order contains information regarding the products (product type and their volume) and their priorities. As shown in Fig. 5, the procedure starts with the highest priority group, followed with the first product type, followed by the first product piece, then by the first operation. System order is completed when the last piece in the run of the last product type in the lowest priority group is assembled.
- **Actual/Target state**—disturbances (assembly station failures, breaking of tools, robot shut downs…) cause deviations from the planned schedule [11]. There is always a difference between planned and realized working scenarios as shown in Fig. 6. Depending on the size of the difference, there are three main modes: normal working mode, transition mode and disturbance mode. In a normal working mode, small differences are compensated by the automatic control system. However, the automatic control system is limited. Transition mode takes place if the difference increases beyond the automatic control systems ability to compensate it.

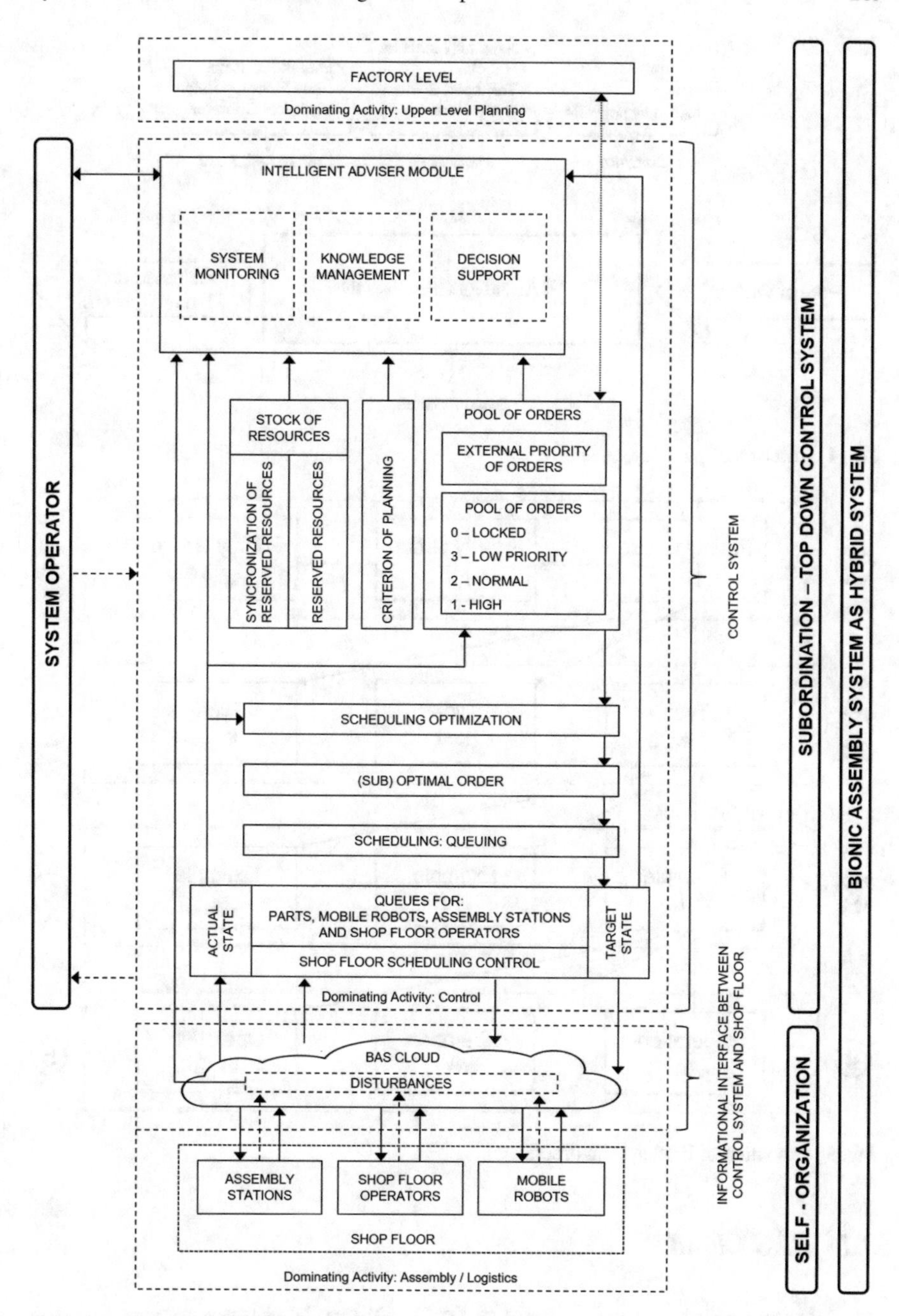

Fig. 3 Bionic assembly system hybrid control structure

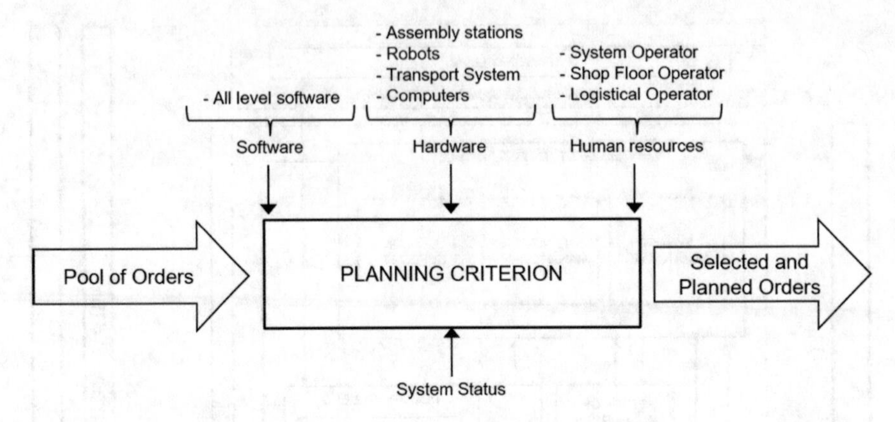

Fig. 4 Criterion of planning

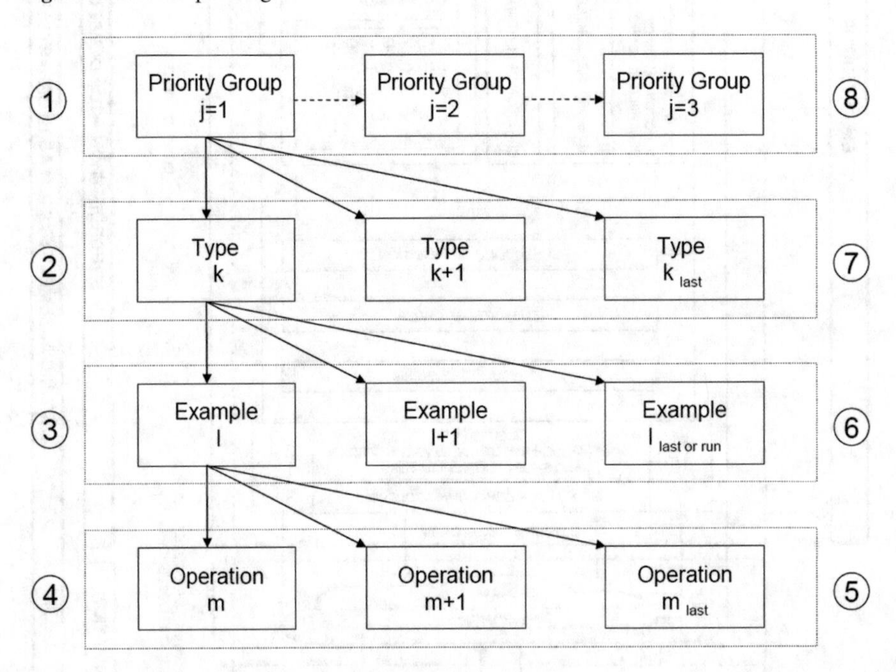

Fig. 5 Execution of BAS system orders

3.1 BAS Cloud

The Bionic Assembly System Cloud (BAS Cloud) [12] is an informational interface between the control system and the shop floor as shown in Fig. 7.

There are two main communication channels in BAS: vertical and horizontal. Vertical communication takes place between the subordinate elements of the control system and is completed with the BAS cloud interface. The information flow from

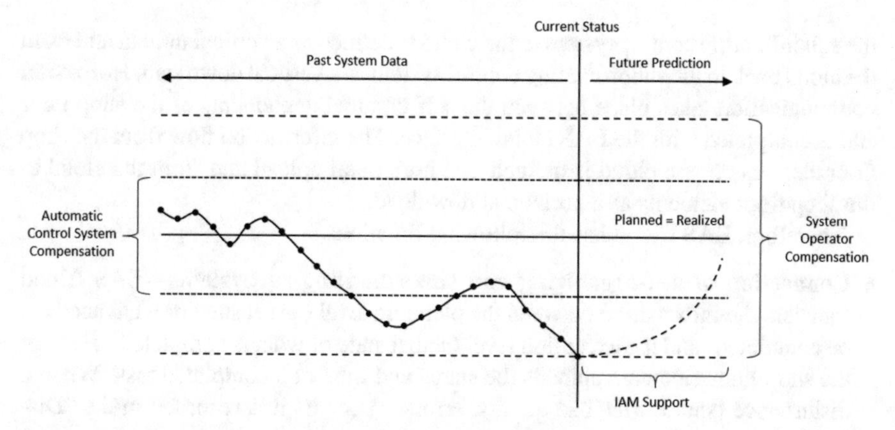

Fig. 6 Difference between planned and realized working scenarios

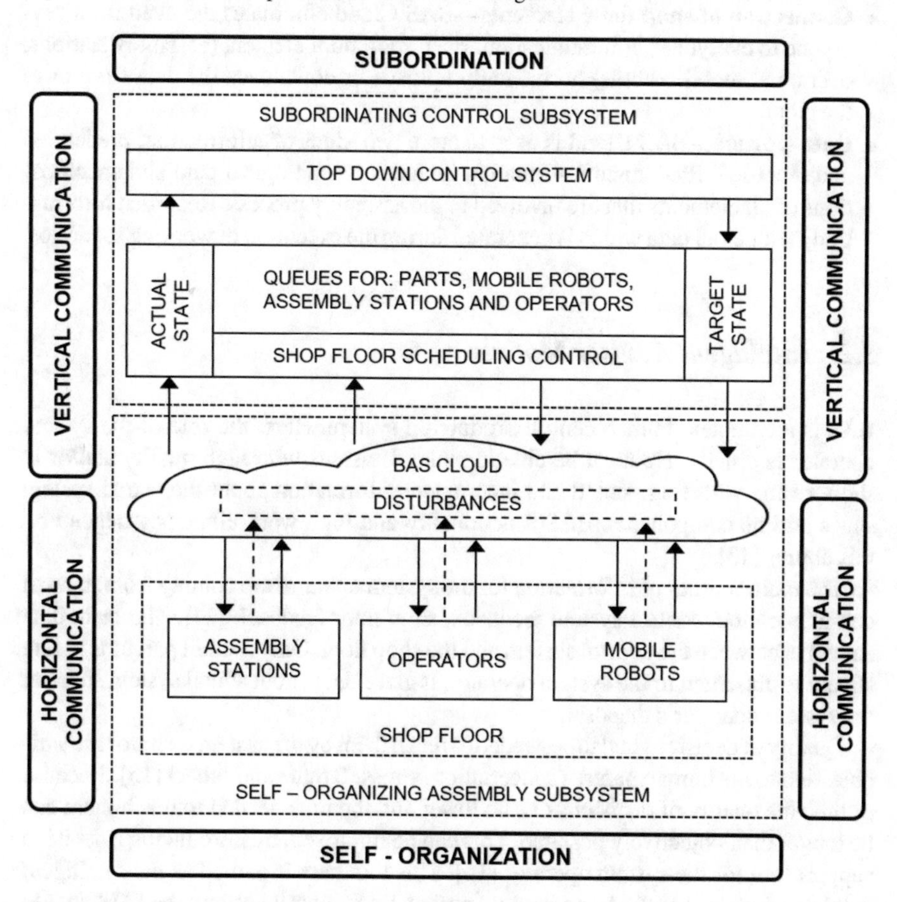

Fig. 7 BAS informational interface

the subordinating control system to the cloud is defined as a vertical upload and from the cloud back to the subordinating control system as a vertical download. Horizontal communication takes place between the self-organizing elements of the shop floor and is completed with the BAS cloud interface. The information flow from the shop floor elements to the cloud is defined as a horizontal upload and from the cloud to the shop floor elements as a horizontal download.

Therefore, BAS Cloud has the following functions:

- **Connection of self-organizing and subordinating subsystems**—BAS Cloud transfers the information between the planning level (target state or what needs to be completed) and the execution level (actual state or what is completed). Each of the shop floor elements uploads the status and time of a completed task. When a disturbance (shutdowns, bad quality, errors…) occurs it is recorded in the "Disturbances" module.
- **Connection of shop floor elements**—BAS Cloud eliminates the need for a "everyone to everyone" communication. Each shop floor element (assembly stations, operators, mobile robots) horizontally uploads or downloads the data from or to the cloud.
- **Data storage**—BAS Cloud is used to store two kinds of information: predefined and recorded. Predefined information includes technological data and specifications of all elements that are involved in the assembly process. Recorded information includes all data which is generated during the execution of working scenarios.

3.2 *Intelligent Adviser Module*

BAS is a complex, human centric production system. Here, the role of the system operator is crucial. He must be able to make decisions with high quality and/or in shorter time with fragmented and incomplete information about the actual system states and its components. BAS functionality and high work efficiency depend on this ability [13].

The main sources of information for the system operator are coming from: human communication, control system feedback, shop floor feedback [14]. The main data stream is between the control system and the shop floor. Only a small part of this data stream is presented to the system operator. It gives data about actual system states at the system operator's disposal.

Quality of decisions and time needed to reach them by the system operator are variable, due to his human nature (concertation, stress, fatigue and other) [15]. Because of this, the quality of decisions can be lower and the time needed to reach them can be longer than objectively possible. This can be improved, by introducing a decision support tool for the system operator [16]. This tool here is named as the Intelligent Adviser Module (IAM). As an integral part of BAS control system, the IAM should take into consideration actual system states, past system states, external data from manuals and other documentation, human experts and past system behaviour [17].

As a result, IAM should be able to learn and to improve the accuracy of its proposals over time [18]. The IAM proposals answer to the question: What to do here and now? The IAM is continuously updating its proposals. They are not mandatory for the system operator.

Therefore, the work of the IAM should be based on:

- Actual system state data from the interface between the control and the controlled system.
- Digitally recorded data from a significant period of past working time.
- Extraction of expert knowledge and expertise from humans directly involved with the system.
- Forecast of the execution of working scenarios for a short time horizon.
- Constant generation of IAM proposals according to the situation. IAM proposals should always be available. The final decision is made exclusively by the system operator. He decides if he will accept, partially accept or ignore the proposals [19].
- Accumulated "situation-decision-results" cases from the past as shown in Fig. 8. A situation represents the problem (type, origin, time, elements) for which the adviser is used. Decision represents the system operator's final solution and reasoning. Result represents if the system operator's decision was positive, neutral or negative. This information can be useful if a similar problem repeats.

4 BAS Reconfigurations

One of the defining BAS characteristics is its ability to reconfigure and adapt to internal and external disturbances. The number of active shop floor elements (assembly stations, mobile robots, operators) constantly changes. It can increase (automatic mobile robot activation, new stations introduced in the system, additional operators…), or decrease (station malfunction, mobile robot low battery, operators missing…). If one unit fails, the execution of working scenarios has to continue. If a new unit becomes available, the workload needs to be distributed.

BAS has the ability to perform:

- Queue rearrangement (mobile robots' queue rearrangement according to available assembly stations)
- Shop floor layout reconfiguration (physical repositioning and/or reorientation of stations on the shop floor).

4.1 Queue Rearrangement

During high volume of assembly, queues of mobile robots will start to form as shown in Fig. 9. The basic principle of waiting in queues is as follows. If we have a station x which is able to perform operations i, j, k on products m, n, l and there is a first

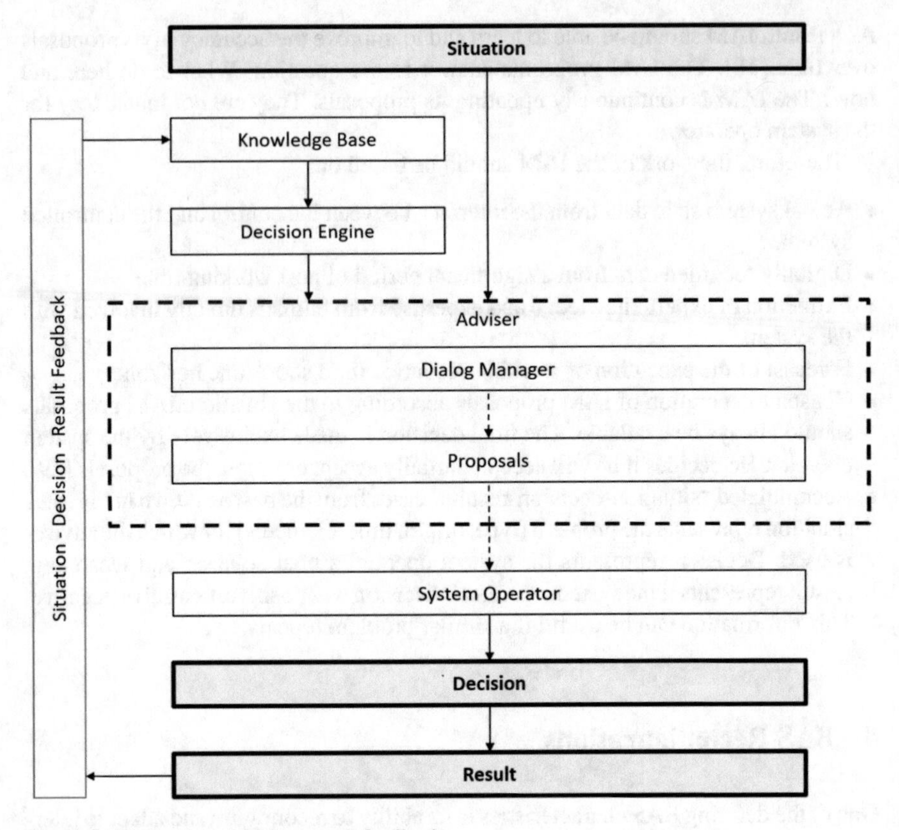

Fig. 8 IAM situation–decision–result feedback

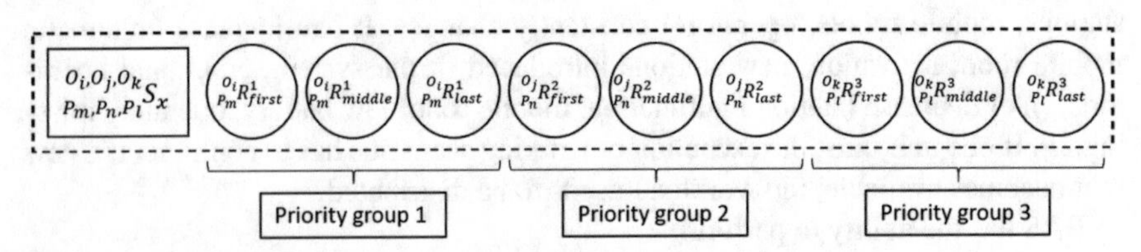

Fig. 9 Queue of mobile robots

group of mobile robots transporting product m waiting for operation i, second group of mobile robots transporting product n waiting for operation j and a third group of mobile robots transporting product l waiting for operation k. These 3 group of mobile robots are grouped according to the priority rule.

Mobile robots carrying the higher priority products have an advantage over mobile robots carrying lower priority products. The main target of BAS is to reach the highest productivity in the given working conditions. This is realized with queue rearrangements which take place when:

- A current station becomes unavailable
- A new station becomes available.

4.1.1 Queue Rearrangement After a New Station Becomes Available

Queue rearrangement after a new station becomes available is shown in Fig. 10. In state A, in front of assembly station S_1, which is capable to perform operation i, j, k on products m, n, l there is a queue of mobile robots. There are mobile robots with the highest priority product m, followed by mobile robots with the priority level 2 product n and lastly mobile robots with the lowest priority product l. In state B, a new station S_2 which is capable to perform the same operations for the same products becomes available. Each priority group is divided into two halves. In state C the second half of each priority group is rearranged in front of the new station S_2 whilst keeping the ordering according to the priorities. This demonstrates BAS robustness against disturbances. It allows to continue with the assembly through flexibility and task redistribution.

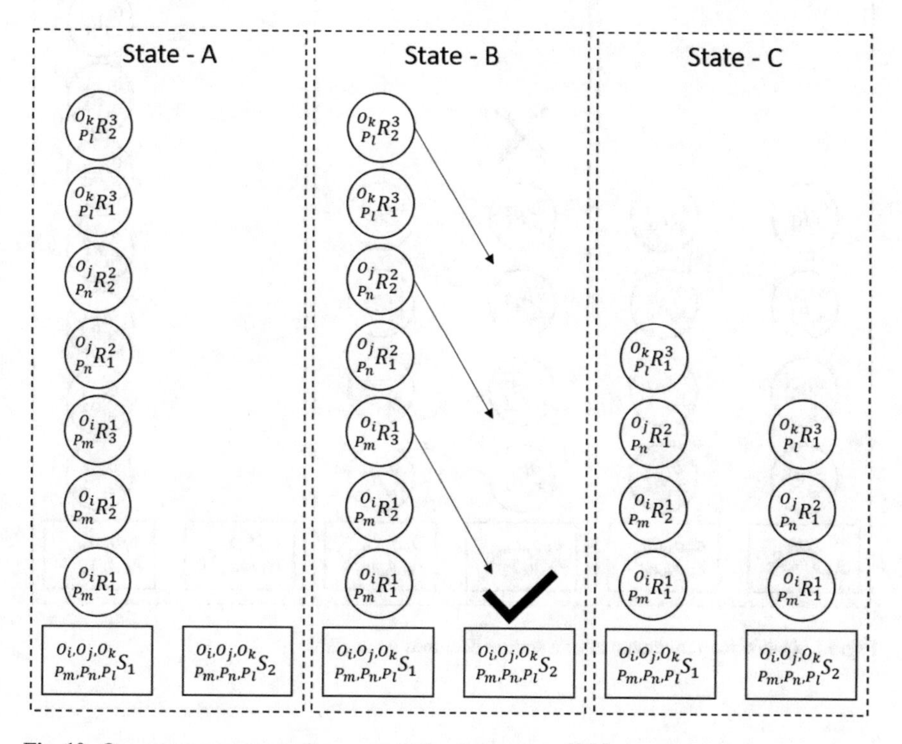

Fig. 10 Queue rearrangement after a new station becomes available

4.1.2 Queue Rearrangement After a Station Becomes Unavailable

Queue rearrangement after a station fails is shown in Fig. 11. In state A, there are stations S_1 and S_2 and both are able to perform same operations on the same products.

In state B, station S_1 fails, and the mobile robots from that failed station need to rearrange themselves in front of the still available station S_2, by going behind the appropriate priority group of mobile robots which are waiting in front of the available station S_2. In state C, all the mobile robots are in a rearranged queue in front of station S_2, whilst keeping the ordering according to the priorities. The described queue rearrangements demonstrated BAS ability to dynamically adapt to different working scenarios. This includes different workloads and variable hardware functionality.

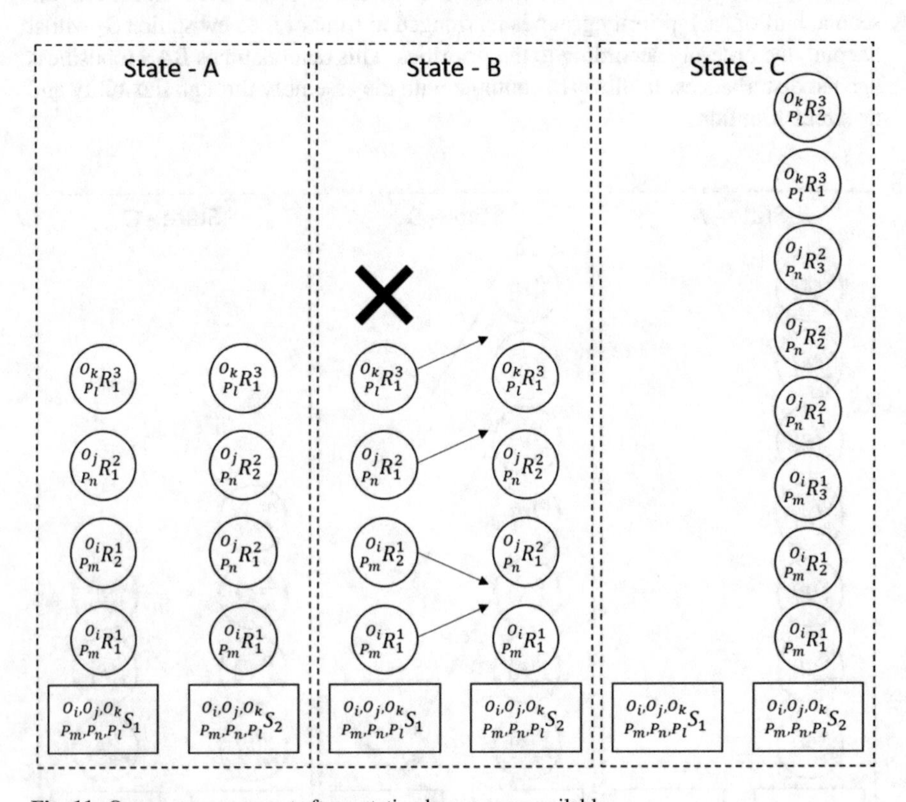

Fig. 11 Queue rearrangement after a station becomes unavailable

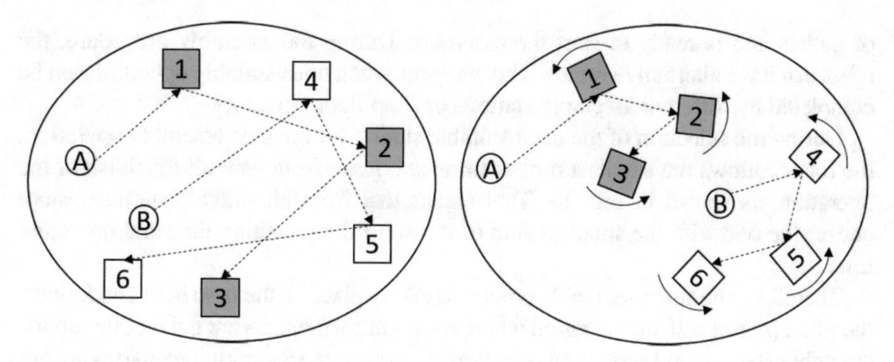

Fig. 12 Layout reconfiguration

4.2 *Shop Floor Layout Reconfiguration*

In certain situations, during the normal execution of working scenarios and when there is a small number of active mobile robots [20], the system could complete layout reconfigurations as shown in Fig. 12. Here, the shop floor can contain a certain number of movable assembly stations which are able to change their position and rotation.

Operators and mobile robots are movable as they can change their position according to the current task. Various stations can be movable, semi-movable and non-movable. Movable stations have the ability to change their position and rotation on the shop floor. Semi-movable stations cannot change their position but can rotate. Non-movable stations are fixed in their place and cannot rotate due to their dimensions, weight and operating demands.

Suppose we have a mobile robot transporting product A between stations 1, 2, 3 and a mobile robot transporting product B between stations 4, 5, 6. After the layout reconfiguration, it is possible to achieve the following:

- Movable assembly stations/change of position and/or orientation
- Optimization of mobile robot trajectories/distance reduction between stations/grouping.

5 BAS Normal Working Scenario

BAS normal working scenarios are realized with the uninterrupted execution of all activities which are needed to assemble a continuous stream of products. Continuous stream of products is made with assembly orders which are formed by the subordinating control subsystem. These orders are vertically uploaded to BAS cloud. All standby mobile robots check if there are any available orders on the cloud. When the robot finds and horizontally downloads an order, it takes the pallet from pool of pallets and is ready to start the assembly. During the assembly procedure, the robot can have alternative routes. This happens when one assembly operation can be completed by different assembly stations or shop floor operators.

During the selection of the most suitable station for the next assembly operation, the robot follows the smallest time resistance criteria from now till the finish of the operation, as shown in Fig. 13. That means that from all suitable stations, robot choses the one with the smallest sum of transport time, waiting time and operation time.

Therefore, the entire assembly process is taking place on the shop floor and follows the principles of self-organization where the main participants are the mobile robots, assembly stations and shop floor operators. However, BAS working scenarios are not always realized with the uninterrupted execution of all activities which are needed to assemble a continuous stream of products [21]. As a result, specific BAS scenarios can occur. These will be described with further research.

6 Conclusion

The research presented in this paper focused on the investigation of working scenarios and efficiency of next generation of modern assembly systems. These systems are known as BAS.

The main focus of the investigation was the hybrid control structure, reconfiguration capabilities, main system layout and its elements. The research was limited to normal working scenarios with idealized working conditions. Here, all activities which are needed to assemble a continuous stream of products were uninterrupted.

The main results of the investigation are that by introducing biologically inspired principles of self-organisation, reduced centralized control, networking between units and natural parallel distribution of processes, it is possible to increase system efficiency and robustness.

As a human centric system, BAS has the possibility to be highly automated on one side and the ability to integrate workers on the other side. Reconfiguration of queues and layout during assembly introduces additional possibilities for optimization of working scenario and system parameters.

This represents a promising direction of development of future modern assembly systems. One possible direction of further development is the analysis of specific BAS scenarios as well as a verification of BAS concept in an industrial application with high technical similarity with BAS.

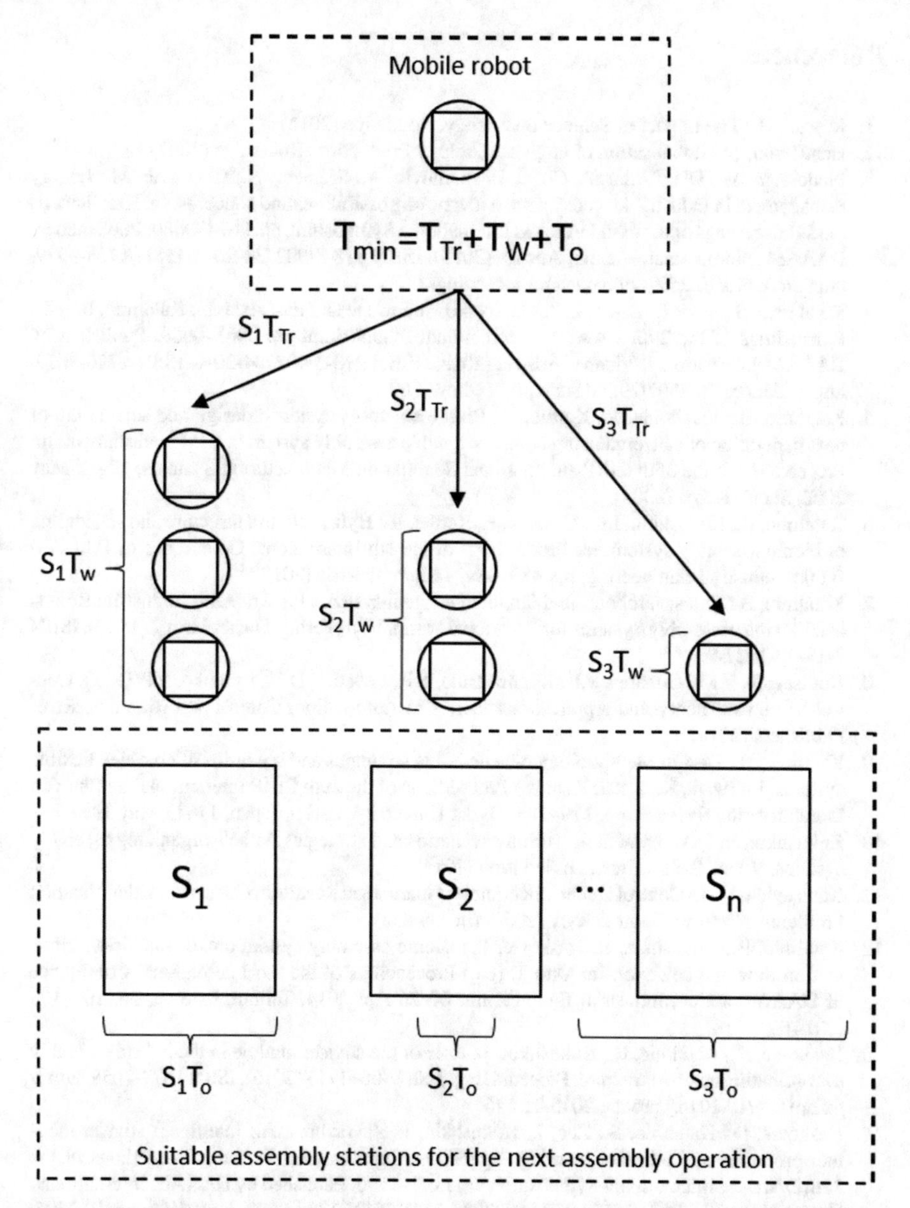

Fig. 13 Smallest time resistance criteria: T_{min} is smallest time resistance criteria; T_{Tr} is transport time; T_w is waiting time; T_O is operating time

References

1. Russell, B.: The Impact of Science on Society. Routledge (2016)
2. Henderson, J.: Globalisation of High Technology Production. Routledge (2002)
3. Medojevic, M., Diaz Villar, P., Cosic, I., Rikalovic, A., Sremcev, N., Lazarevic, M.: Energy management in Industry 4.0 ecosystem: a review on possibilities and concerns. In: Katalinic, B. (ed.) Proceedings of the 29th DAAAM International Symposium, pp. 0674–0680. Published by DAAAM International, Vienna, Austria (2018). ISBN 978-3-902734-20-4. ISSN 1726-9679. https://doi.org/10.2507/29th.daaam.proceedings.097
4. Takakuwa, S., Veza, I., Celar, S.: "Industry 4.0" in Europe and East Asia. In: Katalinic, B. (ed.) Proceedings of the 29th DAAAM International Symposium, pp. 0061–0069. Published by DAAAM International, Vienna, Austria (2018). ISBN 978-3-902734-20-4. ISSN 1726-9679. https://doi.org/10.2507/29th.daaam.proceedings.009
5. Katalinic, B., Visekruna, V., Kordic, V.: Bionic assembly systems: design and scheduling of next generation of self-organising complex flexible assembly system in CIM environment. In: Proceedings of the 35th CIRP-International Seminar on Manufacturing Systems, 12–15 May 2002, Seoul, Korea (2002)
6. Katalinic, B., Kukushkin, I.K., Cesarec, P., Kettler, R.: Hybrid control structure and scheduling of bionic assembly system. In: Proceedings of the 8th International Conference of DAAAM Baltic, Industrial Engineering, pp. 483–489, Tallinn, Estonia (2012)
7. Katalinic, B.: Industrieroboter und flexible Fertigungsysteme für Drehteile (Industrial Robots and Flexible Assembly Systems for Rotational Parts). VDI, Verlag, Duesseldorf 2 (1990). ISBN 3-18-40-1027-9
8. Kuntsevich, V.M., Gubarev, V.F., Kondratenko, Y.P., Lebedev, D.V., Lysenko, V.P. (eds.): Control Systems: Theory and Applications. Series in Automation, Control and Robotics, River Publishers (2018)
9. Katalinic, B.: Design Methodology of scheduling strategies and scenarios of complex flexible systems. In: Iwata, K., Ueda, K. (eds.) Proceedings of the 29th CIRP International Seminar on Manufacturing Systems, pp. 179–185, Osaka University, Osaka, Japan, 11–13 May 1997
10. Kukushkin, I.: Development of working scenarios and strategies for self-organizing assembly systems. Wien, Techn. Univ., Dissertation (2014)
11. Kuntsevich, V.M.: Control Under Uncertainty: Guaranteed Results in Control and Identification Problems. Naukova Dumka, Kyiv (2006) (in Russian)
12. Katalinic, B., Kukushkin, I., Haskovic, D.: Bionic assembly system cloud: functions, information flow and behavior. In: Otto T. (ed.) Proceedings of the 9th International Conference of DAAAM Baltic, Industrial Engineering, 24–26 Apr 2014, Tallinn, Estonia, pp. 103–108 (2014)
13. Haskovic, D., Katalinic, B., Kukushkin, I.: Role of the adviser module in the hybrid assembly subordinating control structure. Procedia Eng. **100**, 1706–1713 (2015). ISSN 1877-7058. http://dx.doi.org/10.1016/j.proeng.2015.01.546
14. Haskovic, D., Katalinic, B., Zec, I., Kukushkin, I., Zavrazhina, A.: Intelligent adviser module: proposals and adaptive learning capabilities. In: Katalinic, B. (ed.) Proceedings of the 28th DAAAM International Symposium, pp. 1191–1196. Published by DAAAM International, Vienna, Austria (2017). ISBN 978-3-902734-11-2. ISSN 1726-9679. https://doi.org/10.2507/28th.daaam.proceedings.165
15. Getty, D.J., et al.: System operator response to warnings of danger: a laboratory investigation of the effects of the predictive value of a warning on human response time. J. Exp. Psychol. Appl. **1**(1), 19 (1995)
16. Haskovic, D., Katalinic, B., Kildibekov, A., Kukushkin, I.: Intelligent adviser module for bionic assembly control system: functions and structure concept. In: Katalinic, B. (ed.) Proceedings of the 26th DAAAM International Symposium, pp. 1158–1165. Published by DAAAM International, Vienna, Austria (2016). ISBN 978-3-902734-07-5. ISSN 1726-9679

17. Haskovic, D., Katalinic, B., Zec, I., Kukushkin, I., Zavrazhina, A.: Structure and working modes of the intelligent adviser module. In: Katalinic, B. (ed.) Proceedings of the 27th DAAAM International Symposium, pp. 0866–0875. Published by DAAAM International, Vienna, Austria (2016). ISBN 978-3-902734-08-2. ISSN 1726-9679. https://doi.org/10.2507/27th.daaam.proceedings.125
18. Pryanichnikov, V., Aryskin, A., Eprikov, S., Kirsanov K., Khelemendik, R., Ksenzenko, A., Prysev, E., Travushkin, A.: Technology of multi-agent control for industrial automation with logical processing of contradictions. In: Katalinic, B. (ed.) Proceedings of the 28th DAAAM International Symposium, pp. 1202–1207. Published by DAAAM International, Vienna, Austria (2017). ISBN 978-3-902734-11-2. ISSN 1726-9679. https://doi.org/10.2507/28th.daaam.proceedings.167
19. Pryanichnikov, V., Helemendik, R.V.: Information technology IGEC and pentalogy. In: Computer Science and Information Technology: Materials of International Science Conference—Saratov: The Publication. Center "Science", pp. 331–333 (2016)
20. Pryanichnikov, V.E., Chernyshev, V., Arykantsev, V., Aryskin, A.A., Eprikov, S., Ksenzenko, A., Petrakov, M.S.: Enhancing the functionality of the groups of autonomous underwater robots. In: Katalinic, B. (ed.) Proceedings of the 29th DAAAM International Symposium, pp. 1319–1325. Published by DAAAM International, Vienna, Austria (2018). ISBN 978-3-902734-20-4. ISSN 1726-9679. https://doi.org/10.2507/29th.daaam.proceedings.190
21. Haskovic, D.: Working Scenarios of Hybrid Self-Organizing Assembly System. Wien, Techn. Univ., Dissertation (2018)

A Flatness-Based Approach to the Control of Distributed Parameter Systems Applied to Load Transportation with Heavy Ropes

Abdurrahman Irscheid, Matthias Konz and Joachim Rudolph

Abstract The cooperative transport of a rigid body carried by multiple heavy ropes that are suspended by tricopters is used as an example for flatness-based trajectory planning for distributed-parameter systems. At first, the load-side ends of the ropes are parametrized by a flat output of the nonlinear boundary system that describes the motion of the rigid body. This parametrization is then used to express the solution of the linearized partial differential equations for the heavy ropes by the means of operational calculus. Evaluating the derived parametrization at the controllable ends yields the desired trajectories for the tricopter positions. A position controller is used to track the reference trajectories of the tricopters. This control can be used to realize a desired transition of the load position and orientation in finite time. Experimental results of the introduced method are presented for validation.

1 Introduction

Rope-suspended transport by multiple members allows for greater payload capacities and the configuration of the load orientation. While these ropes are often modeled as massless rigid links as in [8], the rope mass distribution plays a significant role in describing the overall system motion in certain applications. As a result, so-called heavy ropes are considered in the following with motions described by partial differential equations.

Flatness-based methods are suitable for controlling such distributed-parameter systems [10]. In fact, the flatness of a single heavy rope has already been established

A. Irscheid (✉) · M. Konz · J. Rudolph
Chair of Systems Theory and Control Engineering,
Saarland University, Saarbrücken, Germany
e-mail: a.irscheid@lsr.uni-saarland.de

M. Konz
e-mail: m.konz@lsr.uni-saarland.de

J. Rudolph
e-mail: j.rudolph@lsr.uni-saarland.de

Y. P. Kondratenko et al. (eds.), *Advanced Control Techniques in Complex Engineering Systems: Theory and Applications*, Studies in Systems, Decision and Control 203, https://doi.org/10.1007/978-3-030-21927-7_13

Fig. 1 A triangle suspended by three heavy ropes that are each attached to a tricopter for actuation during an example maneuver

in [5, 9]. An extension of this result is sketched below for a rigid body suspended by multiple heavy ropes, which leads to a straightforward design of an open-loop controller.

The experiments carried out for verification purposes showed great results. Each rope was attached to a tricopter in order to realize the required motion of the rope suspension point. The configuration of the overall system during an example maneuver is depicted in Fig. 1.

A step-by-step approach is presented in this chapter. Focusing on a single heavy rope at first makes it much easier to understand the cooperative transport with multiple ropes. Afterwards, a brief discussion of the used tricopters and their position control algorithms is presented. Finally, the chapter ends with a presentation of experimental results that verify this approach to flatness-based cooperative transport.

2 Modeling Heavy Ropes

Considering the mass distribution along the rope leads to partial differential equations governing its motion. In the following, a brief derivation of the heavy rope model is presented. A more detailed description[1] can be found in [1].

[1] An alternative model using redundant coordinates is derived in [2].

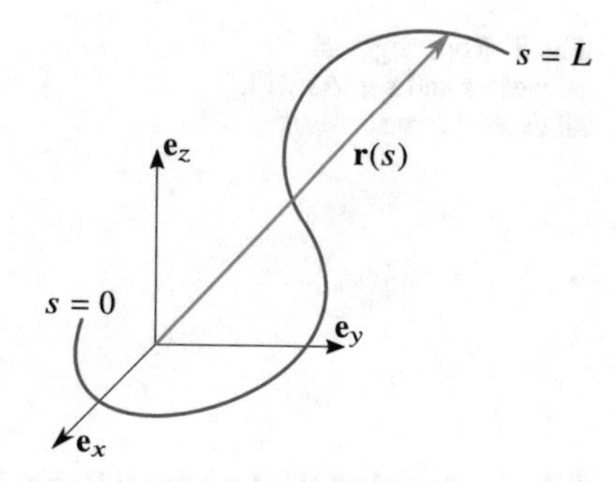

Fig. 2 The vector $\mathbf{r}(s)$ describes a point on the rope corresponding to the arc length s

Instead of directly using equations of (planar) motion of an inelastic[2] homogeneous[3] heavy rope for small angle approximations about a vertical equilibrium configuration as in [9], it is useful to start with the derivation of a nonlinear model in three dimensional space in order to gain more insight.

For fixed time $t \geq 0$ the rope can be described as the image of a curve in three dimensional space. Let the curve parameter s describe the arc length of the curve with a certain orientation that has yet to be defined. This means the curve maps $s \in [0, L]$ onto $\mathbf{r}(s) \in \mathbb{R}^3$ (see Fig. 2), where L is the total length of the rope. Furthermore, let ρ be the mass density of the rope per unit length. Since the dynamic behavior of a rope is of interest, the time dependency of the rope profile

$$\mathbf{r}(s,t) = \begin{pmatrix} x(s,t) \\ y(s,t) \\ z(s,t) \end{pmatrix} \tag{1}$$

is incorporated to describe the curve at each time t. Since s is the arc length of the curve, the derivative of $\mathbf{r}$ with respect to s, i.e. $\partial_s \mathbf{r}$, describes the tangential vector of the curve at the point s for fixed time t with $||\partial_s \mathbf{r}(s,t)|| = 1$.

2.1 The Equations of Motion

Consider a rope segment of length Δs as in Fig. 3 with contact forces $\mathbf{T}(s,t)$ and $\mathbf{T}(s + \Delta s, t)$ exerted by the adjacent rope segments. The additional force $\rho\mathbf{a}$ per unit length accounts for external forces.[4] The force $\mathbf{T}$ is chosen positive in the direction of increasing arc length s, which explains the negative sign in Fig. 3.

[2] An inelastic rope is of constant length.

[3] A rope made of a homogeneous material with a constant cross section is called homogeneous.

[4] Here in particular, accounting for the gravitational force.

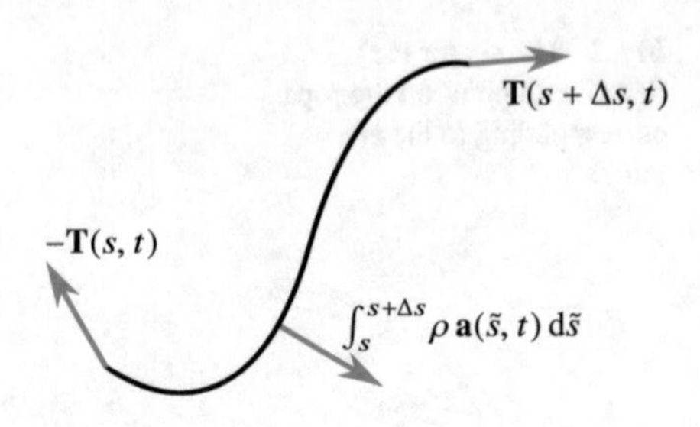

Fig. 3 Rope segment between s and $s + \Delta s$ with all exerted forces

2.1.1 Deriving the Dynamic Behavior

The equations of motion follow by setting the time derivative of the resulting linear momentum of an infinitesimal rope segment equal to the resultant force exerted on that segment:

$$\int_s^{s+\Delta s} \rho \partial_{tt} \mathbf{r}(\tilde{s}, t)\, \mathrm{d}\tilde{s} = \mathbf{T}(s + \Delta s, t) - \mathbf{T}(s, t) + \int_s^{s+\Delta s} \rho \mathbf{a}(\tilde{s}, t)\, \mathrm{d}\tilde{s}\ . \tag{2}$$

Using the mean value theorem for definite integrals yields

$$\rho \partial_{tt} \mathbf{r}(s + \theta_1 \Delta s, t) \Delta s = \mathbf{T}(s + \Delta s, t) - \mathbf{T}(s, t) + \rho \mathbf{a}(s + \theta_2 \Delta s, t) \Delta s \tag{3}$$

with $\theta_i \in [0, 1],\ i \in \{1, 2\}$. Dividing by Δs then letting Δs tend towards zero leads to

$$\rho \partial_{tt} \mathbf{r}(s, t) = \partial_s \mathbf{T}(s, t) + \rho \mathbf{a}(s, t)\ . \tag{4}$$

Modeling a rope without any bending stiffness requires the additional assumption that no torque transmission takes place. Note that this assumption was already used in Fig. 3 since no torques were drawn at both ends.

Using the fact that the time derivative of the resulting angular momentum is equal to the resultant torque[5] on the rope segment leads to

$$\begin{aligned} \frac{\partial}{\partial t} \int_s^{s+\Delta s} \mathbf{r}(\tilde{s}, t) \times \rho \partial_t \mathbf{r}(\tilde{s}, t)\, \mathrm{d}\tilde{s} =& \mathbf{r}(s + \Delta s, t) \times \mathbf{T}(s + \Delta s, t) - \mathbf{r}(s, t) \times \mathbf{T}(s, t) \\ &+ \int_s^{s+\Delta s} \mathbf{r}(\tilde{s}, t) \times \rho \mathbf{a}(\tilde{s}, t)\, \mathrm{d}\tilde{s}\ . \end{aligned} \tag{5}$$

With (4) the following relation immediately follows:

$$\int_s^{s+\Delta s} \partial_s \mathbf{r}(\tilde{s}, t) \times \mathbf{T}(\tilde{s}, t)\, \mathrm{d}\tilde{s} = \mathbf{0}\ . \tag{6}$$

[5]The origin was used as a reference point for both the angular momentum and the torque.

Using the mean value theorem again together with $\Delta s \to 0$ results in

$$\partial_s \mathbf{r}(s,t) \times \mathbf{T}(s,t) = \mathbf{0} , \tag{7}$$

which means that the rope force $\mathbf{T}$ must always be tangential to the rope, i. e.

$$\mathbf{T}(s,t) = T(s,t)\partial_s \mathbf{r}(s,t) . \tag{8}$$

Finally, the motion of a heavy rope is governed by

$$\rho \mathbf{r}_{tt}(s,t) = \partial_s \left(T(s,t)\mathbf{r}_s(s,t) \right) + \rho \mathbf{a}(s,t) , \tag{9}$$

where the subscripts denote derivatives w. r. t. time t and arc length s. If the only external force acting on the rope is gravitation, then $\mathbf{a}$ is equal to the gravitational acceleration $\mathbf{g}$.

2.1.2 Initial Conditions and Boundary Values

The equations of motion (9) are of order two with respect to arc length s and time t. Therefore, two initial conditions and two boundary values are required. An initial profile and its time derivative are obviously suitable initial conditions:

$$\begin{aligned} \mathbf{r}(s,0) &= \mathbf{r}_0(s) \\ \mathbf{r}_t(s,0) &= \mathbf{r}_{t,0}(s) \end{aligned} \tag{10}$$

For instance, one could assign forces applied at both ends of the rope or velocities of the end points as boundary values. For the time being, it is assumed that at $s = L$ the position of the suspension point

$$\mathbf{r}(L,t) = \mathbf{u}(t) \tag{11}$$

is given by an input $\mathbf{u}$.[6] The other end at $s = 0$ is connected to a load, which results in an ordinary differential equation[7]

$$m \left(\mathbf{r}_{tt}(0,t) - \mathbf{g} \right) = T(0,t)\mathbf{r}_s(0,t) . \tag{12}$$

---

[6]This boundary condition does not match the given problem of cooperative transport using tricopters. However, this will be considered in Sect. 5.

[7]The load is modeled here as a point mass m for simplicity. The cooperative transport of a rigid body with multiple heavy ropes in Sect. 4 will result in a system of ordinary differential equations at the load-side boundary.

2.2 Equilibrium Configurations

Before planning transitions between set points, it is important to examine the underlying model to find out which equilibrium configurations are possible. Therefore, let

$$\tilde{\mathbf{r}}(s) = \begin{pmatrix} \tilde{x}(s) \\ \tilde{y}(s) \\ \tilde{z}(s) \end{pmatrix}, \qquad s \in [0, L] \tag{13}$$

describe the profile of a heavy rope in an equilibrium configuration with the constant force

$$\tilde{\mathbf{T}}(0) = \begin{pmatrix} F_x \\ 0 \\ F_z \end{pmatrix}, \qquad F_x \in \mathbb{R}, F_z \geq 0 \tag{14}$$

at its lower end. This force is motivated by the fact that the heavy rope is connected to a load.[8] For the sake of simplicity, the coordinate system has been chosen in such a way that $\tilde{\mathbf{T}}(0)$ is entirely contained in the x-z-plane and that the gravitational acceleration $\mathbf{g}$ points in the negative z-direction, i. e. $\mathbf{g} = -g\mathbf{e}_z$. Assume further that the load-side end is located at the origin: $\tilde{\mathbf{r}}(0) = \mathbf{0}$. At this point we can already predict that $\tilde{y}(s) = 0$.

Evaluating (9) at this equilibrium configuration leads to an ordinary differential equation

$$\frac{\mathrm{d}\tilde{\mathbf{T}}}{\mathrm{d}s}(s) = \rho g \mathbf{e}_z \tag{15}$$

for the rope force. Its solution immediately follows by integration:

$$\tilde{\mathbf{T}}(s) = \begin{pmatrix} F_x \\ 0 \\ F_z + \rho g s \end{pmatrix} \tag{16}$$

Then the differential equations

$$\begin{aligned} \tilde{x}_s(s) &= \frac{F_x}{\sqrt{F_x^2 + (F_z + \rho g s)^2}} \\ \tilde{z}_s(s) &= \frac{F_z + \rho g s}{\sqrt{F_x^2 + (F_z + \rho g s)^2}} \end{aligned} \tag{17}$$

for the rope profile can be derived using the relations $\tilde{\mathbf{T}}(s) = \tilde{T}(s)\partial_s\tilde{\mathbf{r}}(s)$ and $||\partial_s\tilde{\mathbf{r}}(s)|| = 1$. The integration of these differential equations by separation of variables is quite straight forward. The trivial case $F_x = 0$ results in $\tilde{x}(s) = 0$ and

---

[8] Obviously $F_x \neq 0$ only makes sense for the case of at least two ropes carrying a load.

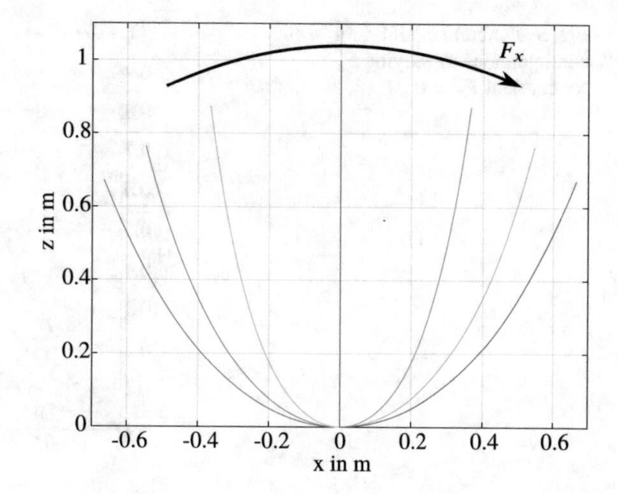

Fig. 4 Different equilibrium configurations by varying F_x

$\tilde{z}(s) = s$, which will be called *vertical equilibrium configuration* throughout the chapter. The other case leads to

$$
\begin{aligned}
\tilde{x}(s) &= \int_0^s \frac{F_x}{\sqrt{F_x^2 + (F_z + \rho g \tilde{s})^2}} \, d\tilde{s} = \left[\frac{|F_x|}{\rho g} \operatorname{arsinh} \frac{F_z + \rho g \tilde{s}}{F_x} \right]_0^s \\
&= \frac{|F_x|}{\rho g} \left(\operatorname{arsinh} \left(\frac{F_z + \rho g s}{F_x} \right) - \operatorname{arsinh} \left(\frac{F_z}{F_x} \right) \right)
\end{aligned}
\tag{18}
$$

and

$$
\begin{aligned}
\tilde{z}(s) &= \int_0^s \frac{F_z + \rho g \tilde{s}}{\sqrt{F_x^2 + (F_z + \rho g \tilde{s})^2}} \, d\tilde{s} = \left[\frac{1}{\rho g} \sqrt{F_x^2 + (F_z + \rho g \tilde{s})^2} \right]_0^s \\
&= \frac{1}{\rho g} \left(\sqrt{F_x^2 + (F_z + \rho g s)^2} - \sqrt{F_x^2 + F_z^2} \right) .
\end{aligned}
\tag{19}
$$

Different equilibrium configurations are depicted in Figs. 4 and 5.

2.3 Linearized Model

The control method presented in Sect. 3 is based on a small angle approximation about a vertical equilibrium configuration of the heavy rope model. The proposed open-loop controller only requires the linearized partial differential equations.

The approximation of the equations of motion

$$
\begin{aligned}
\rho \left(\mathbf{r}_{tt}(s,t) - \mathbf{g} \right) &= \frac{\partial}{\partial s} \left(T(s,t) \mathbf{r}_s(s,t) \right) \\
||\mathbf{r}_s(s,t)|| = 1 &\Longleftrightarrow \mathbf{r}_s^T(s,t)\, \mathbf{r}_s(s,t) = 1
\end{aligned}
\tag{20}
$$

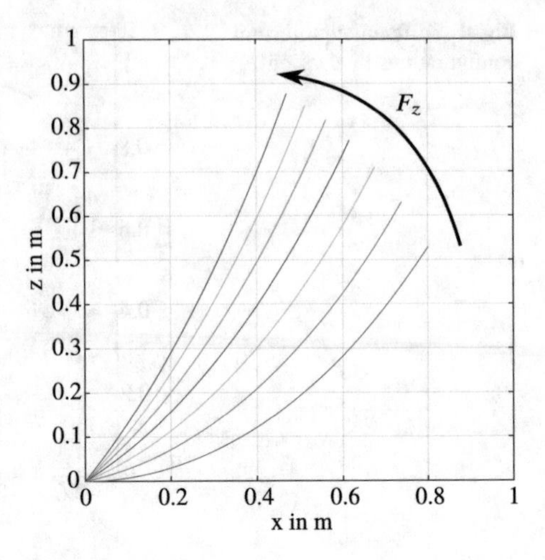

Fig. 5 Different equilibrium configurations by varying F_z for constant $F_x > 0$

about the trivial equilibrium configuration

$$\begin{aligned} \tilde{\mathbf{r}}(s) &= s\mathbf{e}_z \\ \tilde{T}(s) &= F_z + \rho g s \end{aligned} \tag{21}$$

leads to the linear equations

$$\begin{aligned} \rho \Delta \mathbf{r}_{tt}(s,t) &= \Delta T_s(s,t)\mathbf{e}_z + \frac{\partial}{\partial s}\left(\tilde{T}(s)\Delta \mathbf{r}_s(s,t)\right) \\ \Delta z_s(s,t) &= 0\,, \end{aligned} \tag{22}$$

where $\Delta w = w - \tilde{w}$ denotes the difference between a quantity w and its value $\tilde{w}$ in the vertical equilibrium configuration. The latter means that $\Delta z(s,t)$ is only a function of time. Note that $\Delta z(s,t)$ at the boundary $s = L$ is equal to $\Delta u_z(t) = \mathbf{e}_z^T \mathbf{u}(t) - L$. Combining these equations results in

$$\begin{aligned} \rho \Delta x_{tt}(s,t) &= \frac{\partial}{\partial s}\left(\tilde{T}(s)\Delta x_s(s,t)\right) \\ \rho \Delta y_{tt}(s,t) &= \frac{\partial}{\partial s}\left(\tilde{T}(s)\Delta y_s(s,t)\right) \\ \Delta z(s,t) &= \Delta u_z(t) \\ \Delta T_s(s,t) &= \rho \Delta \ddot{u}_z(t)\,. \end{aligned} \tag{23}$$

This linear model consists of two partial differential equations of the same structure, namely the ones for Δx and Δy, in addition to a simple relation between Δz and the vertical movement Δu_z of the suspension point at $s = L$. If needed, we have also gained an ordinary differential equation for the rope force ΔT driven by $\Delta \ddot{u}_z$.

3 Flatness-Based Parametrization

Based on the linear model derived above, there exist flatness-based methods for parameterizing the solution with a flat output of the load-side boundary system [9]. Instead of restricting ourselves to the case of a point mass load, these results will be extended by slightly modifying the boundary conditions.

For the sake of brevity, we will only discuss the parametrization of Δx and Δy. It is also obvious that results obtained for Δx below can be directly used for Δy. If needed, methods for linear systems of ordinary differential equations may be used for Δz and ΔT.

Under the assumption of homogeneous initial conditions for Δx it has been shown in [9], that a parametrization of the form

$$\begin{aligned}\frac{\pi^2}{2}\Delta x(s,t) &= \frac{1}{2}\int_0^\pi\int_0^\pi \Delta x(0, t-\delta(s,\psi,\phi))\,\mathrm{d}\psi\,\mathrm{d}\phi \\ &+ \sqrt{\frac{F_z}{\rho g^2}}\int_0^\pi\int_0^\pi G(s,\psi,\phi)\cos\psi\,\Delta x_t(0, t-\delta(s,\psi,\phi))\,\mathrm{d}\psi\,\mathrm{d}\phi \\ &+ \frac{1}{\rho g}\int_0^\pi\int_0^\pi G(s,\psi,\phi)F_z\Delta x_s(0, t-\delta(s,\psi,\phi))\,\mathrm{d}\psi\,\mathrm{d}\phi\end{aligned} \tag{24}$$

with

$$\begin{aligned}\delta(s,\psi,\phi) &= \frac{2}{\sqrt{\rho g^2}}\left(\sqrt{F_z}\cos\psi + \sqrt{F_z+\rho g s}\cos\phi\right) \\ G(s,\psi,\phi) &= \ln\left(\frac{\sqrt{F_z+\rho g s}\sin^2\phi}{\sqrt{F_z}\sin^2\psi}\right)\end{aligned} \tag{25}$$

can be derived for point mass loads with $F_z = mg$ using operational calculus. Note that the parametrization depends on distributed delays and advances of Δx at the load-side end. This parametrization can be extended to the case of a general load, i.e. a rigid body, by considering the following steps.

Let $\tilde{T}(s) = F_z + \rho g s$ be the rope force in the vertical equilibrium configuration as in Sect. 2.2. Equation (8) implies

$$\mathbf{e}_x^T \Delta\mathbf{T}(s,t) = \tilde{T}(s)\Delta x_s(s,t)\,. \tag{26}$$

Evaluating this expression at $s = 0$ leads to $\mathbf{e}_x^T \Delta\mathbf{T}(0,t) = F_z\Delta x_s(0,t)$, which can be used to replace the expression $F_z\Delta x_s(0, t-\delta(s,\psi,\phi))$ in (24). Now, by further replacing

$$\Delta\mathbf{T}(0,t) = \mathbf{T}(0,t) - F_z\mathbf{e}_z, \quad \Delta x(0,t) = x(0,t) - \tilde{x} \tag{27}$$

in (24) we can use the modified parametrization

$$\frac{\pi^2}{2}x(s,t) = \frac{1}{2}\int_0^{\pi}\int_0^{\pi} x(0, t-\delta(s,\psi,\phi))\,\mathrm{d}\psi\,\mathrm{d}\phi + \sqrt{\frac{F_z}{\rho g^2}}\int_0^{\pi}\int_0^{\pi} G(s,\psi,\phi)\cos\psi\, x_t(0, t-\delta(s,\psi,\phi))\,\mathrm{d}\psi\,\mathrm{d}\phi + \frac{1}{\rho g}\int_0^{\pi}\int_0^{\pi} G(s,\psi,\phi)\mathbf{e}_x^T\mathbf{T}(0, t-\delta(s,\psi,\phi))\,\mathrm{d}\psi\,\mathrm{d}\phi \tag{28}$$

instead. This overcomes the restriction of having to linearize the load-side boundary system.

4 Cooperative Load Transportation

At first, the load is modeled as a rigid body K on which N external forces $\mathbf{F}_i$ are applied at suspension points ${}^b\mathbf{l}_i$ in a body-fixed frame with $i = 1, \ldots, N$. (In the following, all quantities described in the body-fixed frame are denoted by a left superscript b.) The position of the load and its orientation w.r.t. an inertial frame are denoted by the minimal coordinates $\mathbf{y}_K$ and $\boldsymbol{\varphi}$ respectively. For instance, Euler angles can be used as entries of $\boldsymbol{\varphi}$. The attitude R_b of the load, which is a function of $\boldsymbol{\varphi}$, can be used to compute the suspension points

$$\mathbf{l}_i = \mathbf{y}_K + R_b\,{}^b\mathbf{l}_i, \quad i = 1, \ldots, N \tag{29}$$

in the inertial frame. Furthermore, the angular velocity ${}^b\boldsymbol{\omega}$ follows directly from

$$\dot{R}_b = R_b \begin{pmatrix} 0 & -{}^b\omega_z & {}^b\omega_y \\ {}^b\omega_z & 0 & -{}^b\omega_x \\ -{}^b\omega_y & {}^b\omega_x & 0 \end{pmatrix}. \tag{30}$$

Let m be the mass of the load and J its inertia matrix (w.r.t. the body-fixed frame). Thus, the equations of motion of the rigid body read

$$\begin{aligned} m\left(\ddot{\mathbf{y}}_K - \mathbf{g}\right) &= \sum_{i=1}^{N} \mathbf{F}_i \\ J\,{}^b\dot{\boldsymbol{\omega}} + {}^b\boldsymbol{\omega} \times \left(J\,{}^b\boldsymbol{\omega}\right) &= \sum_{i=1}^{N} {}^b\mathbf{l}_i \times {}^b\mathbf{F}_i\,. \end{aligned} \tag{31}$$

4.1 Flatness of the Load System

So far, no assumptions about the quantity N of rope forces or their positions on the load have been made. It was argued in [3] that a suitable choice of rope count N and attachment point locations ${}^b\mathbf{l}_i$ is required for flatness. This observation is going to be briefly explained under the assumption[9] of $N > 2$.

Let $\mathbf{y}_K$ and $\boldsymbol{\varphi}$ be components of a flat output candidate $\mathbf{y}_0$. Then the quantities $R_b, \boldsymbol{\omega}$ and $\mathbf{l}_i$ can be easily parametrized by $\mathbf{y}_0$. The remaining quantities to be parametrized are the N forces $\mathbf{F}_i$. With (31) there are only two (vector valued) equations for the N forces. Now assume that the attachment points ${}^b\mathbf{l}_i$ and the rope count N are chosen such that it is possible to define $(N-2)$ independent quantities $\boldsymbol{\psi}_i$ as functions of the forces $\mathbf{F}_i$ in order to obtain a system of equations that can be used for parameterizing the forces. The new quantities $\boldsymbol{\psi}_i$ are then the remaining components of the flat output $\mathbf{y}_0$. Flatness of the system has thereby been shown.

4.2 Heavy Ropes

For the sake of simplicity, all ropes are assumed to be homogeneous, inelastic and identical with length L and mass density ρ per unit length. Their profiles $\mathbf{r}_i(s,t) \in \mathbb{R}^3, i = 1, \ldots, N$ describe the trajectory of every point on each rope. All profiles are a function of time $t \geq 0$ and the same curve parameter, namely the arc length $s \in [0, L]$. Let $\mathbf{u}_i(t)$ be the position of the controlled end and $\mathbf{T}_i(s,t)$ the internal rope tension force of the i-th rope.

The orientation of the arc length is chosen identically for each rope such that $s = 0$ is at the load-side boundary and $s = L$ at the controlled ends. Thus, at the boundaries the relations

$$\mathbf{r}_i(L,t) = \mathbf{u}_i(t), \quad \mathbf{r}_i(0,t) = \mathbf{l}_i(t), \quad \mathbf{T}_i(0,t) = \mathbf{F}_i(t) \tag{32}$$

hold, where the latter two describe the coupling of each rope with the rigid body.

4.3 Open-Loop Control

It is indirectly assumed that the suspension points $\mathbf{u}_i$ serve as inputs. Of course, this assumption does not hold, especially since the controlled ends are attached to tricopters that exhibit their own dynamics. However, the here presented approach will use this assumption for the open-loop control, and afterwards use the computed suspension points as a reference for the position controllers that are implemented on the tricopters.

[9]This restriction is only necessary for proving flatness with the simple argument used here.

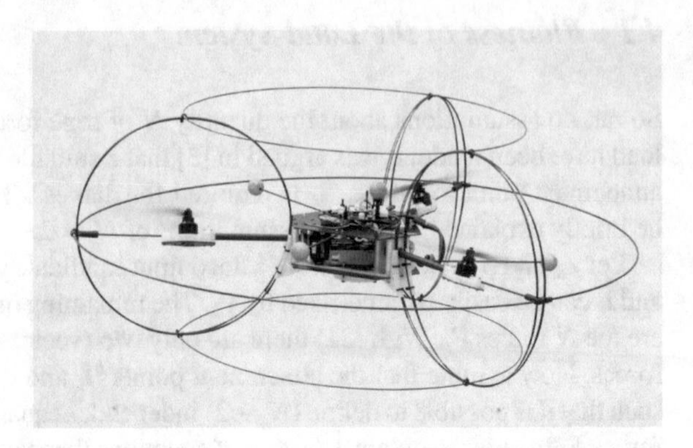

Fig. 6 The LSR Tricopter

Since the load-side boundary system is flat, we can parametrize each rope at its lower end by using the flat output of the boundary system. The main idea of the open-loop controller is taking advantage of the fact that the ropes are only coupled at the load. Therefore, their partial differential equations are not coupled, which allows using the results of the parametrization in Sect. 3.

After parameterizing the rope profile using a suitable[10] reference trajectory for the flat output, the last step is evaluating the expression (28) for each rope at $s = L$ in order to compute the reference trajectories $\mathbf{u}_{i,d}(t)$ of the suspension points.

5 Tricopters

The LSR tricopter design and a pilot-supporting control were first presented in [4]. The most recent realization, shown in Fig. 6, and full configuration tracking control is discussed in [7]. Some key aspects of the tricopter control and adaptations for the heavy rope transport are summarized in the following.

5.1 *Realization*

The developed tricopter is characterized by the three independently tiltable, propeller supporting arms. The tilting mechanism is driven by a standard hobby servo motor and allows for tilt angles of $\pm 75°$. The three 10 inch propellers are each driven by a speed-controlled brushless DC-motor (BLDC) capable of angular velocities up to

[10]The underlying model for the parameterization resulted by linearizing the rope model about a vertical equilibrium configuration. Hence, a suitable choice for a reference trajectory consists of a set point transition for the flat output between two stationary regimes, which results in a transition between two vertical equilibrium configurations for the ropes. The conditions for this kind of configuration have been discussed in detail in Sect. 2.

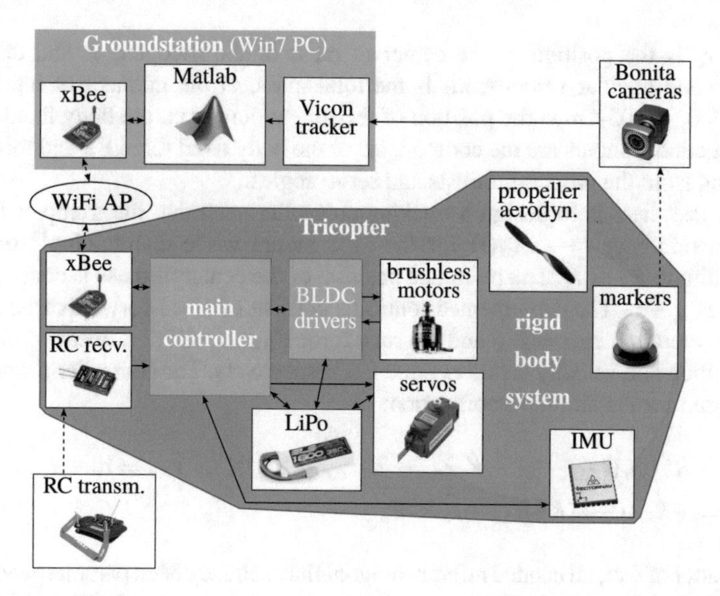

Fig. 7 Realization overview

120 Hz, corresponding to a maximum thrust of 8 N. The vehicle has an outer diameter of about 0.8 m and a mass of about 1.2 kg. It has been tested to carry payloads of up to 0.5 kg. Depending on the payload, the onboard 14.8 V, 2.2 Ah lithium-ion polymer battery (LiPo) allows for up to 15 min of autonomous flight.

The relevant hardware components are summarized in Fig. 7. Crucial for control implementation are the sensors: An inertial measurement unit (IMU) VN100s from VECTORNAV is mounted on the tricopter to measure its angular velocity and acceleration. An external camera-based motion capture system from VICON measures position and attitude by way of reflective markers on the tricopter. The external measurements are processed on a groundstation PC and sent to the controller, which is implemented on the main processor onboard the tricopter. This results in a significant time delay, which is compensated by a model based sensor fusion with the inertial measurements.

5.2 *Configuration Control*

Neglecting the dynamics of the servo motors and the propellers, the tricopter can be regarded as a fully actuated rigid body. Taking into account the rope force $\mathbf{F}_S(t) = \mathbf{T}(L, t)$ at the mount position $\mathbf{r}_S = \mathbf{r}(L, t)$ the model equations are

$$\begin{aligned} \mathbf{r}_S &= \mathbf{r}_c + R\,\mathbf{l}_S, & m_t(\ddot{\mathbf{r}}_c - \mathbf{g}) &= R\mathbf{F}_P - \mathbf{F}_S, \\ \dot{R} &= R\widehat{\omega}, & \Theta\dot{\boldsymbol{\omega}} + \widehat{\omega}\Theta\boldsymbol{\omega} &= \boldsymbol{\tau}_P - \widehat{\mathbf{l}}_S R^T \mathbf{F}_S, \end{aligned} \tag{33}$$

where $\mathbf{r}_c$ is the position of the center of mass of the tricopter, R and $\boldsymbol{\omega}$ are its attitude and angular velocity, m_t is the total mass, Θ the moment of inertia and $\mathbf{l}_S = [0, 0, -0.05]^T$ m is the position of the rope mount w.r.t. the body fixed frame. The six control inputs are the components of the body fixed force $\mathbf{F}_P$ and torque $\boldsymbol{\tau}_P$ resulting from the propeller thrusts and servo angles.

The task here is to design a tracking controller such that the tricopter follows a given reference $t \mapsto \mathbf{r}_{S,d}(t)$ for the rope mount while maintaining[11] the constant attitude $R_d = I_3$. The reference position of the center of mass is consequently $\mathbf{r}_{c,d} = \mathbf{r}_{S,d} - \mathbf{l}_S$. The implemented controller consists of a feed-forward of the desired rope acceleration $\ddot{\mathbf{r}}_{S,d} = \ddot{\mathbf{r}}_{c,d}$ and the rope force $\mathbf{F}_{S,d}$ as well as a linear feedback of the position and velocity errors $\mathbf{r}_{c,e}$ and $\dot{\mathbf{r}}_{c,e}$ respectively. The controller proposed in [6] is used for the attitude stabilization:

$$\mathbf{F}_P = R^T\big(m_t(\ddot{\mathbf{r}}_{S,d} - \mathbf{g}) - K_v\dot{\mathbf{r}}_{c,e} - K_r\mathbf{r}_{c,e} + \mathbf{F}_{S,d}\big), \qquad \mathbf{r}_{c,e} = \mathbf{r}_c - \mathbf{r}_{c,d}, \tag{34a}$$

$$\boldsymbol{\tau}_P = -K_\omega\boldsymbol{\omega} - 2(K_R R)^\vee + \widehat{\mathbf{l}_S} R^T\mathbf{F}_{S,d} \tag{34b}$$

As a matter of fact, all needed reference signals have already been parametrized by the flat output of the load-side boundary system. Therefore, using reference trajectories for the flat output in these parametrizations results in reference trajectories for all other quantities, especially the controlled ends of the ropes. By taking the first and second time derivatives of the reference trajectories after evaluating (28) at $s = L$ we can obtain the reference velocity and reference acceleration of the controlled ends.

As for the reference trajectory for the rope force, i. e. $\mathbf{F}_{S,d}$, one has to take the derivative of (28) w. r. t. s, evaluate the expression at $s = L$ and then multiply the result with $\tilde{T}(L)$ as evident from (26). Of course, the same procedure applies to the y-component of the desired force.

6 Experiments

Experimental results[12] for planned transitions of a single heavy rope, a rod suspended by two heavy ropes and a triangle suspended by three heavy ropes have been used for validation. Ropes of length 1.5 m with a mass density of 0.2 kg/m were used for this purpose. The length of the rod is about 2 m and it weighs roughly 0.15 kg. Each side of the triangle has the same parameters as the rod.

An example maneuver for transporting the triangle is depicted in Fig. 8 with measurements of the tricopters and the load. Note that the tracking behavior of the load position and orientation is at least as accurate as the tricopters. This remarkable result, despite the use of a simple linear rope model for relatively fast transitions, validates the here presented flatness-based approach.

---

[11] Tracking a given attitude $t \mapsto R_d(t)$ at the same time would work just as fine, see [7], but is dropped for the sake of brevity.

[12] Video available at https://youtu.be/epLOAFtv_xA and http://www.uni-saarland.de/?trikopter.

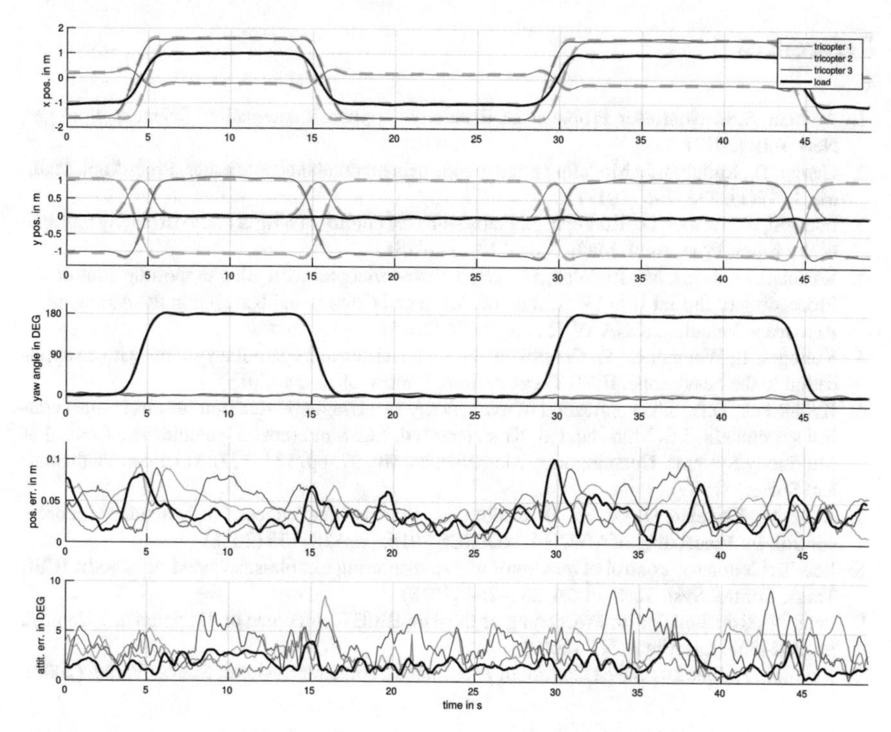

Fig. 8 Example maneuver for the cooperative transport of a rigid body

7 Conclusion

A flatness-based open-loop control based on the linearized partial differential equations of a heavy rope has been combined with position control algorithms for tricopters, which are used to achieve the desired motion of the actuated rope ends. Note that this simple approach does not require measurements of the rope profiles or the load. Nonetheless, experimental results demonstrate a great control performance.

Yet, this approach for cooperative load transportation can be further developed to overcome arising problems. For instance, the realized feedback intrinsically does not directly act on errors or external disturbances along the rope profile. In order to address such problems, controllers can be designed based on the underlying distributed-parameter system. So-called PDE-Backstepping methods can be deployed for instance. These controllers will definitely require additional measurements such as the rope profiles, the load position and its attitude. Thus, combining the controller with a suitable observer is of interest as well. The flatness property of the system can be exploited in both the controller and the observer design.

References

1. Antman, S.S.: Nonlinear Problems of Elasticity. Applied Mathematical Sciences. Springer, New York (2013)
2. Gerbet, D., Rudolph, J.: Modeling a heavy rope using redundant coordinates. Proc. Appl. Math. Mech. **17**(1), 795–796 (2017)
3. Irscheid, A., Gerbet, D., Rudolph, J.: Flatness-based control of a rigid body carried by multiple heavy ropes. Proc. Appl. Math. Mech. **18**(1) (2018)
4. Kastelan, D., Konz, M., Rudolph, J.: Fully actuated tricopter with pilot-supporting control. In: Proceeding of the 1st IFAC Workshop on Advanced Control and Navigation for Autonomous Aerospace Vehicles ACNAAV 15, pp. 79–84 (2015)
5. Knüppel, T., Woittennek, F.: Control design for quasi-linear hyperbolic systems with an application to the heavy rope. IEEE Trans. Autom. Control **60**, 5–18 (2015)
6. Koditschek, D.E.: The application of total energy as a Lyapunov function for mechanical control systems. In: J.E. Marsden, P.S. Krishnaprasad, J.C. Simo (eds.) Dynamics and Control of Multibody Systems. Contemporary Mathematics, vol. 97, pp. 131–157. American Mathematical Society (1989)
7. Konz, M., Kastelan, D., Gerbet, D., Rudolph, J.: Practical challenges of fully-actuated tricopter control. In: Proceeding of MECHATRONICS 2018, pp. 128–135 (2018)
8. Lee, T.: Geometric control of quadrotor uavs transporting a cable-suspended rigid body. IEEE Trans. Control Syst. Technol. **26**, 255–264 (2018)
9. Petit, N., Rouchon, P.: In: Proceeding of the 41st IEEE Conference on Decision and Control, pp. 362–367. Las Vegas, NV (2002)
10. Rudolph, J.: Flatness Based Control of Distributed Parameter Systems. Shaker Verlag (2003)

Part IV
Modern Applications for Management and Information Processing in Complex Engineering Systems

Toward a Secure IoT Architecture

Robert E. Hiromoto, Michael Haney, Aleksandar Vakanski and Bryar Shareef

Abstract The design of a cyber-secure, Internet of Things (IoT), supply chain risk management architecture is proposed. The purpose of the architecture is to reduce vulnerabilities of malicious supply chain risks by applying machine learning (ML), cryptographic hardware monitoring (CHM), and distributed network coordination (DNC) techniques to guard against unforeseen hardware component failures and malicious attacks. These crosscutting technologies are combined into an Instrumentation-and-Control/Operator-in-the-Loop (ICOL) architecture that learns *normal* and *abnormal* system behaviors. In the event that the ICOL detects possible *abnormal* system-component behaviors, an ICOL network alert is triggered that requires an operator verification-response action before any control parameters can affect the operation of the system. The operator verification-response is fed back into the ML systems to recalibrate the classification of *normal* and *abnormal* states of the system. As a consequence, this proposal adheres to the notion of an Operator-in-the-Loop strategy that combines DNC, ML, and CHM with the creative problem-solving capabilities of human intelligence.

Keywords IoT · Supply chain risk management · Normal and abnormal behavior · Network attacks · Throttled network · Machine learning

R. E. Hiromoto (✉) · M. Haney · A. Vakanski · B. Shareef
Center for Advanced Energy Studies (CAES), University of Idaho, 1776 Science Center Drive, 83402 Idaho Falls, ID, USA
e-mail: hiromoto@uidaho.edu

M. Haney
e-mail: mhaney@uidaho.edu

A. Vakanski
e-mail: vakanski@uidaho.edu

B. Shareef
e-mail: shar0416@vandals.uidaho.edu

Y. P. Kondratenko et al. (eds.), *Advanced Control Techniques in Complex Engineering Systems: Theory and Applications*, Studies in Systems, Decision and Control 203, https://doi.org/10.1007/978-3-030-21927-7_14

1 Introduction

In recent years, the development of advanced control methods in both theory and their practical industrial applications has gained considerable attention in SCADA systems, space control, robotics, marine systems, and process control in agriculture and food production. Kuntsevich [33, 34] whose work for which these chapters are dedicated is one of the main contributors to its development.

Advances in microelectronic devices, the development of efficient user friendly apps and the seamless access to connect to the Internet has moved control methods to new levels of demand. In this regard, the Internet of Things are rapidly transforming the landscape of traditional industries and institutions, and ushering in an era of smart homes, smart cities, smart factories, and smart grids. The IoT involves the Internet connection of systems and devices encompassing such technologies as wireless sensors, home heating/cooling units, utility metering systems, actuators and the coordination of autonomous land and aerial vehicles using interoperable protocols, often built into embedded systems. The IoT provide services and communication to coordinate and drive data analytics, which facilitates improved performance, better turn-around time, and a more efficient maintenance schedule. As industrial requirements change over time to meet new demands and new markets, the IoT provide the flexibility to maintain in real-time the ability to scale with the addition of new technologies. IoT offer an economically efficient and a convenient wireless interconnection of systems and devices that can seamlessly coordinate industrial manufacturing, day-to-day home and personal activities, and the control of major power stations to deliver adequate and reliable energy. The ubiquitous growth of IoT is, therefore, a consequence of their economic advantages and practical conveniences.

1.1 The IoT Security Challenges

The desire for flexibility, scalability and real-time network connections over the Internet entails a whole host of IoT data vulnerabilities. To secure the IoT, either by securing the application through system-specific software solutions or by requiring a certified hardware design and the installation of proprietary communication protocols, can ultimately limit their flexibility. In such a boot-strapped IoT environment, the acquisition and monitoring of service devices to ensure cybersecurity becomes a difficult problem in itself. Furthermore, once any IoT device is infected with malicious malware, the embedded system must have an alternative security mechanism to overcome or moderate such intrusions. Without a pre-certified [48] characterization of a device-execution profile, detecting normal or abnormal behavior becomes a difficult task when attempting to predict the onset of possible cyber attacks.

It is the open-ended conveniences of the IoT that amplifies the potential for devastating cyber attacks. In this regard and with the emergence of the Marai attacks [66], the lack of IoT security has become a major concern to the commercial and industrial communities. The US Department of Homeland Security (DHS) offers a "strategic principles for securing IoT," [63] to improve the security of IoT that relies on cooperative practices employed by the various vendors, starting from the design phase to the manufacturing and deployment phases that may rely upon trusted third-party vendors. In reality, the proliferation of the international supply chain manufacturing of IoT components represents an unmanageable marketplace for security.

The privacy and security issues surrounding IoT touches the core of personal liberties and industrial regulations. According to the U.S Federal Trade Commission report on privacy and security [64]: "IoT devices may present a variety of potential security risks that could be exploited to harm consumers by: (1) enabling unauthorized access and misuse of personal information; (2) facilitating attacks on other systems; and (3) creating safety risks. Although each of these risks exists with traditional computers and computer networks, they are heightened in the IoT …"

The reality of IoT security is the vulnerabilities that it contributes to the supply chain of inexpensive off-the-shelf devices. The IoT come with neither the guarantees of a secure embedded system environment nor the absence of malicious malware components. As a consequence, IoT security represents a supply chain risk management issue that goes beyond external IoT cyber attacks.

1.2 An Architecture for Secure IoT Supply Chain Management

In this paper, we treat the security of IoT as a supply chain risk management issue that requires an architectural and administrative approach, which relies upon (1) pattern recognition tools to learn and identify 'normal' and, in particular, 'abnormal' IoT component behavior under hardware/software faults and cyber attacks; (2) an architecture that can respond to potential security intrusions through a system of throttled data transmissions; (3) characteristics of hardware behavior under cryptographic attacks; and (4) an operator-in-the-loop [53] design that requires an operator verification-response to challenge or approve the actions to be taken when a questionable processing event (alert) is detected. This latter requirement serves as a criticality strategy to arbitrate decisions about normal/abnormal behavior, which may have been missed or misclassified by the various pattern recognition tools employed. Finally, a feedback loop between the operator and the monitoring (learning) tools is designed to cross-validate and refine the definitions of normal and abnormal device output classes.

At the core of this proposal is the adherence to an Operator-in-the-Loop strategy that explicitly combines the creative problem-solving capabilities of human intelligence. Under this scenario, we conjecture that "on-off" cyber attacks such as the Stuxnet [34] might well have been averted. Had operators been provided with interrupt driven ICOL alerts, such attacks may have been detected in real time with appropriate operator interventions. As new embedded system components are introduced, or if seemly *abnormal* events are detected, the ICOL verification-response criteria can positively enhance the maintenance and on-the-job training of the operator workforce, and provide incentives to revisiting IC policies and procedures, as needed.

The proposed strategies integrate well-understood techniques to create a novel synergistic approach that can isolate, mitigate and protect IoT security concerns against the consequences of cyber threats in supply-chain instrumentation and control environments. The realization of the proposed concepts has the beneficial effects of lowering the occurrences of detecting false positives and false negatives, and provides a practical cost-efficient IoT infrastructure that protects against cyber attacks and hardware failures. One likely security mechanism is a multilevel architecture that involves a cross-correlated strategy for the classification of abnormal IoT behavior; a throttled, synchronous data flow network; and a cryptographic hardware monitoring capability.

2 Components of a Secure IoT Network

Intrusion detection in IoT devices remains a very challenging problem, and a very important aspect for secure operation of IoT systems. The reliability of a derived IoT system depends upon the normal time-dependent interactions between aggregated components of the system. As a consequence, the learning of normal and abnormal behaviors is a crucial first step in securing an on-line environment against malicious attacks or even hardware/software failures. The proposed secure-IoT architecture is made up from three monitoring components that are summarized at a system (operator) interface level. This interface provides a human operator a system-wide status as it tracks real-time activities. The three monitoring components are derived from machine learning algorithms, cryptographic hardware monitoring, and a synchronous dataflow network that throttles (interrupts) the flow of abnormal device outputs before they are allowed to trigger a response or an action from a receiving instrumentation or control device.

2.1 Machine Learning ICOL Assisted Monitoring

Machine learning (ML) algorithms have made advances over the last several years with the promise of attaining autonomous self-sustaining systems. The total reliance on ML, however, is far from maturity and has not achieved 'failsafe' stature. Detecting normal and abnormal events is an important unsolved ML problem.

2.1.1 Machine Learning for Intrusion Detection

Numerous data mining and machine learning algorithms have been employed for intrusion detection [11]. These systems are generally implemented as misuse-based detection, anomaly-based detection, and hybrid detection [10, 35].

The *misuse-based detection* entails availability of a set of known intrusion attacks on a network. Based on feature patterns that are representative for the documented malicious attacks, a machine learning algorithm is employed for classification of new network activity in a supervised learning manner [11]. In a training phase, a selected classifier algorithm is employed to extract a set of salient features in presented network data. The data comprises both normal and attack instances with corresponding class labels, and a classifier is trained to distinguish normal from abnormal data by using the set of identified class features. In a testing phase, unseen outputs from the network are evaluated and classified as normal or attack behavior, based on the probability of belonging to one of the two classes. Alternatively, different attack patterns can be labeled as individual classes, and the algorithm is trained to differentiate not only between normal and abnormal traffic data, but also it can identify the type of attack in the case of intrusion. Methods that have been employed for misuse detection include random forests [67], decision trees [31], Bayesian networks [40], support vector machines [1], and artificial neural networks [13]. Although these systems yield low rate of false negatives, they require periodic updating of the signatures of known attacks.

The *anomaly-based detection* refers to detecting abnormal activity in the network outputs based on known patterns of normal network behavior. Anomaly detection, however, is more challenging since it is aimed at detecting novel attacks that the network has not experienced or learned before where the attributes of the abnormal behavior are not known or labeled. Therefore, this class of detection approaches often entails implementation of unsupervised learning techniques for detection of abnormal behavior [11]. Among the algorithms that have been used for anomaly-based detection are clustering techniques [57], fuzzy logic [43], artificial neural networks [9], and latent variable models [3]. For anomaly-based detection, the features of the normal behavior are learned at a training phase. In the testing phase, unseen network traffic is evaluated against the known normal behavior, and any significant deviation is classified as abnormal behavior. The outcomes of anomaly-based detection can also be applied in the definition and updates of attack signatures for misuse-based detection. However, anomaly-based detection systems are prone to producing high levels of false negatives, especially when the normal behavior of the network changes over time.

Finally, the *hybrid systems* for intrusion detection combine the advantages of both misuse-based and anomaly-based detection. They utilize historical data for both known attacks and models of normal network behavior. This combination of known attacks and models of normal network behavior provides an enhanced capability for classification of previously seen attacks and recognition of unknown attacks. Even with these added enhancements, the capabilities of a hybrid system can still be vulnerable to sequences of normal behavior that form an attack or the occurrence of abnormal behavior that may be a rare sequence of normal behavior.

2.1.2 Deep Learning

The term deep learning refers to machine learning algorithms that employ artificial neural networks (ANNs) with multiple layers of hidden computational units. A neural network with one layer is considered shallow; most deep networks at the present time contain between 5 and 10 hidden layers, although network architectures exist with over 1,000 layers [23].

In recent years, several different structures of deep ANNs have been applied to intrusion detection. Some of the earlier works are based on restricted Boltzmann machine networks and deep belief networks, where the neural networks are utilized for dimensionality reduction of the data [56], and afterward a classifier algorithm (e.g., support vector machines (SVM)) is used to differentiate between the normal and abnormal data traffic. More recently, autoencoder networks emerged as a preferred architecture for dimensionality reduction and feature extraction [41, 50, 60]. Kim et al. [28] implemented a recurrent neural network with long-term short memory (LTSM) units for intrusion detection and achieved an average detection rate of 98.8% of the attack instances and 90% on the normal instances from the KDD Cup 1999 dataset. Networks with fully connected layers of hidden computational units have also been implemented for intrusion detection [35, 59]; although, they produced lower detection rates in comparison to autoencoders and recurrent neural networks.

2.1.3 Data and Datasets

The network data used by the majority of methods for intrusion detection is packet capture (pcap) data. The network packets data are captured at a computer port by application programming interfaces, such as WinPCap (with Windows systems) and Libpcap (with Unix systems). The order and number of data attributes vary depending on the used network protocol, such as Transmission Control Protocol (TCP), User Datagram Protocol (UDP), Internet Control Message Protocol (ICMP), etc. The network packets contain a header and a payload, which in addition may include data related to higher level protocols, e.g., Network File System, Server Message Block, Hypertext Transfer Protocol, Post Office Protocol, and similar.

Besides packet capture, another network data format is NetFlow, which defines a network flow as a unidirectional sequence of packets with a set of predefined attributes. The network packets attributes typically include: protocol type, duration of the connection, network service on the destination, number of data bytes from source and from destination, flag, number of urgent packets, etc. In comparison to full packet capture, this approach offers a recording of summary data or metadata about network communication sessions that provides a statistical approach to review of network-transmitted information. It is much more highly efficient for storage and processing of network data. Based on the original development by Cisco Systems and still commonly referred to by this name, NetFlow is now codified in an IETF Request For Comments open standard in RFC 5101 as the Internet protocol flow information export (IPFIX).

Most of the methods for intrusion detection are implemented on the KDD Cup 1999 dataset [27], or its updated version NSL-KDD dataset. The KDD Cup 1999 have been used in our preliminary research on this topic and it is briefly described in an ensuing section. Other existing datasets for network intrusion detection include DARPA 1998, 1999 and 2000 [25], DEFCON [16], CAIDA [12], Lawrence Berkley National Laboratory (LBNL) [36], UNIBS [62], and TUIDS dataset.

2.2 *Analysis of Network Security*

A cross-correlation strategy is proposed as part of the network security analysis. In this strategy, the machine learned classification of a device's output states is cross-correlated with an interactive operator-in-the-loop approval or rejection of the machine learned state. The output of the system device component is not streamed between connected components but throttled in a synchronous manner so that perceived abnormal states can be monitored and reasoned about by the operator on duty. In this approach, the operator has final control over the real or misclassified nature of the anomaly.

2.2.1 Machine Learning and Cryptographic Hardware Monitoring

One machine learning approach that appears to address our needs employs a deep artificial neural network (DANN) for analysis of the system behavior that consists of two sub-networks: one for learning feature representation of *normal* data, and a second sub-network for the detection of *abnormal* feature patterns caused by intrusion attempts, malware, undetected source code bugs, or hardware failures. The DANN operates on blocks of embedded system component outputs. The DANN is designed to capture *time-evolving* features. Inputs to the DANN ML system are obtained from system component output data. The ML algorithm will monitor these outputs by clustering patterns in terms of operations (specific to each embedded system), the range of variable scale lengths, and the duration of repeated operations. At one level, physical characteristics of hardware modules under cryptographic attack will be monitored and integrated into the input stream. Under such attacks, differential power consumption, temperature fluctuations, heat generation, acoustic noise, component vibration, etc., have been used successfully when analyzing cryptographic hardware modules for effectively extracting crypto keys, called "side channel attacks" [30, 58]. This leads to the development of novel techniques for examining embedded system components in a "black box" manner that relies on physical characteristics of these components during operations.

A second level of hardware feature analysis is an a priori system-information catalog that summarizes the use of similar hardware, software and firmware components that may make up numerous embedded system architectures. With this catalog and related Internet blogs, specific rules can be formulated and incorporated into the ML rule sets.

2.2.2 Convergence of Embedded System Components

A focus of ongoing research in embedded system security is the trend in the industry to standardize on components of hardware and firmware driven by economies of scale. Different vendors use catalog similarities in hardware, software, and firmware components rather than develop single-purpose components in a proprietary manner [21, 39, 52]. Common use of real-time operating systems and platforms such as VxWorks by Wind River Systems [65] or communications integrated circuits such as SnapDragon 800 chips or others developed by Qualcomm stems from the nature of industrial patents and intellectual property licensing and common embedded systems manufacturing techniques. The use of these common components has the potential to facilitate common attack scenarios across different vendor architectures and different industrial sector installations. Thus, the catalog of common component usage can be incorporated as rules or as cross-correlation inputs to the DANN. By examining and recording common functionality in these constituent system components, the DANN can be trained to recognize common features that may be present in a variety of systems that would otherwise appear to be unrelated.

2.2.3 Instrumentation and Data Acquisition

A test bench setting for device testing, data gathering and recording will be constructed. Individual system components will be instrumented with equipment such as temperature sensors, sensitive directional microphones, broad-spectrum photocells, voltage meters, etc., to evaluate effects similar to methods used in analysis of 'side-channel' attacks in cryptanalysis. Individual system component behavior will be recorded under normal operation for some period of time. This data will be fed into the generative neural network model to "learn normal" behavior. An adversarial counterpart to "learn normal" is the development of cyber attack scenarios that tamper or interfere with normal operation of the system components. The measured physical characteristics of these components while "under attack" will be the basis for detecting deviations from normal operations [30]. Both non-destructive and destructive sampling methods for examining constituent components of systems will be employed to catalog systems and determine expected normal behaviors based on these components. For example, by connecting to an embedded system via a communications interface such as JTAG or UART, the firmware of an embedded system may be queried to determine the version of VxWorks or other RTOS.

3 A Synchronous Large-Grain Dataflow Model

Given the growing concerns and evolving capabilities by sophisticated cyber attacks, it becomes critical to have a centralized system for monitoring and analysis of all network activities. Current network intrusion detection systems rely on both network and security teams having access to network flow data (acquired from devices like routers) that provides traffic volume and a packet's origin/destination trace, and packet data extracted from various network mirroring and filtering mechanisms such as SPAN (Switched Port Analyzer) [7] also known as Port Monitoring and a network TAP (Terminal Access Point) hardware device. The main disadvantage to these approaches lies in the postmortem and microscopic levels of intrusion detection that they offer.

Rather than improving upon the above methods, we propose a different approach that involves an interactive, operator-in-the-loop monitoring system that learns normal and abnormal device behavior, and provides the operator with the capability to react to suspicious or threatening activities. In order to realize this novel approach, a synchronous intrusion detection network is proposed. The synchronicity of the network provides 'safety' latches or 'pulsed' traffic flow of device outputs between the sensor and control elements of the network. The proposed synchronous intrusion detection network provides the operator-in-the-loop a dynamic, incremental and controlled access to the state of the network.

To achieve this interaction between data streams and ICOL alerts, a throttled network is designed that tracks data motion and dependences between interacting system components [18, 19]. The architecture will be interrupted (throttled) as alerts are posted, and resumed or halted depending upon the expert ICOL determination of *normal* or *abnormal* conditions. A graphical interface implemented on top of this network, will allow the tracking of offending data streams back to their originating component(s). This provides component isolation and early detection of compromised supply chain components.

3.1 Dataflow

The proposed throttled, interrupt- and data-driven network architecture is based on a dataflow paradigm. In this paradigm, a directed acyclic precedence graph represents the data dependence for correct program execution. In this graph, vertices represent the computational nodes and directed arcs define the order for which nodes can execute (fire). Tokens are data items that flow from one node to another along adjacent arcs. When all tokens are present on the input arcs of a receiving node, that node can fire. This execution model has the advantage of implicitly scheduling concurrent firing of nodes; and because the program execution is controlled by the arrival of data, it is referred to as data-driven [26]. Figure 1 illustrates the input/output states

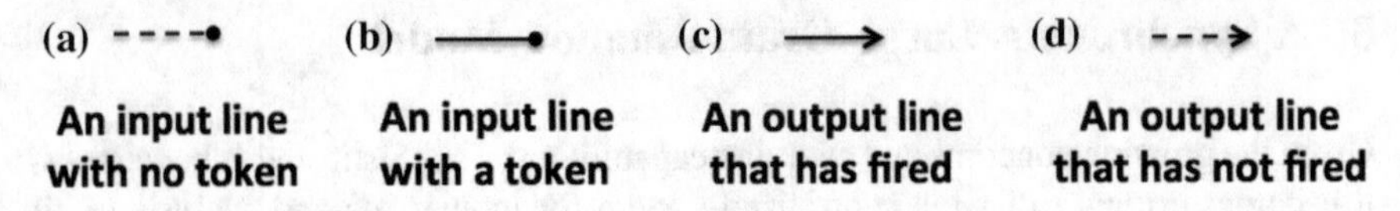

Fig. 1 Dataflow input/output *arc* states

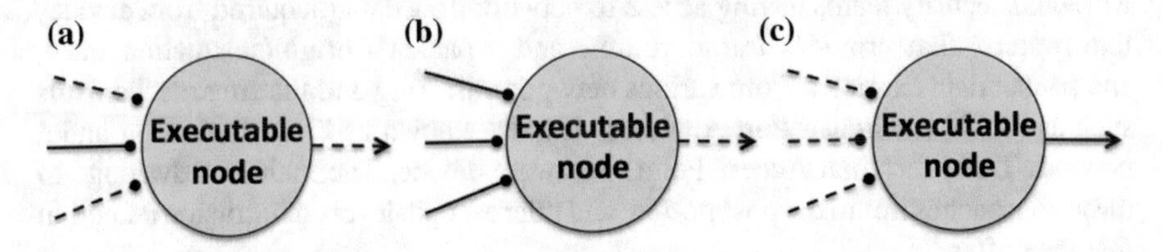

Fig. 2 Dataflow execution rules

of the dataflow *arcs* (paths) between nodes. The dataflow model also assumed that each node executed at most a single instruction.

Using these input/output states, dataflow firing rules are depicted in Fig. 2. Notice in Fig. 2a that all input tokens (data) have not arrived. In Fig. 2b the arrival of all tokens initiates the execution of the node; however, during this time the generation of the output token has not yet occurred. Finally, in Fig. 2c an output token is generated at which point all input tokens are cleared, which creates the synchronous behavior of the dataflow execution graph.

Initially, the dataflow model for program execution was an area of study with the high hopes of extracting massive amounts of parallelism at the instruction level. Although the theoretical results seemed promising, the hardware resources required to maintain maximal processor efficiency was not attainable. However, without abandoning the dataflow framework, another approach that gained some following was the large-grain dataflow (LGDF) model for program execution.

3.2 *Large-Grain Dataflow*

The LGDF model was developed in the late 1980s by Babb [5] and it was demonstrated in the application of large scientific simulations [6]. The LGDF is a computational model that combines sequential programming with dataflow activations and throttling constructs to instantiate modular execution in parallel. These modules (nodes) are large-grain instruction blocks, e.g., functions or entire procedures. The LGDF computational model adheres to the dataflow precedence graph, where each component (node) of the dataflow graph represents a temporally ordered sequence of computations. LGDF applications are constructed using networks of program modules connected by data-paths. Parallel execution is controlled indirectly via the production and consumption of data. Figure 3 illustrates a typical LGDF graph where

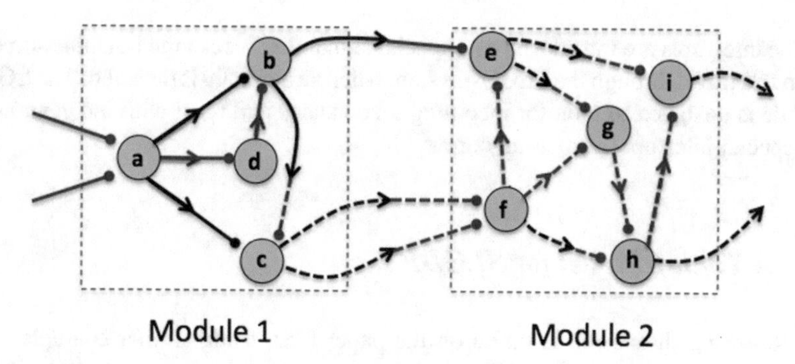

Fig. 3 LGDF execution graph

Module 1 is blocked from completing its execution. Here we note that, although, an output has been generated from node b, the output token has not yet reach node c, and for this reason Module 2 is unable to initiate its execution.

The LGDF model offers the appropriate level of modularity and responsiveness to introduce ICOL alert interrupts throughout the system. These interrupts must be triggered through a throttling mechanism to allow IC operators the option to monitor system behavior under detected abnormal situations or to intervene at a more strategic level. The interrupts can trigger different kinds of events or the selection of specific serial port controller. The dataflow graph also provides a backward-looking anomaly detection capability during a post analysis phase.

3.3 Synchronous Large-Grain Dataflow

The design of the throttling and interrupt mechanisms can be achieved by introducing a Synchronous LGDF (SLGDF) network controller. Lee and Messerschmitt [1987] introduced the concept of SLGDF to automatically extract parallelism in digital signal processing (DSP) algorithms and generate automatic scheduling of processes onto parallel processors. Related work that predates Lee and Messerschmitt is described by Dennis and Misunas [17].

The synchronous data flow execution of a graph and its subgraphs is statically scheduled by balancing the number of activations (or tokens) generated by one node and correspondingly received by a second node within the dataflow graph. The buffer size to hold the number of tokens is also assigned to provide adequate space resources during all repeated data flow iterations. Delay loops are introduced to maintain a block schedule that is periodic, where each cycle must terminate before the next cycle begins. This periodic cyclic scheduling facilitates the deployment of an operator monitoring system without the need for a complicated dynamic trapping mechanism in software or hardware implementation, if so desired.

The interrupts are introduced as a special signal that blocks the transmission of a token and travel through the SLGDF system when an anomaly is detected. The LGDF module is designed to monitor incoming token states and react with the generation of a special interrupt token as an output.

3.4 A Timing Model for SLGDF

The following discussion is based on the paper [38], using similar examples and notations. A "synchronous" **LG**DF block has a predetermined number of inputs that are consumed on each input path (*arc*) and a number of outputs produced on each output arc each time the block is executed. In other words, when a synchronous DF block is executed, it consumes a fixed number of data inputs on each input arc and produces a fixed number of data outputs on each output *arc*.

At the start of execution in Fig. 4a (the execution graph), node *a* can be invoked (fired) because it has no input arcs and hence it needs no input data. After invoking (executing) node *a*, node *b* and node *d* can be invoked, after which node *c* can be fired. This sequence can be repeated, but node *a* produces twice as many outputs on *arc* A_2 as node *c* can consume one at a time. An infinite repetition of this schedule, therefore, causes an infinite accumulation of data in the buffer associated with *arc* A_3. This implies an unbounded memory requirement, which is clearly not practical. Node *c* expects as inputs two signals with the same data rates but instead gets two signals with different or "inconsistent" data rates. In a synchronous DF block, an inconsistent data rate can also cause the execution of a node based on the wrong sequence of input data. For example, if node *a* happens to fire before the second output (token) on arc 2 is consumed then it is possible that node *b* generates a new token that node *c* consumes with the second token, which gives rise to a race condition. The SLGDF graph Fig. 4b does not have this problem.

A periodic admissible sequential schedule (PASS) repeats the invocation {a, d, b, c, c}, where node *a* initiates the execution followed by node *d* then node *b* and finally node c. Notice that node *c* is fired twice as often as the other three. A formalism for automatically checking for consistent data rates and simultaneously determine the relative frequency with which each node must be invoked can be defined in a matrix where (1) a column is assigned to each node, (2) a row is assigned to each arc, (3) the (i, j)th entry in the matrix is the amount of data produced by node j on arc i each time it is fired, (4) if node j consumes data on arc i, the number is negative, and (5) if node j is not connected to arc i then the number is zero. This matrix is called a topology matrix and is illustrated Eq. 1. The topology matrices for SLGDF are displayed in Eqs. 2 and 3.

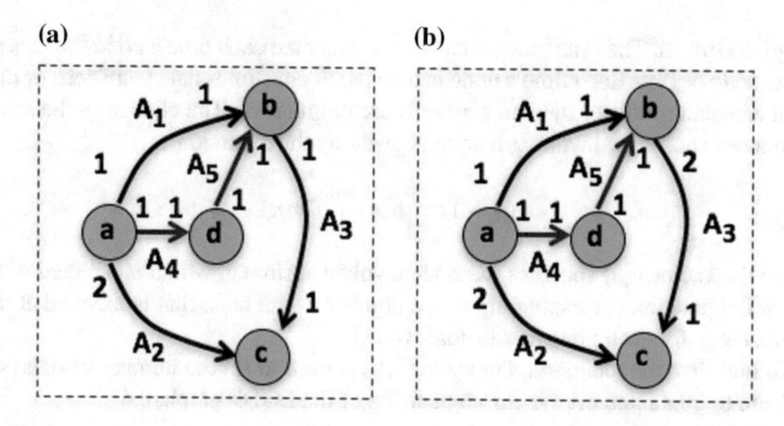

Fig. 4 **a** An example of a defective SLGDF graph. **b** A corrected SLGDF graph

$$\Gamma = \begin{array}{c} \\ \\ \\ \\ \\ \end{array}\begin{bmatrix} (1,a) & (1,b) & (1,c) & (1,d) \\ (2,a) & (2,b) & (2,c) & (2,d) \\ (3,a) & (3,b) & (3,c) & (3,d) \\ (4,a) & (4,b) & (4,c) & (4,d) \\ (5,a) & (5,b) & (5,c) & (5,d) \end{bmatrix} \begin{array}{c} A_1 \\ A_2 \\ A_3 \\ A_4 \\ A_5 \end{array} \quad (\text{columns } a\ b\ c\ d) \tag{1}$$

$$\Gamma_a = \begin{bmatrix} 1 & -1 & 0 & 0 \\ 2 & 0 & -1 & 0 \\ 0 & 1 & -1 & 0 \\ 1 & 0 & 0 & -1 \\ 0 & -1 & 0 & 1 \end{bmatrix} \tag{2}$$

$$\Gamma_b = \begin{bmatrix} 1 & -1 & 0 & 0 \\ 2 & 0 & -1 & 0 \\ 0 & 2 & -1 & 0 \\ 1 & 0 & 0 & -1 \\ 0 & -1 & 0 & 1 \end{bmatrix} \tag{3}$$

A necessary condition for the existence of a PASS with bounded memory is that the rank $(\Gamma) = s - 1$, where s is the number of nodes. This is proved in [38]. In our example (Eqs. 2 and 3), it can be shown that rank $(\Gamma_a) = 4$ and rank $(\Gamma_b) = 3$ for which Γ_b satisfies the necessary condition for the existence of an allowable PASS.

In order to achieve a practical implementation, an *arc* is treated as a FIFO queue. The size of the queue varies at different times in the execution. The size of this queue can be defined by a vector $b(n)$, which contains the number of tokens in each queue at a given time n. The time index n can be incremented each time a node finishes and a new node begins. Each time a node is invoked, it consumes data from zero or more input arcs and produces data on zero or more output arcs. The change in the size of the queues caused by invoking a node is given by the recursion

$$b(n+1) = b(n) + \Gamma v(n)$$

where the vector v(n) specifies the node invoked at time n, which is denoted with a one in the position corresponding to the number of the node that is invoked at time n and zeroes for each node that is not invoked.

To initialize the recursion, the vector $b(0)$ is used to set the number of delays on each arc to guarantee the synchronous firing of the LGDF graph, e.g.,
with $b(0)$ set to

$$b(0) = \begin{bmatrix} 1 \\ 2 \\ 1 \end{bmatrix},$$

which corresponds to the SLGDF graph (see Fig. 5.) with delays on the arcs. Delays, therefore, affect the way the system starts up. Clearly, every directed loop must have at least one delay, or the system cannot be started.

A Periodic Admissible Parallel Schedule (PAPS) is described by Lee and Messerschmitt [38]. In their discussion, the PAPS is the collection of independent sequential SLGDF modules that can be scheduled to optimally execute in parallel. The goal is to ensure that all concurrently executing modules exhibit the same amount of runtime before the next collection of SLGDF modules are invoked. They show how the scheduling problem can be reduced to a related problem in operations research. The details and their proposed scheduling algorithms are left to the readers to workout.

A slightly less restricted synchronous DF approach can be achieved by allowing internal nodes to execute using a "non-strict" execution paradigm.

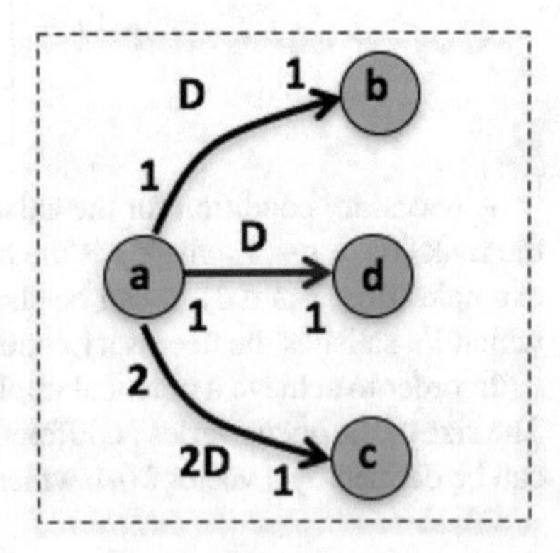

Fig. 5 Delay on arcs to guarantee correct SLGDF execution

3.5 Strict and Non-strict Semantics

The SLGDF represents an execution paradigm referred to as "strict", where all inputs to the LGDF node must arrive on the input arcs before the node can be invoked. This is not, however, the only execution semantics available. A less restrictive semantics is a "non-strict" form of execution. The "non-strict" semantics does not demand that all inputs (tokens) be present on all input arcs before an LGDF sub-expression (sub-node) is invoked [22]. Figure 6 is an illustration of a "strict" execution semantics. Notice in Module 1 that node *c* has not yet sensed the presence of node *b*'s output token on its input arc. For this reason, node c is not invoked and no output token has yet to be generated to instantiate the firing of Module 2. In contrast, Fig. 7 illustrates an example that employs "non-strict" semantics. Here Module 1 experiences the same scenario described in Fig. 6, however, due to the "non-strict" execution semantics node *e* is invoked by the output from node *b* and allowed to execute even thought Module 2 has not received all of its tokens on all of its input arcs. In other words, a "non-strict" execution semantics can evaluate an expression to a value even if some of their other subexpressions do not. Clearly a non-strict implementation can improve the throughput and performance of the LGDF paradigm by introducing an asynchronous scheduling between modules.

In an asynchronous execution semantics, no fixed amount of data consumed or produced on the input or output arcs is specified statically. In order to accommodate an LGDF node, the graph can be divided into synchronous subgraphs connected by asynchronous arcs. These subgraphs can then be scheduled on different processing elements using an asynchronous interprocessor communication protocol. The asynchronous links are then handled by the dataflow scheduler. Functional LGDF node with more than one parameter may be strict or non-strict in each parameter independently, as well as jointly strict or non-strict in several parameters simultaneously.

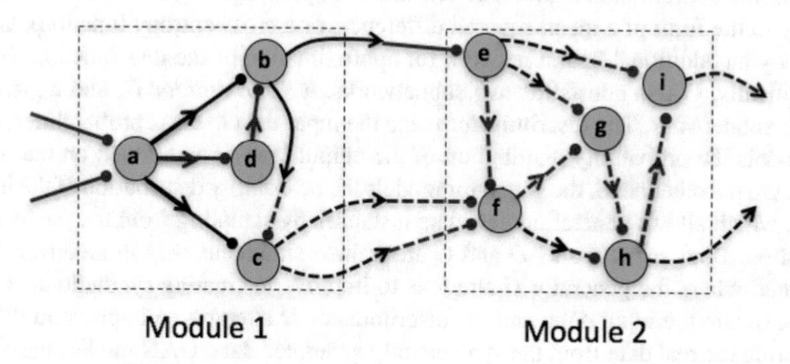

Fig. 6 A strict LGDF execution graph

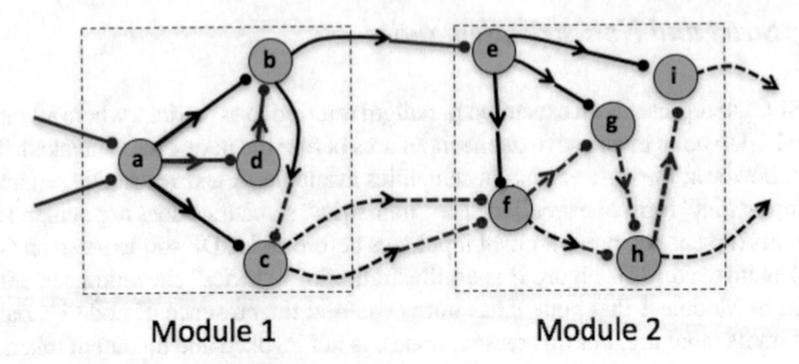

Fig. 7 A non-strict LGDF execution graph

4 Preliminary Results

The proposed approach employs the neural networks architecture Generative Adversarial Networks (GAN), proposed by Goodfellow et al. [20]. In the domain of cyber security, the proposed approach employs a GAN architecture for detecting intrusions in communication networks from unauthorized users. The objective is to initially train the neural network to detect intrusions in a supervised manner based on labeled data, and over time the network to learn detecting new intrusion patterns that were not explicitly taught and labeled.

4.1 Generative Adversarial Networks

Unlike the conventional neural networks trained by minimizing a loss function (typically in the form of a mean-squared difference, or a cross-entropy function), GAN employ an additional neural network for approximation of the loss function. More specifically, GANs consist of two subnetworks: a *discriminator D*, and a *generator G* subnetwork. The discriminator maps the input data to class probabilities, i.e., it models the probability distribution of the output labels conditioned on the input data. On the other hand, the generator models the probability distribution of the input data, which allows generating new data instances by sampling from the model distribution. Both subnetworks D and G are trained simultaneously in an adversarial manner, where the generator G attempts to improve in creating synthetic data that approximate the input data, and the discriminator D attempts to improve in differentiating the real data from the synthetically generated data. GAN models have had a tremendous success in the domain of image processing, e.g., for generating super resolution photo-realistic images from text [37], face aging images in entertainment [2], blending of objects from one picture into the background of another picture, as well as in other applications, such as generating hand-written text, and music sequence generation [24].

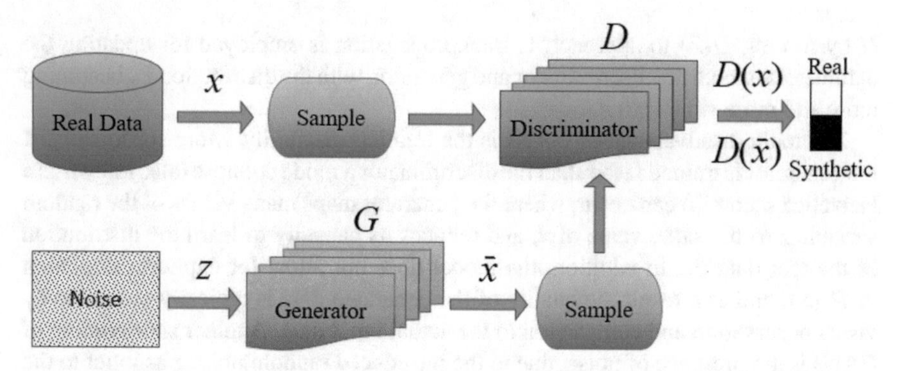

Fig. 8 A GAN model consists of a generator and a discriminator

Let's use x to denote the inputs to the network, where $x \sim \mathbb{P}_r$ and $\mathbb{P}_r$ denotes the probability distribution of the real input data. The goal of the generator in GANs is to learn a model distribution $\mathbb{P}_g$ that approximates the unknown distribution of the real data $\mathbb{P}_r$. For that purpose, a random variable z sampled from a fixed (e.g., uniform or Gaussian) probability distribution is used as the input to the generator, as illustrated in Fig. 8. During the training phase, the parameters of the generator are iteratively varied in order to reduce the distance, or divergence, between the distributions $\mathbb{P}_g$ and $\mathbb{P}_r$. The output of the generator is denoted $\bar{x}$ here, i.e., the generator mapping is $G : z \mapsto \bar{x}$. To solve the described problem, a network loss function H is introduced in the form of a cross-entropy,

$$H(D, G) = \mathop{\mathbb{E}}_{x \sim \mathbb{P}_r} \big[\log(D(x))\big] + \mathop{\mathbb{E}}_{\bar{x} \sim \mathbb{P}_g} \big[\log(1 - D(\bar{x}))\big]. \tag{4}$$

The discriminator is trained to maximize the loss function H, and the generator is trained to minimize the loss function H, i.e.,

$$\min_G \max_D H(D, G). \tag{5}$$

In the game theory this is called a minimax game. The two subnetworks are trained in a competitive two-player scenario, where both the generator and discriminator improve their performance until a Nash equilibrium is reached. One can note that minimizing the function in (4) is equivalent to minimizing the Jensen-Shannon (JS) divergence between the real data distribution $\mathbb{P}_r$ and the model distribution $\mathbb{P}_g$.

In the case of a binary classification, the discriminator is trained to maximize H by forcing $D(x)$ to approach 1 and $D(\bar{x})$ to approach 0 (Fig. 8). The generator takes random noise as input and attempts to produce synthetic data that resemble the real data. The discriminator attempts to discriminate real data from the synthetic data produced by the generator. Contrarily, the generator is trained to minimize

H by forcing $D(\bar{x})$ to approach 1. Backpropagation is employed for updating the parameters of both the discriminator and generator, with the distribution $\mathbb{P}_g$ becoming more and more similar to $\mathbb{P}_r$.

The main disadvantage of GANs is the training instability. More specifically, if the generator is trained faster than the discriminator a mode collapse (also known as a Helvetica scenario) can occur, where the generator maps many values of the random variable z to the same value of x, and reduces its capacity to learn the distribution of the real data $\mathbb{P}_r$. In addition, the model does not allow for explicit calculation of $\mathcal{P}_g(x)$, and as a result the quality of the generated data is typically evaluated by visual observation and comparison to the actual input data. Another shortcoming of GANs is the presence of noise, due to the introduced random noise z as input to the generator.

A number of variants of GANs have been proposed since the original work, which have addressed some of the above shortcomings [4, 8, 45, 54], or have been designed for domain-specific solutions [15, 47]. E.g., deep convolutional GANs (DCGANs) [54] introduce several constraints and modifications to the original GAN architecture for improved stability and performance. As the name implies, the generator and discriminator subnetworks are composed of multiple layers of convolutional computational units, as opposed to the multilayer perceptron (MLP) networks proposed in the original GAN paper [20]. The modifications in DCGANs include replacing pooling layers with strided convolutions, elimination of fully connected layers, batch normalization applied to all layers, and use of ReLU and Tanh activation function in particular layers in the network. By applying the above recommendations, the authors have demonstrated improved classification performance on various datasets of images, and capabilities of generating complex and visually realistic images. Then, Wasserstein GANs (WGANs) [4] introduced a new loss function for training the generator and discriminator subnetworks. The loss function is based on the Wasserstein distance (also known as Earth Mover distance) between the real data distribution $\mathbb{P}_r$ and the model distribution $\mathbb{P}_g$ learned by the generator. Such distance function induces a weaker topology than the Jensen-Shannon (JS) divergence used in the original GANs and given in (4), and the Kullback-Leibler (KL) divergence commonly used in maximum likelihood estimation. The weaker topology provides a lever for the convergence of the probability distribution of the model $\mathbb{P}_g$ to the real distribution of the data $\mathbb{P}_r$. If the discriminator $D(x)$ is a K-Lipschitz function, it was proven that the proposed loss function is continuous and differentiable, and produces stable gradients during training, thereby improving the problem of training instability in GANs. Other forms of loss functions in GANs include least-squares loss for the discriminator [45], lower bound of the Wasserstein distance and use of an autoencoder as a discriminator [8], and introduction of gradient norm penalty [29], among others.

4.2 Application of the KDD Cup 1999 Dataset

For a preliminary evaluation of the proposed approach, the KDD Cup 1999 Dataset [27] was employed. We used 10% of the entire dataset, which consists of a train set with 494,021 data packages (97,278 normal data packages and 396,743 attack data packages) and a test set with 311,029 data packages (60,593 normal data packages and 250,536 attack data packages). The data packages include 41 different features, such as duration of the connection, type of the protocol, network service on the destination, number of data bytes from source to destination, number of data bytes from destination to source, etc. All 41 features are described on the web site hosting the data. The train dataset has 24 types of attacks, and 14 additional types of attacks occur only in the test set, and not in the train set.

The architecture of the used GAN network consists of a discriminative subnet with three hidden layers of 20, 10 and 5 computational units, and an output layer of 1 unit. The discriminative subnet classifies the data packages into a normal or attack class. The generative network has an input layer of 200 computational units, three hidden layer of 150, 100 and 60 computational units, and an output layer of 41 computational units. A dropout rate of 0.5 was set for the hidden layers to avoid overfitting. The architecture of the employed GAN network is shown in Fig. 9.

The input data were scaled between 0 and 1. We used Adam optimization method for training the subnets, with a learning rate of 0.01 for the discriminative subnet, and a learning rate of 0.001 for the generative subnet and for the GAN network. The gradient information for the training of generative and discriminative models was calculated with backpropagation. The network was trained for 3,000 epochs,

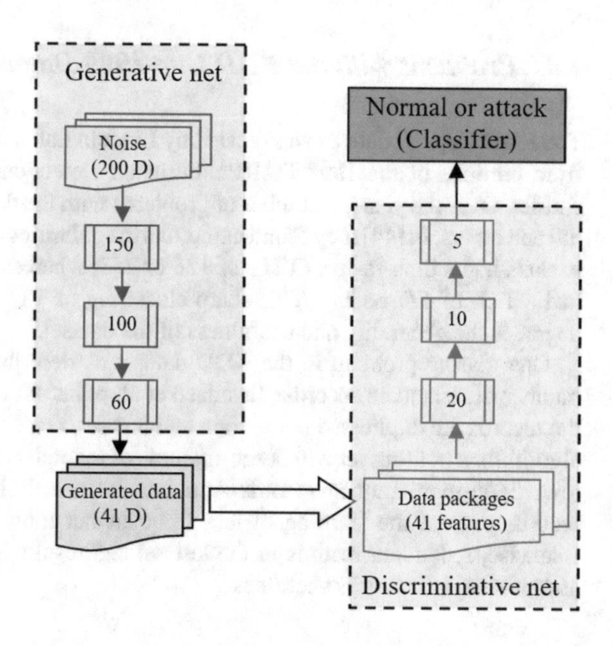

Fig. 9 GAN network architecture

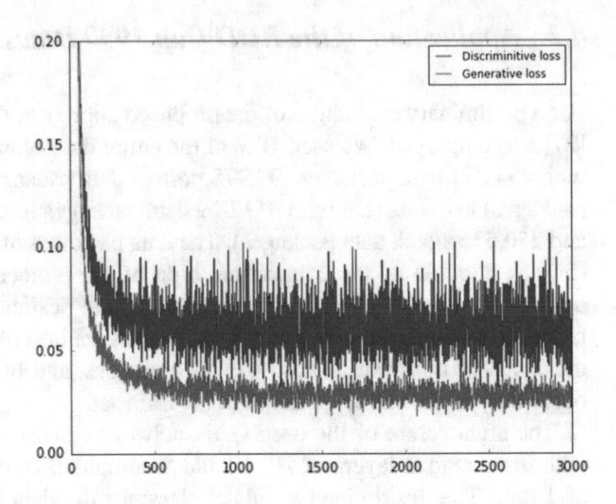

Fig. 10 Loss of the two subnets

with a batch size of 1,000 randomly selected data packages per epoch. Both normal and attack packages with their known labels were used to train the discriminative network. The generative network was trained to generate data packages that were drawn from the same distribution as the normal data packages in the training set. The loss of the two subnets is shown in the Fig. 10.

The accuracy of the train set was 99.22%, and the accuracy of the test set was 91.83%. For the test set, the network correctly classified 98.57% of the normal packages, and 90.20% of the attack packages.

4.3 Problems with the KDD Cup 1999 Dataset

The KDD Cup '99 dataset was created by Lincoln Lab under a DARPA contract [42] from portions of the 1998 DARPA Intrusion Detection System (IDS) Evaluation dataset. Over the years, a number of problems with the dataset are described by [46, 55] and others. In [44] they found numerous irregularities and found that all malicious packets had a time-to-live (TTL) of 126 or 253; whereas, almost all benign packets had a TTL of 127 or 254. This sharp clustering of TTL is not only unrealistic but degrades the generality and usefulness of the dataset.

One glaring problem in the KDD dataset as described in [46, 61] is the large number of redundant records. Tavallaee et al. point out that about 78% and 75% of the records are duplicated in the train and test set, respectively. In a typical learning algorithm, a training set with large amounts of redundant records will bias discovery towards the more frequent records. As a consequence, the heavily weighted redundant records prevent the learning of less frequent but more harmful network records. Likewise, redundant records in the test set can results in high validation accuracy and can, thus, distort its usefulness.

Although the KDD Cup '99 dataset is burdened with problems, it is still a convenient, large-scale open source dataset that reveals aspect of a networks' traffic flow. As such, IDS and knowledge discovery researchers continue to demonstrate their new and novel learning algorithms within the network intrusion detection domain. Unfortunately, given the problems with the dataset, it is difficult to project the value of their research.

It should be noted, that the intrinsic problem of KDD data set has motivated the development of new NSL-KDD data set [51]. The NSL-KDD is based on KDD Cup '99 but it has overcome many problems like redundant records.

4.4 A Synthetic Benchmark Using KDD Cup '99

The KDD cup '99 dataset provides aspects of a network's traffic flow that is readily available for open source access and for the verification of corresponding published results. However, rather than developing new algorithms to extract malicious network traffic attacks within the dataset during the learning process, we propose a much different use of the KDD cup '99 dataset. In this approach, the training set is void of malicious traffic packets. Instead we clean (scrape) the training set by removing all non-benign traffic records. The resulting scraped training set is used to learn "normal" or non-benign traffic patterns. In using this approach, an artificial neural network (ANN) is trained to learn "normal" network traffic behavior. Once trained, the ANN is now able to identify all normal traffic and reject all unclassified traffic using a Softmax or other recognition approach.

The designed network is a deep learning network with an input layer, four hidden layers and one output layer. The rectified linear unit (RELU) activation function is used for all the layers except the final output layer where the Softmax activation function is used. 41 input features are presented to the input layer. The four hidden layers are populated by 10, 50, 10 and 1 neurons, respectively. The cross entropy function is used in classification. The overfitting problem is addressed by employing "Early Stopping" which is a very simple and effective technique used by most of the people in deep neural networks. Early stopping is performed on a validation dataset. Also the "Adam" optimizer algorithm, an extension to the stochastic gradient descent, is used to update network weights.

The preprocessing of the KDD cup dataset required several steps. It was noted that the dataset had some missing fields of data which required encoding to match the datadet. A final step normalized features contains large data values against others with small data values.

The application of early stopping and the cross entropy loss function was used to minimize the loss and increase the accuracy. By applying the early stopping technique, the network averaged 27 epochs for convergence. The accuracy in detecting abnormal behavior is about 99.79% and the loss is about 0.01644.

4.5 An Industrial Data Network Application

One of the most stringent industrial IC networks are defined by the nuclear power industry with input and guidance from the United States Nuclear Regulatory Commission. As a result, the Code of Federal Regulations (CFR) [14] requires that licensees provide high assurance that digital computer, communication systems and networks are adequately protected against cyber attacks, and that each facility currently licensed to operate a nuclear power plant is required to submit a cyber security plan for Commission review and approval. The approval of such plans is outlined by the defense-in-depth protective strategies describe in NEI 08-09 [49]. These strategies are designed to ensure the capability to detect, delay, respond to, and recover from cyber attacks on Critical Digital Assets (CDAs). The realization of these strategies are embodied in the defensive security architecture that provides for cyber security defensive levels that are separated by secure data boundaries (such as firewalls and diodes), where digital communications are monitored and restricted. In addition, data flow rules are defined and enforced between defensive levels that include the criteria that the device (firewall, data diode) will apply to enforce the data flow rule.

The following is an example to illustrate the network data flow layout from an actual commercial nuclear power plant that complies with the NEI 08-09 defensive architecture. The data layout describes an actual process data and cyber security monitoring data from the commercial power plant. The process data results from two monitoring systems and three control systems. The cyber security monitoring data is from two separate cyber security monitoring systems. The real-time systems in the plant are grouped into multiple cyber security layers as defined by NEI 08-09. The data layout is partitioned into four layers. Layer 1 represents the Internet, Layer 2 represents the local plant LAN, Layer 3 represents the monitoring systems in the nuclear power plant while Layer 4 includes the automatic and manual control systems. Layers 3 and 4 provide the cyber security levels dictated by the defensive architecture.

The two monitoring systems are located in Cyber Security Layer 3 while the three control systems are located in Cyber Security Layer 4. The two monitoring systems are monitor-only and collects real-time data from the plant control systems and other plant sensors and present the information to the control room operators. The three control systems collect real-time input data from various plant sensors and provide automatic and manual control actions to plant equipment. The cybersecurity equipment located in each layer provide time synchronization for all of the equipment in that cybersecurity layer via the Network Time Protocol (NTP). The equipment also monitors the equipment in the cybersecurity layer by collecting operational log data and collecting the traffic from all networking equipment located in that cybersecurity layer. This data is available to the cybersecurity monitoring systems via raw network packet traces.

The commercial nuclear power plant's defensive architecture meets the approved criteria of the NEI 08-09 specifications. Specifically, layers 3 and 4, employ important cybersecurity devices to defend against attacks on the collection of monitored data and against the disruption of automatic and manual control actions during the plant's operation. These layers represent the cyber-hardened network layers of the defensive architecture. The architecture, as mentioned above, is partitioned between cyber security levels (Layer 3 and 4) by the use of data diodes (i.e., a unidirectional flow of network data traffic), which is referred to as a 'deterministic' boundary; or the use of a firewall, which is referred to as a 'non deterministic' boundary. Time synchronization for all of the equipment within each cybersecurity layer (via the Network Time Protocol (NTP)) provides, in part, a chronological ordering of events for post forensic analysis. Additionally, data flow rules are used to govern the enforcement of data motion across these boundaries.

As inferred throughout this paper, artificial intelligence (AI) through machine learning has not reached a level of maturity to defend against new and provocative real-time cyber attacks. Data flow rules, defined and enforced between defensive levels, are subject to an aggregation of historical classification of events and procedures. Although extensive, the data flow rules are limited, for this reason, in scope and inflexible in addressing real-time, dynamic changes that can affect the data flow within the network layers.

The defensive architecture or 'fortress' provides the structural framework in an attempt to ensure reasonable resilience against 'nominal' cyber attacks. What the fortress lack, however, is an infrastructure that provides operational feedback from the command center (expert human operators) to 'optimally' manage the resources provided by the defensive architecture. In a metaphorical sense, the success of warding off a malicious attack on a fortress or castle depends not only upon the pre-planned structural resources of the fortress but also (and most importantly) upon the skills of the commanding hierarchy. The human operator-in-the-loop is not a new concept but is one that should not be ignore within a critical infrastructure.

The NEI 08-09 criteria provide a systematic approach to the design of a defensive architecture that adheres to the principles of the defense-in-depth protective strategies. Even if these strategies are properly realized in the deployed architecture, and even if 99.9% security assurance is achievable, it is that 0.1% uncertainty if exploited that can have a devastation consequence on the economic, financial and political institutions of a nation. So that the protection from a one-time rare cyber attack is crucial to the nation's survival.

5 Conclusion

The proposed cyber-secure, operator-in-the-loop architecture offers a logical and reasonable approach to supply-chain IoT management. The proposed strategies integrate various learning techniques that offer capabilities to isolate or mitigate the consequences of cyber threats in an instrumentation and control environments. The realization of the proposed concepts will have the effect to lower the costs of protection and compliance, and provide a collaborative pathway for industry and research partnerships.

The use of dataflow semantics and in particular the Large Grain Data Flow concept, provides self-synchronizing network throttling to trap system anomalies before they entire their intended Instrumentation and Control targets. Alerts generated by these anomalies are intercepted by an operator-monitoring system for a human operator to assess the likelihood that the anomaly is a potential treat. The operator, upon cross evaluation of the threat, may designate the anomaly as a real threat, a potential threat marking it for future occurring detection, or as an anomaly that is an unclassified yet 'normal' behavior. Based on the operator's insight, the appropriate updates to the classification database will be made.

The article presents preliminary evaluation results using a machine learning approach based on the GAN neural network. The implemented architecture was able to correctly classify 91.83% of the test data set into normal and abnormal, i.e., attack data packages. However, a synthetic benchmark was constructed to demonstrate the advantages in knowledge discovery when first the ANN is confined to classify 'normal' behavior (events) and then the resulting architecture is applied to differentiate between 'normal' and 'abnormal' inputs. The KDD Cup '99 training set, which is scraped (cleaned) of malicious data packets, is used in this synthetic example. Without a change in the algorithm, the recognition of abnormal or malicious network data packets was correctly classified at 99.7%.

Although, the use of a GAN neural network is proposed, the more important lesson to be learned is not the need for new algorithms to solve KDD-like problems but the rethinking of what simplification of the input training set is more directly appropriate for the desired solution.

One shortcoming of using the KDD Cup dataset for the proposed approach is that the data packages are not temporally ordered, i.e., the data is static. For the future work, we will consider temporally ordered data packages, where the generative network will learn the distribution of the normal data, and will attempt to generate data samples that correspond to the normal network traffic. The goal is that the discriminative network learns to distinguish between the normal and attack data packages. As the data continues to flow and evolves over time, the generative network will learn the new normal data and will generate samples that emulate the current normal data, and the discriminative network will learn to distinguish between the evolving normal data packages and the attack packages.

Acknowledgements We gratefully acknowledge the support of NVIDIA Corporation with the donation of the Titan XP GPU used for this research.

References

1. Amiri, F., Mahdi, M., Yousefi, R., Lucas, C., Shakery, A., Yazdani, N.: Mutual information-based feature selection for IDSs. J. Netw. Comput. Appl. **34**(4), 1184–1199 (2011)
2. Antipov, G., Baccouche, M., Dugelay, J.L.: Face aging with conditional generative adversarial networks. In Proceedings of IEEE International Conference on Image Processing (ICIP), pp. 2089–2093 (September 2017)

3. Ariu, D., Tronci, R., Giacinto, G.: HMMPayl: an intrusion detection system based on hidden Markov models. Comput. Secur. **30**(4), 221–241 (2011)
4. Arjovsky, M., Chintala, S., Bottou, L.: Wasserstein generative adversarial networks. In: Proceedings of International Conference on Machine Learning (ICML) (2017)
5. Babb II, R.G.: Parallel processing with large-grain data flow techniques. Computer **17**(7), 55–61 (1984)
6. Babb II, R.G., Storc, L., Hiromoto, R.E.: Developing a parallel Monte Carlo transport algorithm using large-gramin dataflow. Parallel Comput. **7**(2), 187–198 (1988)
7. Bejtlich, R.: The Practice of Network Security Monitoring: Understanding Incident Detection and Response, 1st edn. No Starch Press (5 August 2013). ISBN-10:1593275099, ISBN-13:978-1593275099
8. Berthelot, D., Schumm, T., Metz, L.: BEGAN: boundary equilibrium generative adversarial networks (2017). arXiv:1703.10717
9. Bivens, A., Palagiri, C., Smith, R., Szymanski, B., Embrechts, M.: Network-based intrusion detection using neural networks. Intell. Eng. Syst. Artif. Neural Netw. **12**(1), 579–584 (2002)
10. Bhuyan, M.H., Bhattacharyya, D.K., Kalita, J.K.: Network anomaly detection: methods, systems and tools. IEEE Commun. Surv. Tutor. **16**(1), 303–336 (2014)
11. Buczak, A.L., Guven, E.: A survey of data mining and machine learning methods for cyber security intrusion detection. IEEE Commun. Surv. Tutor. **18**(2), 1153–1176 (2016)
12. CAIDA: The Cooperative Analysis for Internet Data Analysis (2011). http://www.caida.org
13. Cannady, J.: Artificial neural networks for misuse detection. In: Proceedings of 1998 National Information Systems Security Conference, Arlington, VA, USA, pp. 443–456 (1998)
14. Title 10 of the Code of Federal Regulations (10 CFR), Last Reviewed/Updated Friday, July 06, 2018
15. Chen, X., Duan, Y., Houthooft, R., Schulman, J., Sutskever, I., Abbeel, P.: InfoGAN: interpretable representation learning by information maximizing generative adversarial nets (2016). arXiv:1606.03657v1 [cs.LG]
16. Defcon: The Shmoo Group (2011). http://cctf.shmoo.com/
17. Dennis, J.B., Misunas, D.P.: A computer architecture for highly parallel signal processing. In: Proceedings of 1974 National Computer Conference, pp. 402–409 (1974)
18. Fakhraei, S., Foulds, J., Shashanka, M., Getoor, L.: Collective spammer detection in evolving multi-relational social networks. In: 21th ACM SIGKDD International Conference on Knowledge Discovery and Data Mining (KDD) (2015)
19. Fakhraei, S., Sridhar, D., Pujara, J., Getoor, L.: Adaptive neighborhood graph construction for inference in multi-relational networks. In: 12th KDD Workshop on Mining and Learning with Graphs (MLG) (2016)
20. Goodfellow, I.J., Pouget-Abadiey, J., Mirza, M., Xu, B., Warde-Farley, D., Ozairz, S., Courville, A., Bengio, Y.: Generative adversarial nets (2014). arXiv:1406.2661
21. Gupta, R.K.: Co-synthesis of hardware and software for digital embedded systems, vol. 329. Springer (2012)
22. A brief introduction to Haskell. https://wiki.haskell.org/A_brief_introduction_to_Haskell. Last modified 29 Oct 2011
23. He, K., Zhang, X., Ren, S., Sun, J.: Deep residual learning for image recognition (2015). asXiv:1512.03385
24. Hong, Y., Hwang, U., Yoo, J., Yoon, S.: How generative adversarial nets and its variants work: an overview of GAN (2018). arXiv:1711.05914v6 [cs.LG]
25. I. S. T. G. MIT Lincoln Lab: DARPA Intrusion Detection Data Sets. http://www.ll.mit.edu/mission/communications/ist/corpora/ideval/data/2000data.html, March 2000
26. Johnston, W.M., Paul Hanna, J.R., Millar, R.J.: Advances in dataflow programming languages. ACM Comput. Surv. **36**(1), 1–34 (March 2004)
27. KDD Cup 1999. http://kdd.ics.uci.edu/databases/kddcup99/
28. Kim, J., Kim, J., Thu, H.L.T., Kim, H.: Long short term memory recurrent neural network classifier for intrusion detection. In: International Conference on Platform Technology and Service, pp. 1–5 (2016)

29. Kodali, N., Abernethy, J., Hays, J., Kira, Z.: On convergence and stability of GANs (2017). arXiv:1705.07215
30. Köpf, B., Basin, D.: An information-theoretic model for adaptive side-channel attacks. In: Proceedings of the 14th ACM conference on Computer and communications security, pp. 286–296. ACM (October 2007)
31. Kruegel, C., Toth, T.: Using decision trees to improve signature based intrusion detection. In: Proceedings of the 6th International Workshop Recent Advances in Intrusion Detection, West Lafayette, IN, USA, 2003, pp. 173–191
32. Kuntsevich, V.M., Gubarev, V.F., Kondratenko, Y.P., Lebedev, D.V., Lysenko, V.P. (eds.): Control systems: theory and applications. In: Series in Automation, Control and Robotics. River Publishers (2018)
33. Kuntsevich, V.M.: Control Under Uncertainty: Guaranteed Results in Control and Identification Problems. Naukova Dumka, Kyiv (2006). (in Russian)
34. Kushner, D.: The real story of stuxnet. In: IEEE Spectrum (26 February 2013)
35. Kwon, D., Kim, H., Kim, J., Suh, S.C., Kim, I., Kim, K.J.: A survey of deep learning-based network anomaly detection. In: Cluster Computing, The Journal of Networks, Software Tools and Applications, pp. 1–13. Springer US (2017). https://doi.org/10.1007/s10586-017-1117-8
36. LBNL: Lawrence Berkeley National Laboratory and ICSI, LBNL/ICSI Enterprise Tracing Project (2015). http://www.icir.org/enterprisetracing/
37. Ledig, C., Theis, L., Huszar, F., Caballero, J., Cunningham, A., Acosta, A., et al.: Photo-realistic single image super-resolution using a generative adversarial network. In: Proceedings of IEEE Conference on Computer Vision and Pattern Recognition (CVPR), pp. 105–114 (July 2017)
38. Lee, E.A., Messerschmitt, D.G.: Static scheduling of synchronous data flow programs for digital signal processing. IEEE Trans. Comput. **C-36**(2), 24–35 (1987)
39. Lee, E.A.: What's ahead for embedded software? Computer **33**(9), 18–26
40. Jemili, F., Zaghdoud, M., Ben, A.: A framework for an adaptive intrusion detection system using Bayesian network. In: Proceedings of IEEE Intelligence and Security Informatics, pp. 66–70 (2007)
41. Li, Y., Ma, R., Jiao, R.: Hybrid malicious code detection method based on deep learning. Int. J. Secur.Appl. **9**(5), 205–216 (2015)
42. Lippmann, R.P., Fried, D.J, Graf, I.: Evaluating intrusion detection systems: the 1998 DARPA off-line intrusion detection evaluation. In: Proceedings of the 2000 DARPA Information Survivability Conference and Exposition (DISCEX'00) (2000)
43. Luo, J., Bridges, S.: Mining fuzzy association rules and fuzzy frequency episodes for intrusion detection. Int. J. Intell. Syst. **15**(8), 687–703 (2000)
44. Mahoney, M.V., Chan, P.K.: An analysis of the 1999 DARPA/Lincoln Laboratory evaluation data for network anomaly detection. In: Vigna. G., Jonsson, E., Krugel, C. (eds.) Proceedings of 6th International Symposium on Recent Advances in Intrusion Detection (RAID 2003), Lecture Notes in Computer Science, Pittsburgh, vol. 2820, PA, pp. 220–237. Springer (8–10 September 2003)
45. Mao, X., Li, Q., Xie, H., Lau, R., Wang, Z., Smolley, S.P.: Least squares generative adversarial network (2016) arXiv:1611.04076
46. McHugh, J.: Testing intrusion detection systems: a critique of the 1998 and 1999 darpa intrusion detection system evaluations as performed by lincoln laboratory. ACM Trans. Inf. Syst. Secur. **3**(4), 262–294 (2000)
47. Mirza, M., Osindero, S.: Conditional generate adversarial nets (2014). arXiv:1411.1784v1 [cs.LG]
48. Munson, J., Krings, A., Hiromoto, R.E.: The architecture of a reliable software monitoring system for embedded software systems. In: ANS 2006 Winter Meeting and Nuclear Technology Expo, Albuquerque New Mexico, November 12–16 (2006)
49. Cyber Security Plan for Nuclear Power Reactors [Rev. 6], Nuclear Energy Institute, 1776 I Street N. W., Suite 400, Washington D.C. (202.739.8000) (April 2010)
50. Niyaz, Q., Sun, W., Javaid, A.Y., Alam, M.: A deep learning approach for network intrusion detection system. In: International Conference on Bio-Inspired Information and Communications Technologies, pp. 1–11 (2016)

51. Nsl-kdd data set for network-based intrusion detection systems. http://nsl.cs.unb.ca/NSL-KDD/, November 2014
52. Ota, N., Wright, P.: Trends in wireless sensor networks for manufacturing. Int. J. Manuf. Res. **1**(1), 3–17 (2006)
53. Pinto, R., Mettler, T., Taisch, M.: Managing supplier delivery reliability risk under limited information: foundations for a human-in-the-loop DSS. Decis. Support Syst. **54**(2), 1076–1084 (2013)
54. Radford, A., Metz, L., Chintala, S.: Unsupervised representation learning with deep convolutional generative adversarial networks (2016). arXiv:1511.06434v2 [cs.LG]
55. Revathi, S., Malathi, A.: A detailed analysis of KDD cup99 dataset for IDS. Int. J. Eng. Res. Technol. (IJERT) **2**(12) (December 2013)
56. Salama, M.A., Eid, H.F., Ramadan, R.A., Darwish, A., Hassanien, A.E.: Hybrid intelligent intrusion detection scheme. Soft Comput. Ind. Appl., 293–303 (2011)
57. Sequeira, K., Zaki, M.: ADMIT: anomaly-based data mining for intrusions. In: Proceedings of 8th ACM SIGKDD International Conference on Knowledge discovery and data mining, pp. 386–395 (2002)
58. Standaert, F.-X., Malkin, T.G., Yung, M.: A unified framework for the analysis of side-channel key recovery attacks. In: Annual International Conference on the Theory and Applications of Cryptographic Techniques, pp. 446–461. Springer, Berlin, Heidelberg (2009)
59. Tang, T.A., Mhamdi, L., McLernon, D., Zaidi, S.A.R., Ghogho, M.: Deep learning approach for network intrusion detection in software defined networking. In: International Conference on Wireless Networks and Mobile Communications, pp. 1–6 (2016)
60. Tao, X., Kong, D., Wei, Y., Wang, Y.: A big network traffic data fusion approach based on Fisher and deep auto-encoder. Information **7**(20), 1–10 (2016)
61. Tavallaee, M., Bagheri, E., Lu, W., Ghorbani, A.A.: A detailed analysis of the KDD CUP 99 data set. In: Proceedings of the IEEE Symposium on Computational Intelligence in Security and Defense Applications (CISDA 2009), pp. 1–6 (2009)
62. UNIBS: University of Brescia dataset (2009). http://www.ing.unibs.it/ntw/tools/traces/
63. U.S. Department of Homeland Security: Strategic Principles for Securing the Internet of Things (IoT), Version 1.0 (15 November 2016)
64. U.S. Federal Trade Commission Report: Internet of Things: Privacy & Security in a Connected World (2016)
65. VxWorks Programmers Guide: Wind River Systems, Almeda, CA (1997)
66. Whittaker, Z.: Mirai botnet attack hits thousands of home routers, throwing users offline. ZDNet (29 November 2016)
67. Zhang, J., Zulkernine, M., Haque, A.: Random-forests-based network intrusion detection systems. IEEE Trans. Syst. Man Cybern. C Appl. Rev. **38**(5), 649–659 (2008)

Formal Concept Analysis for Partner Selection in Collaborative Simulation Training

Marina Solesvik, Yuriy P. Kondratenko, Igor Atamanyuk and Odd Jarl Borch

Abstract In this paper, we propose to apply formal concept analysis (FCA) technique to partner selection for joint simulator training. Simulator training is an important tool for preparation of seafarers in a classroom. However, the costs to acquire simulator centers is high. That is why the educational institutions and centers can only afford limited number of simulators. Though the price for simulator classes are quite high, it still more advantageous to train sailors in class than offshore. New cloud-based technologies allow to connect simulators situated in different places, including cross-country communication. This makes it possible to carry out joint training of different simulator types. In the study, we elaborate the approach of formal concept analysis is used to facilitate partner selection from the pool of potential collaborating institutions.

Keywords Partner selection · Formal concept analysis · Fuzzy logic · Simulation technology

M. Solesvik (✉) · O. J. Borch
Nord University Business School, Nord University, Bodø, Norway
e-mail: marina.solesvik@hvl.no

O. J. Borch
e-mail: odd.borch@nord.no

Y. P. Kondratenko
Petro Mohyla Black Sea National University, Mykolaiv, Ukraine
e-mail: yuriy.kondratenko@chmnu.edu.ua

I. Atamanyuk
Mykolaiv National Agrarian University, Mykolayiv, Ukraine
e-mail: atamanyuk@mnau.edu.ua

Y. P. Kondratenko et al. (eds.), *Advanced Control Techniques in Complex Engineering Systems: Theory and Applications*, Studies in Systems, Decision and Control 203, https://doi.org/10.1007/978-3-030-21927-7_15

1 Introduction

Modern ICT technologies assist to manage and control different functional areas in various industries [1–3]. Amongst the key priority areas of the Norwegian maritime industry, is digitalization [4]. Modern ICT technologies can help to open unique possibilities for a new era of functional solutions in the maritime sector from remotely operated vessels to autonomous ships. The interaction between shore and sea-based services (coordination and control, including planning, testing and training the elements of such services) is an area of innovation [5]. Simulator technology can strongly connect pure virtual simulations to real-time remote control over the vessels [6]. There are shipping companies that cope through different challenges, such as operational planning and continuous training [7]. Long distances to the education infrastructure, together with extra demands to planning, competences, and the renewal of certificates call for flexible planning and training platforms. The goal of this study is to explore the challenges and opportunities related to partner selection for joint training using virtual maritime simulators.

An open, secure and collaborative simulator platform is created to assist shipping companies and educational institutions to carry out digital training. The added value is related to fast and flexible competence development through training opportunities and simulator possibilities that may facilitate safe navigation and increase operational efficiency. In the case of emergencies, the solutions may also serve as a platform for joint operations. Shore, air and sea-based units are required to cooperate closely between each other when responding to crisis and emergencies. Different and more efficient solutions that may help during multi-service operations may be found using the simulator resources. These may also benefit in training for both shore and sea-based personnel. Joint digital simulation platforms allow the faculty at universities, training centres and real vessels in operation to work together. This will facilitate shared simulation activities. The solution will help various education systems and commercial simulator centres, promoting university-business cooperation [4]. This may also increase the amount of available assets for shipping companies and rescue services, in case of emergencies through integrating computer systems and this may deliver optimal emergency response solutions. The key research question of this research is: How can formal concept analysis methodology be used for partner selection for joint digital simulator training?

Selection of the best suitable partner university might be tough since responsible people might not be know to each other and not know facilities of prospective partners. Under the conditions of incomplete information about possible partners decisions making can be difficult and lead to selection of the wrong partners [8]. Fuzzy logic is a tremendous tool to simplify decision-making under the conditions of legal, environmental, market, and political uncertainty that confront organizations [9–15]. Hence, the aim of this study is to use a novel mathematical method of formal concept analysis (FCA) to partner selection for joint virtual simulator training connecting educational institutions worldwide. The novel contribution of this study is utilization of mathematical method of formal concept analysis that can be used by university managers to select partners with right facilities and staff for joint training.

The paper is structured as follows. In the next section, we elaborate the theoretical background of the paper related to joint digital simulation and partner selection. Then we present the formal concept analysis approach. In the next section, we illustrate the proposed approach with an illustrative example that shows how the FCA can be applied to facilitate a selection of partner for joint simulator training. The paper ends with discussion of results and conclusions.

2 Theoretical Background

2.1 Collaborative Digital Simulation

Maritime firms wish to have highly skilled crews on board of their fleet. The human factor is the main cause of catastrophes in the sea. A number of shipping firms have established links with maritime universities and controls the quality of navigational education and preparation during the whole cycle of the maritime training and provide trainee places for the best students [16]. Navigational simulators became an important element of basic education of young sailors, as well as update and new course preparation of experienced vessel crews [7, 17, 18]. Notably, navigational simulators are quite expensive to buy for universities. The price of one simulator class for 20–25 computers starts from USD 1 million. Furthermore, this is the price of only one simulator type (for example, navigational). Other simulators include engine room simulators, fire fighting simulators, oil spill recovery simulators, dynamic positioning simulators, and others. A single university might not (a) have demand for all possible simulator types; (b) cannot afford the whole range of simulators. An innovative digital platform can permit connecting independent simulation equipment situated in different locations.

The first virtual simulator campus is built recently in the Northern Norway [6, 7]. The simulator park connects simulators of two navigation schools and one university situated close to each other. The three educational institutions have long-term collaboration relations and know each other for a long time. The University and navigation schools have different maritime simulators related to navigation and emergency modules. One navigation school bought the oil spill preparedness simulator. The second navigation school acquired a dynamic positioning simulator for operations in emergency situations and the University has a brand new ship's bridge simulator with visual scenes, command and control systems as well as log systems. This set is fundamental in crisis management education and training. These simulator facilities are linked via LAN. The next step is to connect them via cloud-based facilities. This will enable data gathering and calculation that will be used to connect simulators for joint training related to navigation, safety, security and other decision-support systems on board.

The joint simulator platform can be also used for cross-industry collaboration [19, 20], as well as a platform for joint search and rescue (SAR) operations. It might work at all stages starting from data collection to data storage and processing. A centralized data collection and computation will be made via the cloud-based system.

The virtual simulator system will enable:

(1) Safe navigation planning through the simulation of itineraries employing integrated simulator-navigation scheduling systems.
(2) Efficient shore-vessel coordination and decision-making by connecting different subsystems in the cloud.
(3) SAR and oil spill response operations will be coordinated via cloud-based simulation systems.

Currently, LAN-based system is tested and integrated into the educational systems at three educational institutions situated in the same region. The next step will be to build and test cloud-based system that will facilitate the connection of all simulators worldwide interested in joint training. The virtual digital platform can also be used by vessel crews to acquire new skills related to offshore operations [21]. The collaborative digital simulation platform will also facilitate the digital twin ship development.

There are many maritime educational institutions worldwide, and many of them are willing to participate in joint training. This will enable (a) cross-cultural communications; (b) open access to other types of simulators not available to single universities; (c) train different types of operations not available for a single university (like oil recovery joint operations, SAR operations, anti-terror operations, etc.). The development of the joint digital platform will be a step forward to a broader usage of artificial intelligence in maritime operations, as well as for the transfer of steering and full control over the vessel to a remote operator. Since there are many different maritime simulators around the world, the selection of collaborative university can be a time consuming procedure, and the established partner selection procedures are necessary.

2.2 Partner Selection

The partner selection is quite popular research area. Geringer [22] divided partner selection criteria to partner-related and task-related dimensions. The partner-related criteria embrace reputation, strategic fit, trust between the top management teams, financial stability of the partner, position within the industry, and enthusiasm for the project. The task-related criteria embrace knowledge of the local and international markets, competence in new product development, knowledge of partner's culture and internal standards, links with major buyers, suppliers and distribution channels, product-specific knowledge, capital and finance, local regulatory knowledge, political influence and other criteria relating to the industry and alliance goals.

Scholars offered both qualitative and quantitative approaches to partner selection process. However, the decision-makers still use mainly intuitive and qualitative methodology to collaborative partner selection, and often neglect quantitative approaches to partner evaluation and selection [23–25]. The main advantages of quantitative approach are unbiased ranging of candidates for collaboration and possibility to easy transfer of quantitative algorithms to computer-based platforms. Quantitative procedures assisting partner selection for various collaborative arrangements in firms and organizations are increasingly more dominant [10, 26].

Among the most widespread quantitative procedures are fuzzy set logic [27] the mixed integer linear programming [28, 29], optimization modeling [30, 31], the analytic hierarchical process [27, 32], and the analytic network process approaches [33–35].

Broad research relating to partner selection for collaboration has been done in different industrial contexts [36]. However, there is still little research relating to the maritime [37]. Notably, the accurate choice of collaborative partners is one of the essential success aspects of cooperation [38].

Formal concept analysis applied in this study is aimed to facilitate the selection of best suitable partners for computer-based maritime simulation training among participating Universities and maritime schools worldwide. Particularly, quantitative techniques are not replacements for qualitative methods of decision-making [35, 39]. The aim of FCA is to add subjective perceptions of decision-makers when they evaluate an appropriate partner for definite training tasks.

2.3 *Formal Concept Analysis*

Information handling is a complex cognitive procedure. Transformation of information into the concepts enables information investigation and treatment simpler. The lattice theory is complex mathematical theory that can be applied to evaluate concepts. Nevertheless, it was problematic to utilize the lattice theory by non-mathematicians. Wille [40] elaborated the formal concept analysis in order to make application of the lattice theory simpler. Wormuth and Becker [41, p. 4] offered the following definition of FCA—it is "a method to visualize data and its inherent structures, implications and dependencies".

FCA is based on an idea of a *concept* which comprises two fragments *extension* and *intension*. A concept is identified as "a set of specific objects, symbols, or events which are grouped together on the basis of shared characteristics and which can be referenced by a particular name or symbol" [42, p. 3]. The *extension* includes all objects related to the concept, whereas the *intension* includes all attributes associated with particular objects [43]. A *context* is a triple (G, M, I) where G and M are sets

Table 1 A formal context

	m_1	m_2	m_3
G_1	×		×
G_2		×	
G_3	×		
G_4		×	

and $I < G \times M$. The elements of G and M are termed as *objects* and *attributes* respectively. I is a relation between G and M. Sporadically the relation I is termed the incidence relation of the context [44]. If the sets are finite, the context can be identified with a help of a cross-table. An example of a simple context is presented in Table 1.

A *concept* of a context (G, M, I) is described as a set (G', M'), where a group of objects which members of M' have in common is G' and the group of attributes that the members of G' have in common is M' [45]. From Table 1, ($\{g_1, g_3\}$, $\{m_1\}$) is a concept of the context, where g_1 and g_3 are the elements of the extension and m_1 is an element of intension.

A lattice is an association-like ordering composition that can be shaped automatically from a term-document indexing relationship. Such an association construction surpasses hierarchical classification construction since the first facilitates many paths to a particular node while the second limits each node to keep only one parent [45]. Therefore, the lattice routing offers an alternate browsing-based methodology that can overpower the limitation of hierarchical classification browsing. The group of all concepts of the context (G, M, I) is a complete lattice and it is recognized as the *concept lattice* of the context (G, M, I). The concept lattice that corresponds to the context in Table 1 is demonstrated in Fig. 1.

The main disadvantage of the previous mathematical methods employed in partner selection is that their application requires experienced computer programmers [46]. FCA leaves behind other mathematical methods used in previous research on collab-

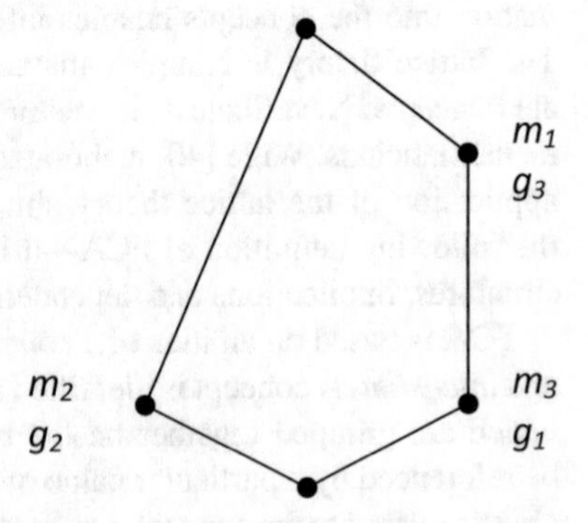

Fig. 1 A concept lattice corresponding to a context in Table 1

oration partner selection by the comparative simplicity for practitioners, opportunity to be incorporated into ICT products and to visualise the analysis.

3 Illustrative Example

In this illustrative example, we will demonstrate how the FCA can be utilized to assist decision-making in the process of partner university selection for joint simulator training. University A's simulator managers determined to invite partners having certain simulators for joint training. University's A's managers investigated potential partners. There are four possible collaborators for joint training (i.e. universities 1, 2, 3, and 4). In order to construct a lattice, it is required to have a context that is a binary link between objects and attributes. The attributes that prospective university partners possessed are listed below and showed in Table 2:

S cultural fit
O organizational fit
T time-zone fit
F knowledge of foreign language (English)
R experience with oil spill recovery training and possessing oil spill recovery simulator
D competence in dynamic positioning (DP) training and possessing the DP simulator
A access to cloud-based technology, good Internet connection
H competence in anti-terror training

Table 2 is a cross-table where cells contain crosses (×) or empty cells. Value (×) means that a university has a certain attribute, and an empty cell means that it does not possess needed attribute.

For further analysis, we portray data in visual way. We apply a Hasse diagram to visualize a lattice. Nodes characterize formal concepts and edges display the subconcepts. Using a concept lattice, we can discover relations between concepts, objects, and attributes. The concepts portrayed hierarchically (i.e. the closer a concept is to the supremum (the peak node), the more attributes it has). Moving from one

Table 2 Context—characteristics of potential collaborators

	S	O	T	F	R	D	A	H
U 1	×	×		×		×		×
U 2		×	×			×		×
U 3			×		×		×	×
U 4	×				×			×

U1, U2, U3, U4 (objects) universities 1, 2, 3, and 4 correspondingly, *(×)* a university has an attribute, *an empty cell* a university does not have and attribute

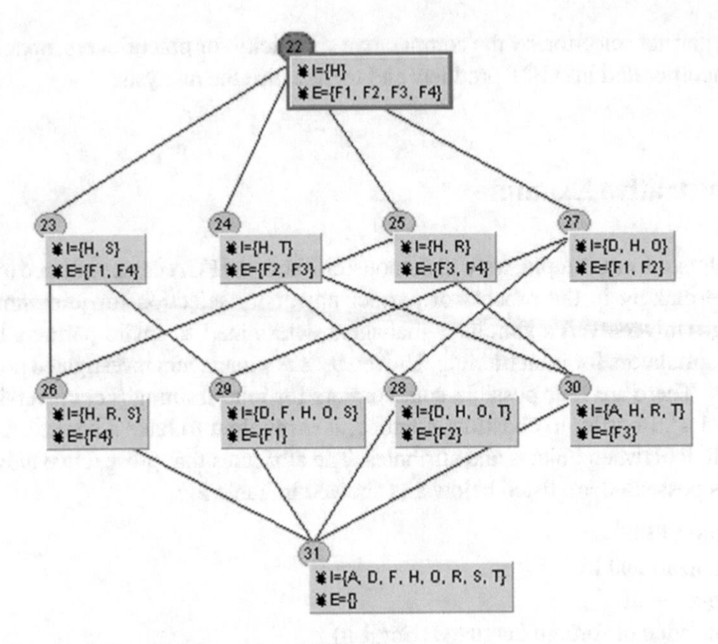

Fig. 2 Concept lattice matching to data in Table 2

vertex to a connected vertex that is nearer to the supremum indicates moving from a more general to a more accurate portrayal of the attributes, if an object appears in both concepts.

Concepts are displayed by the tags attached to the nodes of the lattice (Fig. 2). The connotation of the utilised tags is as follows:

- Node number 23 has a label *I = {H, S}*, *E = {U1, U4}*. This indicates that universities *U1* and *U4* possess two shared traits, *H* and *S* (i.e. competencies in anti-terror training and have cultural fit with a university that looks for partners).
- Node number 27 has a label *I = {D, H, O}*, *E = {U1, U2}*. This indicates that universities *U1* and *U2* have three shared traits, *D, H, O* (i.e. experience with dynamic positioning training, anti-terror training and organizational fit).

4 Conclusions

In this study we aimed to elaborate FCA approach to a partner selection for joint simulation training of sailors. Navigational simulators are situated in maritime educational different countries around the world. The new cloud-based platform allows to connect simulators situated in different countries with each other and carry out joint training. This opens completely new possibilities starting for training for emergency situations when actions of several vessels, police, ambulance, and other services are necessary, and ending with the new possibilities for digital twin development.

Quantitative methods for partner selection are steadily more popular. FCA is a novel approach to data analysis, an important technique of the graphical representation of knowledge and information management. Utilization of FCA has increased in different areas. Numerous applications of FCA are described in research studies [47–50]. This study demonstrated that the FCA could be also applied for selection of partners worldwide for collaborative digital simulation training of seafarers. Further research can build on previous research on ICT applications in maritime sector [51–60], and look for new possibilities to further utilization of modern ICT techniques in this important industry, as well as new ways of practical application of non-classical and paraconsistent logic [61–69].

References

1. Kuntsevich, V.M.: Control in Uncertain Conditions: Garanteed Results in Control and Identification Tasks. Naukova Dumka, Kyiv (2006)
2. Kuntsevich, V.M., Gubarev, V.F., Kondratenko, Y.P., Lebedev, D.V., Lysenko, V.P. (eds.): Control Systems: Theory and Applications. Automation, Control and Robotics. River Publishers, Gistrup (2018)
3. Atamanyuk, I., Kondratenko, V., Kondratenko Y., Shebanin, V., Solesvik, M.: Models and algorithms for prediction of electrical energy consumption based on canonical expansions of random sequences. In: Kharchenko, V., Kondratenko, Y., Kacprzyk, J. (eds.) Green IT Engineering: Social, Business and Industrial Applications. Studies in Systems, Decision and Control, vol. 171, pp. 397–421. Springer, Cham (2019). https://doi.org/10.1007/978-3-030-00253-4_17
4. Gulbrandsen, M., Solesvik, M.: Comparing university-industry linkages in two industrial clusters in Norway. In: Academy of Management Proceedings, vol. 2015, no. 1, p. 18826. Academy of Management (2015)
5. Solesvik, M.: A collaborative design in shipbuilding: two case studies. In: 2007 5th IEEE International Conference on Industrial Informatics, vol. 1. IEEE (2007)
6. Solesvik, M., Kondratenko, Y.: Architecture for Collaborative Digital Simulation for the Polar Regions. In: Kharchenko V., Kondratenko Y., Kacprzyk J. (eds.) Green IT Engineering: Social, Business and Industrial Applications. Studies in Systems, Decision and Control, vol. 171, pp. 517–531. Springer, Cham (2019). https://doi.org/10.1007/978-3-030-00253-4_22
7. Solesvik, M., Borch, O.J., Kondratenko, Y.: Joint Digital Simulation Platforms for Safety and Preparedness. In: Luo Y. (eds.) Cooperative Design, Visualization, and Engineering. CDVE 2018. Lecture Notes in Computer Science, vol. 11151. Springer, Cham, pp. 118–125. https://doi.org/10.1007/978-3-030-00560-3_16
8. Gil-Aluja, J.: Elements for a Theory of Decision in Uncertainty, vol. 32. Springer Science & Business Media, Berlin (1999)
9. Kondratenko, Y., Kondratenko, V.: Soft computing algorithm for arithmetic multiplication of fuzzy sets based on universal analytic models. In: Ermolayev, V. et al. (eds.) Information and Communication Technologies in Education, Research, and Industrial Application. Communications in Computer and Information Science, ICTERI'2014, vol. 469, pp. 49–77. Springer International Publishing, Switzerland (2014). https://doi.org/10.1007/978-3-319-13206-8_3
10. Encheva, S., Kondratenko, Y., Solesvik, M.Z., Tumin, S.: Decision Support Systems in Logistics. AIP Conf. Proc. **1060**, 254–256 (2008). https://doi.org/10.1063/1.3037065

11. Zadeh, L.A.: Toward a theory of fuzzy information granulation and its centrality in human reasoning and fuzzy logic. Fuzzy Sets Syst. **90**(2), 111–127 (1997)
12. Kondratenko, Y.P., Sidenko, Ie.V.: Decision-making based on fuzzy estimation of quality level for cargo delivery. In: Zadeh, L.A. et al. (eds.) Recent Developments and New Directions in Soft Computing. Studies in Fuzziness and Soft Computing, vol. 317, pp. 331–344. Springer International Publishing, Switzerland (2014). https://doi.org/10.1007/978-3-319-06323-2_21
13. Solesvik, M., Kondratenko, Y., Kondratenko, G., Sidenko, I., Kharchenko, V., Boyarchuk, A.: Fuzzy decision support systems in marine practice. In: IEEE International Conference on Fuzzy Systems (FUZZ-IEEE), pp. 1–6, Naples, Italy (2017). https://doi.org/10.1109/fuzzieee.2017.8015471
14. Kondratenko, Y.P., Klymenko, L.P., Sidenko, Ie.V.: Comparative analysis of evaluation algorithms for decision-making in transport logistics. In: Jamshidi, M., Kreinovich, V., Kacprzyk, J. (eds.) Advance Trends in Soft Computing, Studies in Fuzziness and Soft Computing, vol. 312, pp. 203–217 (2014). https://doi.org/10.1007/978-3-319-03674-8_20
15. Kondratenko, Y.P., Kozlov, O.V., Topalov, A.M.: Fuzzy controllers for increasing efficiency of the floating dock's operations: design and optimization. In: Kuntsevich, V.M., Gubarev, V.F., Kondratenko, Y.P., Lebedev, D.V., Lysenko, V.P. (eds.) Control Systems: Theory and Applications. Automation, Control and Robotics, pp. 197–232. River Publishers, Gistrup, Delft (2018)
16. Prause, G., Solesvik, M.: University-Business Cooperation in Maritime Sector-The German-Norwegian Experience, pp. 20–37. University-Business Cooperation, Tallinn (2011)
17. Sidenko, I., Filina, K., Kondratenko, G., Chabanovskyi, D., Kondratenko, Y.: Eye-tracking technology for the analysis of dynamic data. In: Proceedings of 2018 IEEE 9th International Conference on Dependable Systems, Services and Technologies, DESSERT 2018, 24–27 May 2018, Kiev, Ukraine, pp. 509–514. https://doi.org/10.1109/dessert.2018.8409181
18. Kondratenko, Y., Simon, D., Atamanyuk I.: University curricula modification based on advancements in information and communication technologies. In: Proceedings of the 12th International Conference on Information and Communication Technologies in Education, Research, and Industrial Application. Integration, Harmonization and Knowledge Transfer, 21–24 June 2016, Kyiv, Ukraine. Ermolayev, V. et al. (eds.), ICTERI'2016, CEUR-WS, vol. 1614, pp. 184–199 (2016)
19. Solesvik, M.: Interfirm collaboration in the shipbuilding industry: the shipbuilding cycle perspective. Int. J. Bus. Syst. Res. **5**(4), 388–405 (2011)
20. Borch, O.J., Solesvik, M.: Collaborative design of advanced vessel technology for offshore operations in Arctic waters. In: Luo, Y. (ed.) International Conference on Cooperative Design, Visualization and Engineering, pp. 157–160. Springer, Berlin (2013)
21. Borch, O.J., Solesvik, M.Z.: Innovation on the open sea: examining competence transfer and open innovation in the design of offshore vessels. Technol. Innov. Manag. Rev. **5**(9), 17–22 (2015)
22. Geringer, J.M.: Strategic determinants of partner selection criteria in international joint ventures. J. Int. Bus. Stud. **22**(1), 41–62 (1991)
23. Solesvik, M.Z., Encheva, S.: Partner selection for interfirm collaboration in ship design. Ind. Manag. Data Syst. **110**(5), 701–717 (2010)
24. Borch, O.J., Solesvik, M.Z.: Partner selection versus partner attraction in R&D strategic alliances: the case of the Norwegian shipping industry. Int. J. Technol. Mark. **11**(4), 421–439 (2016)
25. Solesvik, M., Gulbrandsen, M.: Partner selection for open innovation. Technol. Innov. Manag. Rev. **3**(4), 11–16 (2013)
26. Solesvik, M.: Partner selection in green innovation projects. In: Berger-Vachon, C., Gil Lafuente, A., Kacprzyk, J., Kondratenko, Y., Merigó, J., Morabito, C. (eds.) Complex Systems: Solutions and Challenges in Economics, Management and Engineering, vol. 125, pp. 471–480. Springer, Cham (2018)
27. Wang, T.-C., Chen, Y.-H.: Applying consistent fuzzy preference relations to partnership selection. Omega **35**(4), 384–388 (2007)

28. Jarimo, T., Salo, A.: Multicriteria partner selection in virtual organisations with transportation costs and other network interdependencies, Technical report. Helsinki University of Technology, Helsinki (2008)
29. Wu, N., Su, P.: Selection partners in virtual enterprise paradigm. Robot. Comput. Integr. Manuf. **21**(2), 119–131 (2005)
30. Cao, Q., Wang, Q.: Optimizing vendor selection in a two-stage outsourcing process. Comput. Oper. Res. **34**(12), 3757–3768 (2007)
31. Fuqing, Z., Yi, H., Dongmei, Y.: A multi-objective optimization model of the partner selection problem in a virtual enterprise and its solution with generic algorithms. Int. J. Adv. Manuf. Technol. **28**(11–12), 1246–1253 (2006)
32. Mikhailov, L.: Fuzzy analytical approach to partnership selection in formation of virtual enterprises. Omega **30**(5), 393–401 (2002)
33. Chen, S.-H., Lee, H.-T., Wu, Y.-F.: Applying ANP approach to partner selection for strategic alliance. Manag. Decis. **46**(3), 449–465 (2008)
34. Sarkis, J., Talluri, S., Gunasekaran, A.: A strategic model for agile virtual enterprise partner selection. Int. J. Oper. Prod. Manag. **27**(11), 1213–1234 (2007)
35. Wu, W.Y., Shih, H.-A., Chan, H.-C.: The analytic network process for partner selection criteria in strategic alliances. Expert Syst. Appl. **36**(3), 4646–4653 (2009)
36. Solesvik, M.: Innovation strategies in shipbuilding: the shipbuilding cycle perspective. Shipbuild. Mar. Infrastruct. **5**(1/2), 44–50 (2016)
37. Solesvik, M.Z., Westhead, P.: Partner selection for strategic alliances: case study insights from the maritime industry. Ind. Manag. Data Syst. **110**(6), 841–860 (2010)
38. Reuer, J.J., Ariño, A., Olk, P.M.: Entrepreneurial Alliances. Prentice Hall, Upper Saddle River (2011)
39. Ordoobadi, S.M.: Fuzzy logic and evaluation of advanced technologies. Ind. Manag. Data Syst. **108**(7), 928–946 (2008)
40. Wille, R.: Restructuring lattice theory: an approach based on hierarchies of concepts. In: Rival, I. (ed.) Ordered Sets, pp. 445–470. Reidel, Dordrecht (1982)
41. Wormuth, B., Becker, P.: Introduction to formal concept analysis. In: 2nd International Conference of Formal Concept Analysis, February 2014, vol. 23 (2014)
42. Merrill, M.D., Tennyson, R.D.: Concept Teaching: An Instructional Design Guide. Educational Technology, Englewood Cliffs, NJ (1977)
43. Wolff, K.E.: A first course in formal concept analysis: how to understand line diagrams. In: Faulbaum, F. (ed.) Advances in Statistical Software, pp. 429–438. Gustav Fischer Verlag, Stuttgart (1993)
44. Carpineto, C., Romano, G.: Concept Data Analysis: Theory and Applications. Wiley, Hoboken, NJ (2004)
45. du Boucher-Ryan, P., Bridge, D.: Collaborative recommending using formal concept analysis. Knowl. Based Syst. **19**, 309–315 (2006)
46. Choy, K.L., Lee, W.B., Lo, V.: An enterprise collaborative management system—a case study of supplier relationship management. J. Enterp. Inf. Manag. **17**(3), 191–207 (2004)
47. Ganter, B., Stumme, G., Wille, R.: Formal Concept Analysis: Foundations and Applications. LNAI, vol. 3626. Springer, Heidelberg (2005)
48. Solesvik, M.Z., Encheva, S., Tumin, S.: Lattices and collaborative design in shipbuilding. Int. J. Bus. Inf. Syst. **7**(3), 309–326 (2011)
49. Wille, R.: Conceptual knowledge processing in the field of economics. In: Ganter, B., Stumme, G., Wille, R. (eds.) Formal Concept Analysis: Foundations and Applications, vol. 3626, pp. 226–249. Springer, Heidelberg, LNAI (2005)
50. Borch, O.J., Solesvik, M.: Partner selection for innovation projects. In: International Society for Professional Innovation Management (ISPIM) Americas Innovation Forum (2014)
51. Gausdal, A.H., Czachorowski, K.V., Solesvik, M.Z.: Applying blockchain technology: evidence from Norwegian companies. Sustainability **10**(6), 1–16 (2018)

52. Czachorowski, K., Solesvik, M., Kondratenko, Y.: The Application of blockchain technology in the maritime industry. In: Kharchenko, V., et al. (eds.) Green IT Engineering: Social, Business and Industrial Applications, pp. 561–577. Springer, Cham (2019). https://doi.org/10.1007/978-3-030-00253-4_24
53. Solesvik, M.Z., Encheva, S.: Benefits and drawbacks of software platforms used in cooperative ship design practice. In: Proceedings of the WSEAS International Conference on Applied Computing Conference, pp. 87–90 (2008)
54. Timchenko, V.L., Kondratenko, Y.P.: Robust stabilization of marine mobile objects on the basis of systems with variable structure of feedbacks. J. Autom. Inf. Sci. **43**(6): 16–29 (2011). (Begel House Inc., New York) https://doi.org/10.1615/jautomatinfscien.v43.i6.20
55. Altameem T.A., Al Zu'bi E.Y.M., Kondratenko Y.P.: Computer decision making system for increasing efficiency of ship's bunkering process. In: Annals of DAAAM for 2010 & Proceeding of the 21th International DAAAM Symposium "Intelligent Manufacturing and Automation", pp. 0403–0404, 20–23 October 2010, Zadar, Croatia. DAAAM International, Vienna, Austria (2010)
56. Solesvik, M.Z., Encheva, S.: Offshoring decision making in the logistics of the Norwegian shipbuilding yards. In: WSEAS Conference in Dallas, March 2007, pp. 22–24 (2007)
57. Solesvik, M.Z., Iakovleva, T., Encheva, S.: Simulation and optimization in collaborative ship design: innovative approach. In: International Conference on Cooperative Design, Visualization and Engineering, pp. 151–154. Springer, Berlin, Heidelberg (2012)
58. Solesvik, M.Z.: Collaboration model for ship design. In: International Conference on Cooperative Design, Visualization and Engineering, pp. 245–248. Springer, Berlin, Heidelberg (2008)
59. Encheva, S., Tumin, S., Solesvik, M. Z.: Decision support system for assessing participants reliabilities in shipbuilding. In: Proceedings of the 9th WSEAS international conference on Automatic control, modelling and simulation, pp. 270–275. World Scientific and Engineering Academy and Society (WSEAS), Stevens Point (2007)
60. Kondratenko Y.P., Timchenko V.L.: Increase in navigation safety by developing distributed man-machine control systems. In: Proceedings of the Third International Offshore and Polar Engineering Conference, vol. 2, pp. 512–519, Singapore (1993)
61. Solesvik, M.Z., Encheva, S.: Benefits and drawbacks of software platforms used in cooperative ship design practice. In: Proceedings of the WSEAS International Conference on Applied Computing Conference, pp. 87–90. World Scientific and Engineering Academy and Society (WSEAS), Stevens Point (2008)
62. Parsyak, V.N., Solesvik, M.B.: Integration of marine engineering and information and communication technology. Shipbuild. Mar. Infrastruct. **2**, 144–155 (2014)
63. Encheva, S., Kondratenko, Y., Tumin, Sh., Kumar, K.S.: Non-Classical logic in an intelligent assessment sub-system. In: Gervasi, O., Gavrilova, M.L. (eds.) International Conference on Computational Science and Its Applications—ICCSA 2007, Kuala Lumpur, Malaysia, 26–29 August 2007. Proceedings, Part I. Lecture Notes in Computer Science, vol. 4705, pp. 305–314. Springer, Berlin (2007). https://doi.org/10.1007/978-3-540-74472-6_24
64. Encheva, S., Tumin, Sh., Kondratenko, Y.: Application of paraconsistent annotated logic in intelligent systems. In: Huang, D.-S., Heutte, L., Loog, M. (eds.), Advanced Intelligent Computing Theories and Applications. with Aspects of Theoretical and Methodological Issues, Proceedings of the Third International Conference on Intelligent Computing, ICIC 2007, Qingdao, China, 21–24 Aug 2007. Lecture Notes in Computer Science 4681, pp. 702–710. Springer, Berlin/Heidelberg (2007). https://doi.org/10.1007/978-3-540-74171-8_69
65. Stetsuyk, E.D., Maevsky, D.A., Maevskaya, E.J., Shaporin, R.O.: Information technology for evaluating the computer energy consumption at the stage of software development. In: Kharchenko, V. et al. (eds.), Green IT Engineering: Social, Business and Industrial Applications, Studies in Systems, Decision and Control, vol. 171, pp. 21–40. Springer International Publishing, Berlin (2019). https://doi.org/10.1007/978-3-030-00253-4_2
66. Kudermetov, R., Polska, O.: Towards a formalization of the fundamental concepts of SOA. In: Proceedings of the 13th International Conference on Modern Problems of Radio Engineering, Telecommunications and Computer Science (TCSET). IEEE, Lviv, Ukraine (2016). https://doi.org/10.1109/tcset.2016.7452096

67. Shebanin, V., Atamanyuk, I., Kondratenko, Y., Volosyuk Y.: Application of fuzzy predicates and quantifiers by matrix presentation in informational resources modeling. In: Proceedings of the International Conference on Perspective Technologies and Methods in MEMS Design (MEMSTECH-2016), pp. 146–149. Lviv-Poljana, Ukraine, 20–24 April 2016. https://doi.org/10.1109/memstech.2016.7507536
68. Kondratenko, G.V., Kondratenko, Y.P., Romanov, D.O.: Fuzzy models for capacitive vehicle routing problem in uncertainty. In: Proceedings 17th International DAAAM Symposium "Intelligent Manufacturing and Automation: Focus on Mechatronics & Robotics", Vienna, Austria, pp. 205–206 (2006)
69. Skarga-Bandurova, I., Derkach, M., Kotsiuba, I.: The information service for delivering arrival public transport prediction. In: Proceedings of the 2018 IEEE 4th International Symposium on Wireless Systems within the International Conferences on Intelligent Data Acquisition and Advanced Computing Systems, IDAACS-SWS 2018, Lviv, Ukraine (2018). https://doi.org/10.1109/idaacs-sws.2018.8525787

Printed in the United States
By Bookmasters